Maurice Holt

Numerical Methods in Fluid Dynamics

Second revised edition

With 114 Figures

Springer-Verlag
Berlin Heidelberg New York Tokyo 1984

Maurice Holt

Professor of Aeronautical Sciences
College of Engineering, Mechanical Engineering
University of California, Berkeley, CA 94720, USA

Editors   H. B. Keller

                                                   Applied Mathematics 101-50
                                                   Firestone Laboratory
                                                   California Institute of Technology
                                                   Pasadena, California 91125/USA

Henri Cabannes

Mécanique Théorique   John Killeen
Université Pierre et Marie Curie
Tour 66-4, place Jussieu   Lawrence Livermore Laboratory
F-75005 Paris/France   P.O. Box 808
                                                   Livermore, California 94551/USA

Maurice Holt

College of Engineering   Stephen A. Orszag
Mechanical Engineering
University of California   Department of Mathematics
Berkeley, California 94720/USA   Massachusetts Institute of Technology
                                                   Cambridge, Massachusetts 02139/USA

ISBN-13: 978-3-540-12799-4   e-ISBN-13: 978-3-642-69341-0
DOI: 10.1007.978-3-642-69341-0

Library of Congress Cataloging in Publication Data.
Holt, Maurice. Numerical methods in fluid dynamics.
(Springer series in computational physics)
Includes bibliographies and index.
1. Fluid dynamics—Mathematics. 2. Numerical calculations. I. Title. II. Series.
TA357.H63   1983   620.1'06'01515   83-14674

2153/3020-543210

Springer Series in Computational Physics

Editors:
H. Cabannes   M. Holt   H. B. Keller   J. Killeen   S. A. Orszag

Preface to the Second Edition

During the six years following the appearance of the first edition of this book many new developments have been made in numerical techniques applied to problems in fluid dynamics. Most of these have been described in companion volumes in the series, especially those by Peyret and Taylor, by Thomasset and by Fletcher. In these works special attention is given to finite element techniques, spectral methods and generalized Galerkin methods. There is no need to duplicate the treatment of these topics in the present volume and the revisions made here are concentrated in two areas. Firstly, Chapter 2, in which the Godunov schemes for unsteady gas dynamics problems were presented, is now expanded to cover Glimm's method. This is a broader version of Godunov's first scheme and has been exploited by Chorin and others in application to one and multidimensional unsteady flow problems. The second revision is in Chapter 5, and covers the many recent applications of the Method of Integral Relations (in its orthonormal version) to turbulent boundary layers, internal pipe flows, and turbulent mixing flows.

At the same time a number of errors in the first edition have been corrected and the references have been updated.

The new work described in the revised edition was supported in part by the U.S. Air Force Office of Scientific Research, by the U.S. Office of Naval Research and by the U.S. Department of Energy.

Berkeley, California
November 1983   Maurice Holt

Preface to the First Edition

This monograph is based on a graduate course, Mechanical Engineering 266, which was developed over a number of years at the University of California—Berkeley. Shorter versions of the course were given at the University of Paris VI in 1969, and at the University of Paris XI in 1972. The course was originally presented as the last of a three-quarter sequence on Compressible Flow Theory, with emphasis on the treatment of non-linear problems by numerical techniques. This is reflected in the material of the first half of the book, covering several techniques for handling non-linear wave interaction and other problems in Gas Dynamics. The techniques have their origins in the Method of Characteristics (in both two and three dimensions). Besides reviewing the method itself the more recent techniques derived from it, firstly by Godunov and his group, and secondly by Rusanov and his co-workers, are described. Both these approaches are applicable to steady flows calculated as asymptotic states of unsteady flows and treat elliptic problems as limiting forms of unsteady hyperbolic problems. They are therefore applicable to low speed as well as to high speed flow problems.

The second half of the book covers the treatment of a variety of steady flow problems, including effects of both viscosity and compressibility, by the Method of Integral Relations, Telenin's Method, and the Method of Lines. The objective of all these methods is to eliminate finite difference calculations in one or more coordinate directions by using interpolation formulae, especially polynomials, to represent the unknowns in selected directions. These methods were used originally to solve flow problems connected with re-entry aerodynamics but have subsequently been applied over the whole speed range. They are, in principle, applicable to a broad range of problems governed by elliptic, mixed elliptic-hyperbolic, and parabolic partial differential equations.

Many of the applications described in the book result from research, carried out at Berkeley, sponsored by the Air Force Office of Scientific Research, NASA Ames Research Center and the Office of Naval Research. The support of all these agencies is gratefully acknowledged.

The less familiar methods discussed in the book are illustrated by solutions to model problems worked out by graduate students enrolled in the class and their contributions are recognized in the text. In addition, several students assisted in checking the equations, especially A. Falade, W. K.

Chan, and K. S. Chang. The manuscript was carefully typed by Mrs. Arlene Martin and I am indebted to her for completing this exacting task with cheerful forbearance.

My original venture into the numerical field was encouraged by Sydney Goldstein, who pointed out to me the importance of this approach to problems in Fluid Dynamics many years ago, when non-linear effects first assumed significance. In the present enterprise Victor Rusanov was very helpful not only in providing material for Chapter 3, but also in obtaining less accessible papers for Chapters 2 and 6. Oleg Belotserkovskii has been a steady source of information on the Method of Integral Relations and many of the applications of the method to inviscid problems originate from his group.

I am grateful to Dr. W. Beiglböck for including the monograph in this new Springer series. I wish to express my appreciation to Mrs. Oelschläger and the editorial staff of Springer-Verlag for their assistance and courteous cooperation in the production of the book. Finally, I wish to thank my wife, Eileen, for her patient support during the writing of the manuscript.

Berkeley, California
January 21, 1977   Maurice Holt

Contents

Chapter 4.   The Method of Characteristics for Three-Dimensional Problems
             in Gas Dynamics

Chapter 5.   The Method of Integral Relations

Chapter 6.   Telenin's Method and the Method of Lines

CHAPTER 1

General Introduction

Brief Review of Concepts of Numerical Analysis

1.1 Introduction

At the present time the majority of unsolved problems in Fluid Dynamics are governed by non-linear partial differential equations and can only be treated by a numerical approach. As a consequence, specialists in Fluid Dynamics have recently devoted increasing attention to numerical, as opposed to analytical, techniques. Of course, there is no point in developing a novel numerical method unless it can be applied to actual problems of interest. In the early days of research on numerical analysis the capacity of computing machines was too restricted to permit many applications to be carried out. Today this situation has changed; the machines now available are sufficiently advanced to deal with an almost limitless range of problems; all that is needed is to discover effective numerical methods to attack them.

Although the major advances in construction and development of actual computing machines have taken place in the United States, many of the principal advances in Numerical Methods were made in the Soviet Union. This is especially true of methods applicable to problems in Fluid Dynamics and here the methods can be divided into two categories. The first depend purely on finite difference techniques, while in the second, the number of independent variables in the numerical scheme is reduced by supposing that the unknowns are polynomials or trigonometric functions of one, or more, of these variables.

In the first part of the monograph we shall be concerned mainly with problems in Gas Dynamics. In many of these problems viscosity is unimportant and the equations of motion reduce to a system of partial differential equations of the first order. If the motion is unsteady, this system is always hyperbolic. If the motion is steady the classification of the system depends on the magnitude of the fluid speed, being hyperbolic if it is supersonic and elliptic if subsonic. Thus the problems of Gas Dynamics are of three types, firstly, elliptic for steady flow at low speeds; secondly, hyperbolic for steady supersonic flow and all unsteady flow; and finally, of mixed type, when the flow is steady and subsonic in one region while being supersonic elsewhere. In all discussions of finite difference methods given in this course we shall always regard a steady flow as the asymptotic state of an unsteady flow and shall therefore only consider hyperbolic systems.

We shall describe two recent finite difference methods. The first is due to GODUNOV, originally presented in 1960 and revised in 1970 (see Refs. Chapt. 2). The second method was developed principally by RUSANOV, also in two stages. The original formulation was presented in 1964 in collaboration with BABENKO, VOSKRESENSKII and LIUBIMOV and is familiarly known as the BVLR method (BABENKO et al., Refs. Chapt. 3). This method was extended to three dimensional unsteady flow in 1970 by RUSANOV and LYUBIMOV (Refs. Chapt. 3). Both the Godunov and BVLR methods have their origins in the method of characteristics. In principle one could solve all nonstationary problems of Gas Dynamics by a method of characteristics and there are at present several research workers in the field who rely exclusively on this method. However, the method does not lend itself easily to machine computations. The main difficulty here results from the fact that the system of characteristic coordinates is not rectangular but curvilinear and, frequently, the angle between coordinate lines of opposing families is very small. It is, of course, always preferable to work with a rectangular coordinate system if possible.

The starting point in Godunov's method is the solution of the problem of piston motion in a cylinder. This is a classical problem. When the piston speed is constant the solution is well known; a shock wave propagates into the undisturbed gas, moving ahead of the piston with a larger but still constant speed. The value of this speed, for a perfect gas, is defined by a simple quadratic formula. When the piston motion is nonuniform the problem can be solved numerically by a method of characteristics employing Riemann invariants.

GODUNOV proposes to solve the general piston problem as follows: the region between the piston face and the shock wave is divided into a number of cells of small length (in general the cells are of equal length). If the velocity distribution along the cylinder is given at one time we can calculate the mean value of the velocity (and also the value of the other dependent variables) in each cell. The actual distribution can then be replaced by a sequence of constant values, one per cell. Then, across the boundary between two adjacent cells the values of the velocity, pressure and density are in general discontinuous. To determine the corresponding values at a slightly later instant it is necessary to solve a problem of breakdown at a diaphragm. This problem has an analytical solution defined by algebraic formulae. Thus, at the later instant, the values of the unknowns on each cell boundary are defined (they are in fact constant) and new cell values are determined as the means of values on the left and right boundaries. GODUNOV applies this process at successive times.

To solve a more general problem, for example, that of flow of supersonic gas past a cylinder, the field of flow is divided into strips, the boundaries of which are parallel to the axis of symmetry. The spacing of the strips is constant and each strip is treated as a channel. However, it is now necessary to take account of diaphragm breakdown between adjacent strips as well as between cell boundaries along each strip.

The BVLR method is purely a finite difference method. Three independent variables are considered; namely, the time (or a space variable which plays the role of time) and two coordinates. To advance the calculation in time we must connect the values in two planes representing conditions at successive time intervals by certain finite difference relations along characteristic lines. In the BVLR method these relations are replaced by equivalent conditions along lines running in the time or coordinate directions. Furthermore, an important part of the calculation is the determination of the shape of shock which encloses the given body. To this end, the boundary conditions at the body surface must be connected with conditions satisfied at the shock wave. To carry this out RUSANOV et al. use an extension of the double sweep method originally proposed by GEL'FAND and LOKUTSIEVSKII (see Refs. Chapt. 3).

To make the coverage of finite difference methods for hyperbolic equations complete, a chapter is included on the method of characteristics in three dimensions. In this the different versions of the method are described and particular attention is given to two of these; namely, the bicharacteristics method of BUTLER and the near characteristics method of SAUER.

The later chapters are devoted to techniques based partly on polynomial or other series representations in one (or more) of the independent variables. The first of these, the Method of Integral Relations, was introduced by DORODNITSYN in 1950, principally as a means to solve the problem of high speed flow past a blunt nosed body (see Refs. Chapt. 5). The application of this method carried out by BELOTSERKOVSKII provided the first solution to the problem of reentry of a space vehicle in the earth's atmosphere. In 1960 DORODNITSYN extended the method to apply to viscous flow, especially to boundary layer and wake problems. The method has been used widely in the United States and many applications will be discussed. The method consists in writing the equations of motion in divergence form and then integrating them with respect to one of the independent variables from one side of the disturbed flow field to the other. To calculate the integrals it is assumed that the integrands are polynomial or trigonometric functions of the variable of integration. These functions contain unknown coefficients which satisfy a reduced system of ordinary or partial differential equations.

The second group of methods using function fitting comes under the headings of Telenin's method and the Method of Lines. The Method of Lines has a comparatively long history as an essentially Soviet technique for solving linear partial differential equations—this is reviewed by LISKOVETS (see Chapt. 6 Refs.). Telenin's method was developed in collaboration with GILINSKI, TINYAKOV and LEBEDEV from 1964 onwards. In both methods the unknowns are represented as polynomials or trigonometric functions of one of the independent variables but it is no longer required to integrate the equations of motion with respect to this variable. TELENIN and his collaborators applied this method to the blunt body problem in both two and three dimensions, considering a wide range of body shapes. At Berkeley Telenin's method was applied to the Supersonic Yawed Cone Problem by NDEFO and HOLT and by FLETCHER and HOLT. It was applied by CHATTOT to solve the transonic double wedge flow problem in the hodograph plane.

The Method of Lines differs from Telenin's method in using local polynomial fitting rather than fitting over a whole coordinate range. It has been applied to the Yawed Cone Problem by JONES, SOUTH, and by FLETCHER (see Chapt. 6 Refs.).

In the remainder of this chapter we shall give a brief review of boundary value and initial value problems, followed by a discussion of one dimensional unsteady flow needed as an introduction to the Godunov schemes. We then deal with the method of characteristics for two dimensional steady supersonic flow as necessary background for Chapt. 4. The chapter concludes with an outline of the basic concepts of finite difference methods, such as stability, consistency and convergence. This draws on unpublished course notes of CHORIN and MILLER used at the University of California.

1.2 Boundary Value Problems and Initial Value Problems

The partial differential equations to be solved in problems in Fluid Mechanics are of three main types, elliptic, hyperbolic and parabolic. Steady inviscid flow in an incompressible fluid, or at subsonic speeds in a compressible fluid is governed by elliptic equations. When viscous boundary layer effects are included in such problems the addition of a diffusion type term converts these equations to parabolic form. Problems of steady supersonic or unsteady inviscid flow of a compressible fluid require the solution of hyperbolic equations. Transonic flow problems are governed by equations of mixed type, elliptic in subsonic regions and hyperbolic in supersonic regions. Although many finite difference methods have been developed for such problems they will not be discussed in the present monograph. Other methods for transonic problems will be discussed in Chapts. 5 and 6.

The three types of equations can be identified by their simplest forms, namely, Laplace's equations, the Wave equation, and the Heat Conduction equation. Elliptic problems are associated with values of the unknowns (or their normal derivatives) given on a closed curve and require the solution of boundary value problems. The solution at a general point of an elliptic problem depends on the data at every point of the boundary and a change in values at one boundary point changes the whole solution. In hyperbolic problems either one of the independent variables is the time or it has time like character and their solution requires the specification of values of unknowns on an open line or surface at some initial time. In this case the solution at a general point only depends on data on a part of the line. Parabolic problems require, in addition, that data be prescribed on certain fixed boundaries at all times. Both problems are classified as initial value problems.

We consider now the solution of Laplace's equation in two dimensions in which unknowns are prescribed on a closed curve. The analytical solution of this Dirichlet problem requires finding the conformal mapping transforming the boundary into the unit circle, equivalent to finding the Green's function for the problem. For a complicated boundary curve this is a difficult task and it may often be easier to solve the problem numerically. If we seek a solution by Finite Differences we divide the region bounded by the curve into a network and set up five point difference equations for the unknown at each interior point, supplementing these with difference equations derived from boundary points. These difference equations are coupled and, although in principle their solution is unique, in practice either an iterative or matrix inversion method must be used to find it.

An alternative method to the iterative process of solving Laplace's equation for such problems is to replace the steady boundary value problem by an unsteady initial value problem. In this boundary values are fixed on the closed curve at all times and values of the unknowns are estimated at the interior network points at some initial time. Values of unknowns at interior points at a later time are then found from difference equations in time and space variables. The solutions of these are determined from only local network values at the new and old times and are found directly without any iteration. The process is carried out at successive instants until the difference between the current value of the unknown at a general point and its corresponding value at the previous instant is less than some assigned small quantity. In other words, the solution of the steady state plane potential problem is found as the limiting approach to steady state of the unsteady initial value problem in two space variables.

The equivalent unsteady problem can either be formulated in parabolic or hyperbolic form, depending on whether a first or second order time derivative is added to Laplace's equation.

In discussing solutions of problems in Fluid Mechanics by Finite Difference Methods we shall treat Boundary Value Problems by this technique of unsteady approach to the steady state, normally using equations in wave propagation rather than parabolic form. Thus we shall only discuss finite difference methods for time dependent problems.

We first consider the simplest equations of motion in one-dimensional gas dynamics, defining characteristic lines, Riemann invariants and the role they play in finite difference methods, including the Method of Characteristics. We then generalize to equations of unsteady flow in two space variables so that we can deal with problems of compressible inviscid motion in plane or axi-symmetric flow. We describe techniques of solving these equations directly by the Method of Characteristics (in three independent variables) and then consider finite difference methods based on treating two dimensional flows as interdependent layers of one dimensional flows. Two methods of the latter type have been successfully developed over the past decade or so in the USSR, the first due to GODUNOV (1961, 1970), and the second by RUSANOV and others, usually known as the BVLR method (BABENKO et al., 1964; RUSANOV and LYUBIMOV, 1970).

1.3 One-Dimensional Unsteady Flow Characteristics

The equations of motion of unsteady flow in one dimension are

$$\frac{\partial u}{\partial t} + u \frac{\partial u}{\partial x} + \frac{1}{\varrho} \frac{\partial p}{\partial x} = 0 \qquad (1.3.1)$$

$$\frac{\partial \varrho}{\partial t} + u \frac{\partial \varrho}{\partial x} + \varrho \frac{\partial u}{\partial x} = 0 \qquad (1.3.2)$$

$$\frac{\partial S}{\partial t} + u \frac{\partial S}{\partial x} = 0 \qquad (1.3.3)$$

Here t is the time measured from some initial instant; x is the distance in the direction of motion; u, p, ϱ and S are velocity, pressure, density, and entropy, respectively.

If we define the speed of sound a by

$$a^2 = \left(\frac{\partial p}{\partial \varrho} \right)_S \qquad (1.3.4)$$

Eqs. (1.3.1) and (1.3.2) may be combined to yield

$$\left(\frac{\partial u}{\partial t} \pm \frac{1}{\varrho a} \frac{\partial p}{\partial t} \right) + (u \pm a) \left(\frac{\partial u}{\partial x} \pm \frac{1}{\varrho a} \frac{\partial p}{\partial x} \right) = 0 \qquad (1.3.5)$$

Eqs. (1.3.5) and (1.3.3) are called the equations of one dimensional unsteady flow in characteristic form and have special properties connected with the families of lines defined by

$$\frac{dx}{dt} = u \pm a \qquad (1.3.6)$$

$$\frac{dx}{dt} = u \qquad (1.3.7)$$

called *characteristic lines* or *characteristics*.

Eqs. (1.3.3) and (1.3.5) state that the original equations of motion (1.3.1)–(1.3.3) can be replaced by equivalent equations in characteristic form, namely, that

$$du \pm \frac{1}{\varrho a} dp = 0 \qquad (1.3.8)$$

$$dS = 0 \qquad (1.3.9)$$

along the directions (1.3.6) and (1.3.7), respectively.

Eqs. (1.3.8) and (1.3.9) are in "inner" differential form, i.e., they only contain derivatives along the corresponding directions (1.3.6) and (1.3.7). The characteristics therefore have the property that derivatives normal to them may be discontinuous. Eqs. (1.3.8) and (1.3.9) form the basis for solving problems of one dimensional gas dynamics in finite difference form, called the Method of Characteristics.

For isentropic flow of a perfect gas, (1.3.9) disappears and (1.3.8) simplify. We can now introduce the variable

$$\sigma = \int \frac{dp}{\varrho a} \qquad (1.3.10)$$

and (1.3.8) can be integrated to give

$$u + \sigma = \alpha$$
$$\qquad (1.3.11)$$
$$u - \sigma = \beta$$

where α and β are constant along lines (1.3.6), respectively, and are called *Riemann invariants*.

In general Eqs. (1.3.11) and (1.3.6) state that Riemann invariants α and β are propagated without change along plus and minus characteristic directions, respectively. In problems where disturbances are propagated in one direction only either α or β is constant throughout the whole flow region. Exact solutions can easily be found in such flows, which are called simple waves. Simple waves also can be used in certain flows with propagation in both directions.

We now cite two problems with simple wave solutions resulting from the sudden breakdown of a discontinuity. In both cases the undisturbed gas obeys perfect gas behavior with constant specific heat ratio γ and is uniform.

Problem 1

A semi infinite column of gas is bounded at its right end by a diaphragm to the right of which is a vacuum. At time $t = 0$ the diaphragm is suddenly ruptured. Determine the way in which gas escapes from the column.

Solution

No fundamental length or time enter the problem so the solution is a function of x/t only, where x is the space coordinate measured from the original position of the diaphragm and t is the time measured from the instant of rupture. If γ is the specific heat ratio, and suffix zero refers to undisturbed conditions, it can easily be shown that the solution is

$$u = \frac{2}{\gamma+1}\left(\frac{x}{t} + a_0\right) \tag{1.3.12}$$

$$a = \frac{2}{\gamma+1}\left(a_0 - \frac{\gamma-1}{2}\frac{x}{t}\right) \tag{1.3.13}$$

This gives a linear variation with x/t for both u and a, and has the property that at the original position of diaphragm $x=0$, $u=a=2a_0/(\gamma+1)$ for all time. This is the simplest of the breakdown solutions used in Godunov's method.

Problem 2

A diaphragm separates two semi infinite columns of gas, initially at rest, with pressure p_1, density ϱ_1 on the left and pressure p_5, density ϱ_5 on the right. If $p_1 > p_5$, determine the motion after the diaphragm is instantaneously ruptured.

Solution

This is the shock tube problem. A shock wave is propagated to the right while a centered expansion wave is propagated to the left. The expansion and compression regions are separated by a contact discontinuity which acts like a uniform piston moving to the right.

The significant flow regions after breakdown are shown in Fig. 1.1.

1 undisturbed	2 expansion fan	3 uniform region	4 uniform region	5 undisturbed
	Head Tail	C.D.	Shock	

Fig. 1.1   Shock tube problem

The problem is solved iteratively. One approach is to assume a value for p_4 and hence for $\xi = p_4/p_5$, the pressure ratio across the shock. This then determines the velocity of the contact discontinuity. We then solve in the expansion regions 2 and 3 regarding the contact discontinuity as a withdrawing piston and obtain a value for the velocity of sound a_1, at the head of the wave. This value is compared with the undisturbed value $a_1 = (\gamma p_1/\varrho_1)^{1/2}$ and if the two are unequal, another value of p_4 must be prescribed and the cycle repeated.

In the full breakdown solutions the semi-infinite columns of gas 1 and 2 are not initially at rest but are moving with *different* uniform velocities. These can also be solved iteratively. The full formulae are given in Sect. 2.2.

1.4 Steady Supersonic Plane or Axi-Symmetric Flow. Equations of Motion in Characteristic Form

The equations of motion for steady irrotational flow in a plane or with axial symmetry are

$$u_y - v_x = 0 \tag{1.4.1}$$

$$(a^2 - u^2)u_x - 2uvu_y + (a^2 - v^2)v_y + ja^2 v/y = 0 \tag{1.4.2}$$

Here x, y are Cartesian coordinates (plane flow)

Cylindrical coordinates (axi-symmetric flow with x along the axis of symmetry)

(u, v) are velocity components in the directions (x, y)

a = speed of sound, $j = 0$ plane
$\quad\quad\quad\quad\quad\quad = 1$ axial symmetry.

We wish to write (1.4.1) and (1.4.2) in characteristic form. This means that we seek a new set of coordinates $\xi = \xi(x, y)$, $\eta = \eta(x, y)$ with the following property: When (1.4.1) and (1.4.2) are referred to ξ and η instead of x and y as independent variables, the first equation contains only ξ derivatives and the second equation only η derivatives. Each equation therefore only contains derivatives along the coordinate direction in question (inner derivatives) and derivatives in directions oblique or normal to the coordinate (outer derivatives) are absent.

To determine (ξ, η) we investigate the following problem. Given a curve $x = x(s)$, $y = y(s)$ and values of u, v along the curve $u = u(s)$, $v = v(s)$, under what conditions will (1.4.1) and (1.4.2) determine outer derivatives of u and v?

Without loss of generality we suppose that the given curve is nowhere parallel to the y axis. Then it is sufficient to investigate the conditions under which (1.4.1) and (1.4.2) determine u_y, v_y on the line. We denote the slope of the line by $m = (y_s/x_s)$.

The inner derivatives of u, v on the line are given by

$$u_s = u_x x_s + u_y y_s \tag{1.4.3}$$

$$v_s = v_x x_s + v_y y_s \tag{1.4.4}$$

where $u_s = (du/ds)$, etc., $u_x = (\partial u/\partial x)$, etc.

We can solve (1.4.3) and (1.4.4) for u_x, v_x on the line, in terms of inner derivatives and u_y, v_y. We have

$$u_x = \frac{u_s}{x_s} - mu_y \tag{1.4.5}$$

$$v_x = \frac{v_s}{x_s} - m v_y \tag{1.4.6}$$

Substitute for u_x, v_x in (1.4.1) and (1.4.2) and rearrange as simultaneous equations for u_y, v_y. Then we obtain,

$$u_y + m v_y = v_x / x_s \tag{1.4.7}$$

$$-\{m(a^2 - u^2) + 2uv\} u_y + (a^2 - v^2) v_y = -(a^2 - u^2) \frac{u_s}{x_s} - j \frac{a^2 v}{y} \tag{1.4.8}$$

Eqs. (1.4.7) and (1.4.8) are simultaneous algebraic equations to determine u_y, v_y. The matrix of the pair is

$$\begin{pmatrix} 1 & m & v_s/x_s \\ -\{m(a^2-u^2)+2uv\} & (a^2-v^2) & -(a^2-u^2)u_s/x_s - ja^2 v/y \end{pmatrix}$$

Denote the leading determinant of the matrix by Δ, and the determinant formed from the first and last columns by Δ_1.
    Then

$$\Delta = \begin{vmatrix} 1 & m \\ -\{m(a^2-u^2)+2uv\} & (a^2-v^2) \end{vmatrix}$$

$$\Delta_1 = \begin{vmatrix} 1 & v_s/x_s \\ -\{m(a^2-u^2)+2uv\} & -(a^2-u^2)u_s/x_s - ja^2 v/y \end{vmatrix}$$

In connection with solving (1.4.7) and (1.4.8) for u_y, v_y the following three possibilities arise,
    (i) $\Delta \neq 0$.
In this case u_y, v_y are determined uniquely by the data on the line and by the equations of motion.
    (ii) $\Delta = 0$, $\Delta_1 \neq 0$.
In this case (1.4.7) and (1.4.8) give no solutions for u_y, v_y.
    (iii) $\Delta = 0$, $\Delta_1 = 0$.
In this case solutions for u_y, v_y are finite but are not unique. In fact there is a single infinity of pairs of (u_y, v_y) satisfying the linear relation (1.4.7) (or (1.4.8) which is now the same equation).
    Conditions (iii) correspond to the case of interest. The condition $\Delta = 0$ determines the directions of the given line for which u_y, v_y are not uniquely

determined. The condition $\varDelta_1 = 0$ is a relation which must be satisfied by the inner derivatives along such lines.

The directions defined by $\varDelta = 0$ are called *characteristic directions*. The integral curves of the equation $\varDelta = 0$ define the new coordinates $\xi = \xi(x, y)$. $\eta = \eta(x, y)$ called characteristic coordinates. The conditions $\varDelta_1 = 0$ satisfied along characteristic lines are called *compatibility conditions* or the *second characteristic equations*.

Expanding $\varDelta = 0$ we obtain the following quadratic for m

$$(a^2 - u^2)m^2 + 2uvm + (a^2 - v^2) = 0 \tag{1.4.9}$$

This has roots

$$m = m_1, m_2 = \frac{-uv \pm a(u^2 + v^2 - a^2)^{1/2}}{(a^2 - u^2)} \tag{1.4.10}$$

Eq. (1.4.10) defines two characteristic directions at each point which are real and distinct if $q > a$, coincident if $q = a$, and complex conjugate if $q < a$. Put $u = q \cos\theta$, $v = q \sin\theta$, $a = q \sin\mu$ (definition). (Polar coordinates in velocity plane). Then

$$m = \tan(\theta \mp \mu) \tag{1.4.11}$$

Eq. (1.4.11) shows that the two characteristic directions are equally inclined to the stream line direction at the Mach angle μ. They are called *Mach line* directions.

We call $m_1 = \tan(\theta - \mu)$ the $+$ Mach line
$\quad\quad\quad m_2 = \tan(\theta + \mu)$ the $-$ Mach line

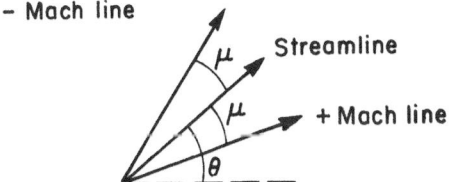

Fig. 1.2 Characteristic directions in steady flow

The velocity component normal to a Mach line (normal direction is obtained by anticlockwise rotation through $\pi/2$ from Mach line direction) is

$+a$ on the $+$ Mach line

$-a$ on the $-$ Mach line

The characteristic curves, lines or coordinates (ξ, η) are the integral curves of the equations

$$\frac{dy}{dx} = \tan(\theta \mp \mu) \tag{1.4.12}$$

The conditions $\Delta_1 = 0$ give us the equations of motion (equivalent to (1.4.1) and (1.4.2)) referred to characteristic coordinates (ξ, η).

Expanding $\Delta_1 = 0$ we have

$$(a^2 - u^2)u_s/x_s - \{(a^2 - u^2)m + 2uv\}v_s/x_s + ja^2 v/y = 0 \tag{1.4.13}$$

Remembering that $u_s = du/ds$, the inner derivative of u, we can write (1.4.13) in differential form (along a characteristic curve)

$$(a^2 - u^2)du - \{(a^2 - u^2)m + 2uv\}dv + ja^2 v\,dx/y = 0$$

Using (1.4.10) for m this becomes

$$(a^2 - u^2)du - \{uv \pm a(u^2 + v^2 - a^2)^{1/2}\}dv + ja^2 v\,dx/y = 0 \tag{1.4.14}$$

Now express (1.4.14) in terms of q, θ and μ.

Then, after some reduction, we obtain

$$\cot \mu \frac{dq}{q} \pm d\theta - j\frac{\sin \mu \sin \theta}{\sin(\theta \mp \mu)} \frac{dy}{y} = 0 \tag{1.4.15}$$

Eq. (1.4.15) are the second characteristic equations satisfied along the $+$ and $-$ Mach lines, respectively.

If we identify the ξ coordinate with the $+$ Mach lines $(\xi = \xi(x, y)$ defined by $(dy/dx) = \tan(\theta - \mu))$ and the η coordinate with $-$ Mach lines, then (1.4.15) written as differential equations are

$$\frac{\cot \mu}{q} \frac{\partial q}{\partial \xi} + \frac{\partial \theta}{\partial \xi} - \frac{j \sin \mu \sin \theta}{y \sin(\theta - \mu)} \frac{\partial y}{\partial \xi} = 0 \tag{1.4.16}$$

$$\frac{\cot \mu}{q} \frac{\partial q}{\partial \eta} - \frac{\partial \theta}{\partial \eta} - \frac{j \sin \mu \sin \theta}{y \sin(\theta + \mu)} \frac{\partial y}{\partial \eta} = 0 \tag{1.4.17}$$

Eq. (1.4.12) define the characteristic coordinates (ξ, η). When we transform the coordinates (x, y) to (ξ, η) (1.4.1) and (1.4.2) are transformed into (1.4.16) and (1.4.17). These, then, are the equations of motion referred to characteristic coordinates. They have the following simple properties

(i) Each equation only contains derivatives in one coordinate direction (inner derivatives).

(ii) Since outer derivatives are absent each equation is valid if derivatives normal to the coordinate direction are discontinuous. Thus in (1.4.16) discontinuities in derivatives normal to the ξ direction are permissible.

Because of (ii), a characteristic curve can be the boundary between a region of uniform flow and a non-uniform region.

We have shown that, in supersonic flow, the equations of motion of steady plane or axi-symmetric flow can be written

Along a +Mach line

$$\frac{dy}{dx} = \tan(\theta - \mu) \tag{1.4.18}$$

$$d\theta + \cot\mu \frac{dq}{q} - j\frac{\sin\mu\sin\theta}{\sin(\theta-\mu)}\frac{dy}{y} = 0 \tag{1.4.19}$$

Along a −Mach line

$$\frac{dy}{dx} = \tan(\theta + \mu) \tag{1.4.20}$$

$$d\theta - \cot\mu \frac{dq}{q} + j\frac{\sin\mu\sin\theta}{\sin(\theta+\mu)}\frac{dy}{y} = 0 \tag{1.4.21}$$

Eqs. (1.4.18)–(1.4.21) are simple to solve by a finite difference process called the *Method of Characteristics*. We illustrate the method for initial data given on a non-characteristic line,

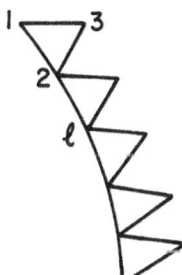

Fig. 1.3   Characteristic network

Let l be such an initial line with u, v prescribed along it. Divide l into a number of small segments bounded by nodes $1, 2, \ldots$.

Draw the + Mach line through 1 (approximated by its tangent at 1) and the − Mach line through 2 to intersect at point 3.

Now write (1.4.18) and (1.4.19) as difference equations along 1,3
(1.4.20) and (1.4.21) as difference equations along 2,3.

Using simple difference equations we first have

$$y_3 - y_1 = \tan(\theta_1 - \mu_1)\{x_3 - x_1\} \tag{1.4.22}$$

$$y_3 - y_2 = \tan(\theta_2 + \mu_2)\{x_3 - x_2\} \tag{1.4.23}$$

Solve (1.4.22) and (1.4.23) for (x_3, y_3).
From (1.4.19) and (1.4.21) we have

$$\theta_3 - \theta_1 + \frac{\cot\mu_1}{q_1}(q_3 - q_1) - j\frac{\sin\mu_1\sin\theta_1}{\sin(\theta_1 - \mu_1)}\frac{y_3 - y_1}{y_1} = 0 \tag{1.4.24}$$

$$\theta_3 - \theta_2 - \frac{\cot\mu_2}{q_2}(q_3 - q_2) + j\frac{\sin\mu_2\sin\theta_2}{\sin(\theta_2 + \mu_2)}\frac{y_3 - y_2}{y_2} = 0 \tag{1.4.25}$$

Now solve (1.4.24) and (1.4.25) for θ_3, q_3. Then the position of point 3 and flow conditions there are determined. We can thus construct data on a new initial data line joining intersections of + and − Mach lines through nodal points on l. We then perform the same calculations on the new line and hence determine the solution of (1.4.1) and (1.4.2) inside the characteristic triangle based on l and bounded by the + Mach line through the upper point of l and the − Mach line through the lower point.

This scheme has first order accuracy. It can be improved to give second order accuracy if, in (1.4.22)–(1.4.25) simple forward differences are replaced by mean differences.

Simplification in Plane Flow: In plane flow $j=0$ and (1.4.19) and (1.4.20) simplify. For a perfect gas they can be integrated.
From Bernoulli's equation

$$q/q_{max} = \{1 + \beta\sin^2\mu\}^{-1/2} \tag{1.4.26}$$

where $\beta = 2/(\gamma - 1)$.
Hence (1.4.19) and (1.4.21) integrate to

$$\theta \pm P(\mu) = \text{constant} \tag{1.4.27}$$

where

$$P(\mu) = -\int^\mu \frac{\beta\cos^2\mu}{1 + \beta\sin^2\mu} d\mu$$

$P(\mu)$ is identical with the Prandtl-Meyer function.

The combinations $\theta + P(\mu)$, $\theta - P(\mu)$ are analogous to Riemann invariants in one dimensional unsteady flow.

If we express $P(\mu)$ in terms of q, using (1.4.26), (1.4.27) define two families of curves in the (q, θ) or hodograph plane. They can be shown to be epicycloids with cusps on the sonic circle and touching the maximum velocity circle.

Extension to Non-Isentropic Flow: In this case the second characteristic equations are modified by the addition of an entropy term $\sin\mu\cos\mu \, T \, dS/a^2$ and a third characteristic relation is satisfied along streamlines, namely, $dS = 0$. This adds a step to the finite difference procedure just described.

1.5 Basic Concepts Used in Finite Difference Methods

We conclude this chapter with an outline of the main ideas, definitions and theorems used in the application of finite difference methods to the solution of partial differential equations. Much of the material presented is derived from Keith Miller's Lecture Notes for the course on Numerical Analysis given at the University of California, Berkeley, Mathematics 228B. These in turn refer to Alexandre Chorin's notes for Mathematics 228 B. These notes are as yet unpublished and the reader is referred to RICHTMYER and MORTON (1967) for a fuller treatment of the concepts presented.

An essential requirement for a given finite difference scheme to be useful is that it be *stable*.

RICHTMYER and MORTON illustrate the importance of stability vividly by means of the following example:

Solve the heat conduction equation in one dimension

$$\frac{\partial u}{\partial t} = \sigma \frac{\partial^2 u}{\partial x^2} \tag{1.5.1}$$

(where σ is a constant, x is the space variable in $(0, \pi)$, t is the time, and u the temperature) under the initial conditions

$$u(x, 0) = \phi(x), \quad 0 \le x \le \pi, \tag{1.5.2}$$

and the boundary conditions

$$u(0, t) = 0, \quad u(\pi, t) = 0, \quad t > 0 \tag{1.5.3}$$

For simplicity, take ϕ as the roof-top function

$$\phi(x) = C(x) \qquad 0 \leq x \leq \frac{\pi}{2}$$

$$= C(\pi - x) \quad \frac{\pi}{2} \leq x \leq \pi$$

Then the exact solution of this problem is

$$u(x,t) = \sum_{-\infty}^{\infty} A_m e^{(imx - m\sigma^2 t)} \tag{1.5.4}$$

where

$$A_m = 0 \qquad\qquad m \text{ even}$$

$$= \frac{2iC}{\pi m^2}(-1)^{(m+1)/2} \quad m \text{ odd} \tag{1.5.5}$$

To solve the finite difference version of the problem we divide $(0, \pi)$ into J equal steps Δx and advance in equal time steps Δt. The value of u at a network point $(j\Delta x, n\Delta t)$ after n time steps is denoted by u_j^n.

Eq. (1.5.1) is approximated by a suitable finite difference equation. If we use a forward time difference on the left and a centered x difference (evaluated at $t = n\Delta t$) the finite difference approximation to (1.5.1) is

$$\frac{u_j^{n+1} - u_j^n}{\Delta t} = \sigma \frac{u_{j+1}^n - 2u_j^n + u_{j-1}^n}{(\Delta x)^2} \tag{1.5.6}$$

u_j^n satisfies the boundary conditions (1.5.3) and the initial condition (1.5.2) at all nodes $(j\Delta x, 0)$ $0 \leq j \leq J$. The solution u_j^n to this problem is

$$u_j^n = \sum_{-\infty}^{\infty} A_m e^{imj\Delta x} [\xi(m)]^n \tag{1.5.7}$$

where

$$A_m = \frac{1}{2\pi} \int_{-\pi}^{\pi} \phi(x) e^{-imx} dx$$

and

$$\xi(m) = 1 - 2\lambda(1 - \cos m\Delta x)$$

where

$$\lambda = \sigma \Delta t / (\Delta x)^2.$$

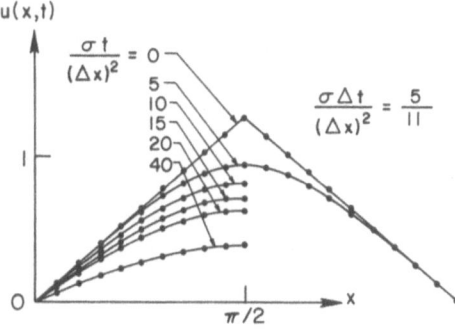

Fig. 1.4 Solution of the one dimensional heat flow problem

Now if $\lambda = 5/11$ the u, x distributions at successive times are as shown in Fig. 1.4 and it is seen that the analytic and finite difference solutions are always in close agreement. On the other hand, if $\lambda = 5/9$, the u, x curves given by the finite difference approximations, shown in Fig. 1.5, deviate more and

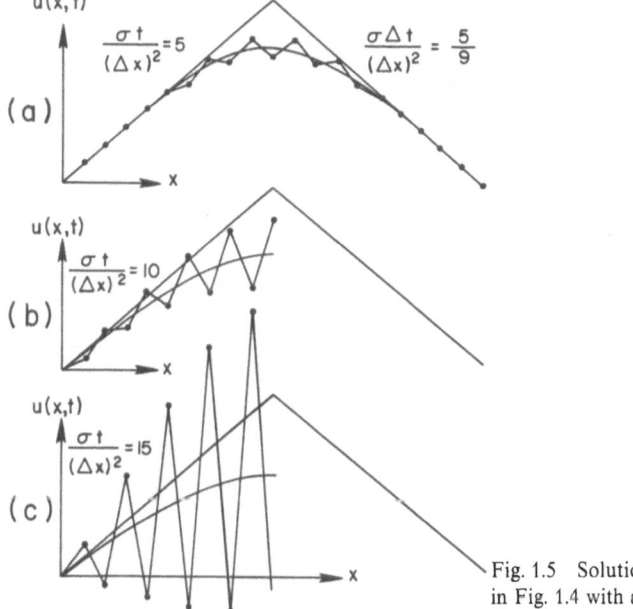

Fig. 1.5 Solution of the problem shown in Fig. 1.4 with a slightly larger time step

more from the exact solution as time increases. This steady worsening of the finite difference approximations is not improved by reducing the time step and is caused by *instability* of the finite difference scheme for this value of λ. For the smaller value of λ the scheme is stable.

Stability requires that u_j^n should remain bounded as $n \to \infty$, for all network points. In the present problem the scheme will be stable if all terms in (1.5.7) remain bounded as $n \to \infty$. Noting the expression for λ, this will be true if

$$\lambda = \sigma \frac{\Delta x}{(\Delta x)^2} < \frac{1}{2} \tag{1.5.8}$$

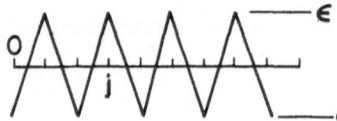

Fig. 1.6   Alternating initial perturbation

MILLER models the instability phenomenon by considering the special initial form of u_j^n with alternating peaks $+\varepsilon, -\varepsilon$, etc., illustrated in Fig. 1.6. Then (1.5.6) gives

$$u_j^1 = (1 - 2\lambda)u_j^0 + \lambda u_{j+1}^0 + \lambda u_{j-1}^0$$

Now $u_j^0 = \varepsilon$, $u_{j+1}^0 = u_{j-1}^0 = -\varepsilon = -u_j^0$, so that $u_j^1 = (1 - 4\lambda)u_j^0$, which is true for all j. Repeating this calculation at successive time steps we find, at the n-th time step,

$$u_j^n = (1 - 4\lambda)^n u_j^0 = (1 - 4\lambda)^n (\pm \varepsilon)$$

This shows that if $\lambda > 1/2$ the peaks are magnified without limit as $n \to \infty$. If $\lambda < 1/2$, on the other hand, the peaks disappear as n increases and the finite difference solution approaches the analytic solution. This confirms the stability criterion (1.58). If we solve the problem by a finite difference scheme which represents $\partial^2 u / \partial x^2$ by a combination of centered differences in x evaluated at $t = n\Delta t$ (old time) and $t = (n+1)\Delta t$ (new time), the stability conditions are less stringent.

A general scheme of this type has the form

$$\frac{u_j^{n+1} - u_j^n}{\Delta t} = \sigma \frac{\theta(\partial^2 u)_j^{n+1} + (1 - \theta)(\partial^2 u)_j^n}{(\Delta x)^2}$$

where

$$(\partial^2 u)_j^n = \{u_{j+1}^n - 2u_j^n + u_{j-1}^n\}$$

and $0 \le \theta \le 1$.

If $0 < \theta < 1/2$ scheme (1.59) is stable, provided that $2\lambda < 1/(1 - 2\theta)$.

If $1/2 < \theta < 1$ the scheme (1.59) is *always* stable.

If $\theta=0$ (scheme (1.5.6)) we call the finite difference scheme *explicit*. If $\theta \neq 0$ the scheme is said to be *implicit*. If $\theta=1$ the scheme is wholly implicit and is always stable.

Implicit schemes either have no restrictions on step size to ensure stability or have less severe restrictions than corresponding explicit schemes. On the other hand, the solution of an implicit finite difference set of equations requires the inversion of a matrix, while explicit equations immediately determine unknowns at all points at the new time.

We now give a more precise definition of stability, as well as of the other properties associated with solving initial value problems in partial differential equations by approximate finite difference methods. For this purpose we frequently work in a functional Banach space, representing the initial value problem.

Banach Space, Norm and Operators: To represent a linear initial value problem we use a function space, called a Banach space $\mathscr{B}$, which has the following properties

(i) The space is linear and elements satisfy the associative, distributive and commutative laws. Also pure multiples of elements belong to the space.

(ii) The space is normed. If u is an element of $\mathscr{B}$ then a non-negative number $\|u\|$ can be defined, called the norm, with the properties

$$\|au\| = |a|\,\|u\|$$

$$\|u+v\| \leq \|u\| + \|v\|$$

$$\|u-v\| = 0 \quad \text{iff} \quad u=v.$$

(iii) $\mathscr{B}$ is a complete space.

A simple example of $\mathscr{B}$ is the class of continuous functions $f(x)$ defined in $a \leq x \leq b$ with $\|f(x)\| = |f(x)|_{\max} \; a \leq x \leq b$.

Norms: In a finite dimensional space for vectors $u=(u_1, u_2, ..., u_m)$ two possible norms are

(a) The maximum norm $\|u\|_{\max} = |u_j|_{\max}$ over all j.
(b) The Euclidean or L_2 norm $\|u\|_L = (\sum u_i^2)^{1/2}$.

The Euclidean norm is the more frequently used of these two in numerical analysis. It can be shown that, in a finite dimensional space, any two norms are equivalent.

If A is a matrix operator, arising, for example, in the difference equation used to solve an initial value problem, the norm of A is written

$$\|A\| = \sup_{\|u\| \neq 0} \frac{\|Au\|}{\|u\|} = \sup_{\|u\|=1} \|Au\|$$

Then

$$\|AB\| \leq \|A\|\,\|B\|.$$

Difference Operators: If f is a grid function $\{f_j\}$ defined at equally spaced points
$\ldots -2h, -h, 0, h, 2h, \ldots$ (the unknowns in the finite difference equation connected with a given initial value problem) we define the following operators.
  Forward shift (or transfer) S_+

$$S_+ f = f_{j+1}$$

  Backward shift S_-

$$S_- f = f_{j-1}$$

  Identity I

$$I f = f_j$$

Clearly $S_+ S_- = I$
Furthermore, $S_+ S_+ = (S_+)^2$, $S_- = (S_+)^{-1}$
  Forward difference D_+

$$D_+ f = \frac{f_{j+1} - f_j}{h} = \frac{(S_+ - I)f}{h}$$

  Backward difference D_-

$$D_- f = \frac{f_j - f_{j-1}}{h}$$

  Central difference D_0

$$D_0 f = \frac{f_{j+1} - f_{j-1}}{2h} = \frac{(S_+ - S_-)f}{2h}$$

$$D_0 = \tfrac{1}{2}(D_+ + D_-)$$

Note that

$$D_+ D_- u = D_- D_+ u = \frac{u_{j-1} - 2u_j + u_{j+1}}{h^2} \quad \text{second centered difference}$$

Fourier Space: Investigations of stability and other properties of finite difference schemes are conveniently carried out in a space derived from the
physical space by applying the Fourier transform.

We recall that if ξ is an integer the set of functions $(1/\sqrt{2\pi})e^{i\xi x}$ is orthonormal in $(-\pi,\pi)$. We represent the range $(-\pi,\pi)$ by points on the circle C shown and ξ by integers on the line C'

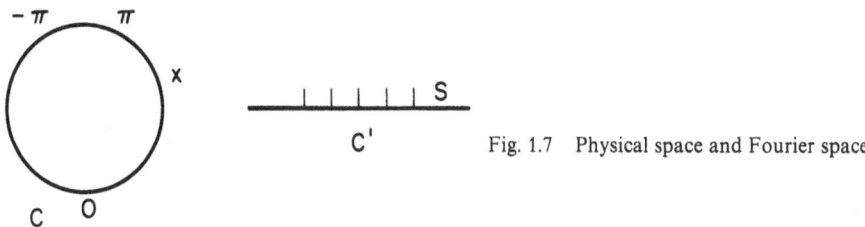

Fig. 1.7   Physical space and Fourier space

If $f(x)$ is a periodic function of bounded variation in C we can represent it by the complex Fourier series

$$f(x) = \frac{1}{\sqrt{2\pi}} \sum_{\xi \in C'} \hat{f}(\xi)e^{i\xi x} \qquad (1.5.9)$$

Then

$$\hat{f}(\xi) = \frac{1}{\sqrt{2\pi}} \int_{x \in C} f(x)e^{-i\xi x}dx \qquad (1.5.10)$$

is the Fourier transform of $f(x)$.
Parseval's relation gives

$$\int_{x \in C} |f(x)|^2 dx = \sum_{\xi \in C'} |\hat{f}(\xi)|^2 \qquad (1.5.11)$$

so that

$$\|f\|_{L^2(C)} = \|\hat{f}\|_{l^2(C')} \qquad (1.5.12)$$

Further, if g is a second function, of bounded variation, periodic in $(-\pi,\pi)$

$$(f,g)_{L^2(C)} = (\hat{f},\hat{g})_{l^2(C')} \qquad (1.5.13)$$

There is $(1,1)$ correspondence between f on C and $\hat{f}$ on C', defined by (1.5.10), with isometry between $L^2(C)$ and $l^2(C')$. The inversion of f to $\hat{f}$ is defined by (1.5.9) and Parseval's relation (1.5.11) still holds.
MILLER points out that the periodic exponentials $e^{i\xi x}$ are eigenfunctions for the shift operator S^+ so that all the linear finite difference operators can be

represented by powers and linear combinations derived from S^+. In fact, if $u(x)$ is the unknown in an IVP

$$\hat{S}_+ u(\xi) = e^{ih\xi} \hat{u}(\xi)$$

Thus the operation S_+, in the transform space is equivalent to multiplication by $e^{i\xi h}$. Similarly $(S^+)^2 \equiv$ multiplication by $e^{i2\xi h}$. Then

$$\hat{D}_+ u(\xi) = \left(\frac{\hat{S}_+ - \hat{I}}{h}\right) u(\xi) = \frac{e^{ih\xi} - 1}{h} \hat{u}(\xi)$$

As $h \to 0$ in the limit

$$\frac{\partial}{\partial x} u(\xi) = \left(\underset{h \to 0}{Lt} \frac{e^{ih\xi} - 1}{h}\right) \hat{u}(\xi) = i\xi \hat{u}(\xi)$$

Then we may call $e^{i\xi h}, (e^{ih\xi} - 1)/h$ and $i\xi$ the *symbols* of S_+, D_+, and $\partial/\partial x$, respectively.

Initial Value Problem in Banach Space: We consider the solution in a Banach space of the Cauchy problem

$$\frac{\partial u}{\partial t} = L(D) u(x, t)$$

with initial conditions

$$u(x, 0) = u_0(x) \tag{1.5.14}$$

where $u = \{u_1, ..., u_k\}$ is a vector representing the unknowns, $x = \{x_1, ..., x_m\}$ is the coordinate vector and $L(D)$ is a matrix linear differential operator $D = \{D_1, ..., D_m\}, D_i \equiv \partial/\partial x_i$.

The analytical solution of this problem can be represented by means of a linear transfer operator S which gives the solution at time t_2 in terms of the solution at an earlier time t_1

$$u(t_2) = S(t_2, t_1) u(t_1) \quad 0 \leq t_1 \leq t_2 \leq T \tag{1.5.15}$$

(This uses the Huygens-Hadamard principle.)

Now consider the solution of (1.5.14) by a two level finite difference scheme (a scheme connecting unknowns at two successive time intervals $t = nh$, $t = (n+1)h$). This can be written

$$\frac{u^{n+1} - u^n}{h} = \Lambda_1(S^+) u^{n+1} + \Lambda_0(S^+) u^n \tag{1.5.16}$$

where Λ_1, Λ_0 are series of positive and negative powers of the shift operator S^+ which approximate the differential operator L. [If the scheme is explicit $\Lambda_1 \equiv 0$.]

Then, solving (1.5.16) for u^{n+1},

$$u^{n+1} = A u^n \tag{1.5.17}$$

where

$$A = (I - h \Lambda_1)^{-1} (1 + h \Lambda_0)$$

defines the finite difference solution of problem (1.5.14).

We now give definitions of *consistency*, *stability*, and *convergence* connecting the solutions of the finite difference equations (1.5.17) with the IVP (1.5.14) in differential form.

Consistency

If u is a smooth solution of the analytical IVP (1.5.14) we say that scheme (1.5.17) is *consistent* with system (1.5.14) if

$$u^{n+1} = A u^n + k \tau^n \quad n = 0, 1, \ldots$$

where $\|\tau^n\| \leq \tau(h)$, k is bounded, and $\tau(h) \to 0$ as $h \to 0$ for all t_n in $0 \leq t_n \leq T$.

Stability

The scheme (1.5.17) is stable if the set of operators A^n are uniformly bounded for all t_n in $0 \leq t_n \leq T$.

Thus $\|A^n\| \leq M(T)$.

Convergence

The solution of the finite difference equation (1.5.17) converges to the solution of the differential IVP (1.5.14) provided that

$$\|u^n(x) - u(x, nh)\| = \|(A^n - S(nh, 0)) u_0\| \to 0$$

uniformly as $h \to 0$ in $0 \leq t \leq T$ for arbitrary $u_0 \in \mathcal{B}$.

Consistency, Stability and Convergence are connected through the following theorem due to Lax.

Lax Equivalence Theorem

If a finite difference approximation for a properly posed initial value is *consistent*, the necessary and sufficient condition for *convergence* is that the scheme is *stable*.

A general criterion for stability was established by VON NEUMANN, stated in the following theorem.

Theorem

The necessary and sufficient condition for the stability of a finite difference scheme approximating a single PDE is that the *symbol* $\varrho(\xi h)$ of the solution operator A satisfy

$$|\varrho(\xi h)| \leq 1 + Ck$$

for all $0 \leq \xi h \leq 2\pi$ and all $k \leq k_0$ (k_0 given) where C is a positive constant.

Errors

The difference between the approximate, finite difference solution of a IVP at a network point and the solution of the differential IVP at the same point is called the *truncation error*. Thus

$$\text{Truncation error} = \|u^n(x_j) - u(x_j, nh)\|.$$

The difference between values of the unknown as defined by a finite difference formula and its actual value as calculated by a computing machine is called the *round off error*.

We now examine these properties of IVP's for a number of examples, usually carrying out our analysis in Fourier space.

Example 1. Hyperbolic equations of the first order

Solve the equation representing wave propagation to the left

$$\frac{\partial u}{\partial t} = c \frac{\partial u}{\partial x} \tag{1.5.18}$$

with

$$u(0, x) = u^0(x).$$

This clearly has the analytical solution

$$u = u^0(x + ct) \tag{1.5.19}$$

To solve by finite differences consider a simple scheme which is forward in time and in distance

$$\frac{u_j^{n+1} - u_j^n}{k} = c \frac{u_{j+1}^n - u_j^n}{h}$$

so $u^{n+1} = (I + kcD_+)u^n$.

Hence, in Fourier space

$$\hat{u}^{n+1} = [(1-\lambda) + \lambda e^{i\xi h}] \hat{u}^n. \tag{1.5.20}$$

where

$$\lambda = \frac{ck}{h}.$$

For stability the spectral radius locus

$$\varrho = 1 - \lambda + \lambda e^{i\xi h}$$

must lie inside the unit circle as shown in the Fig. 1.8.

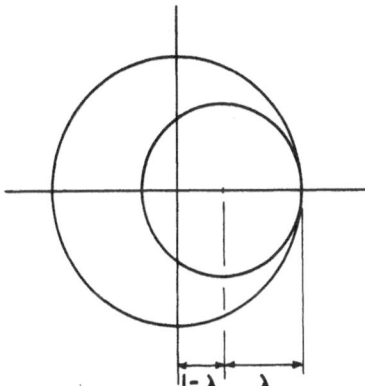

Fig. 1.8 Stability bound for first order hyperbolic equations

This is true if $\lambda \leq 1$.

On the other hand, if $\lambda > 1$ the locus will lie outside the unit circle, and in this case $|\varrho| > 1$.

Hence the stability criterion is

$$\lambda \leq 1.$$

This means that at a given network point the line drawn through diagonal network points back to the initial line parallel to $t/x = k/h$ has a slope exceeding that of the characteristic line through the point parallel to $x = -ct$. Then the finite difference calculation of the unknowns at the network point uses data on the initial line including the segment intersected by the characteristic through the point.

This conforms with the Courant-Friedrichs-Lewy condition for convergence of finite difference schemes applied to hyperbolic equations.

Example 2

Consider the same problem with forward differences in time and centered in space

$$u^{n+1} - u^n = \lambda D_0 u^n$$

$$u^{n+1} = (I + \lambda D_0) u^n.$$

In Fourier space

$$\hat{u}^{n+1} = \left\{ 1 + \lambda \frac{e^{i\xi h} - e^{-i\xi h}}{h} \right\} \hat{u}^n$$

$$\hat{u}^{n+1} = \{ 1 + 2i\lambda \sin \xi h \} \hat{u}^n. \tag{1.5.21}$$

The locus of the spectral radius is a vertical segment between $(1, 2i\lambda)$ and $(1, -2i\lambda)$ and lies outside the unit circle for all λ. Hence this scheme is unstable.

Example 3. Heat conduction equation. Explicit scheme.

The scheme here is defined by (1.5.6).
In Fourier space this becomes

$$\hat{u}^{n+1} = [\lambda e^{i\xi h} + (1 - 2\lambda) + \lambda e^{-i\xi h}] \hat{u}^n$$

so the spectral radius is

$$\varrho(\xi h) = 2\lambda \cos \xi h + (1 - 2\lambda)$$

This is in the range $1 - 4\lambda, 1$
so we require that $1 - 4\lambda \geq -1$
or $\lambda \leq \frac{1}{2}$.

Example 4. Heat conduction equation. Dufort-Frankel scheme.

This is defined by

$$\frac{u_j^{n+1} - u_j^{n-1}}{2k} = \frac{u_{j+1}^n - u_j^{n+1} - u_j^{n-1} + u_{j-1}^n}{h^2} \tag{1.5.22}$$

This is a three level scheme which we convert to a two level scheme by introducing

$$v_j^{n+1} = u_j^n \tag{1.5.23}$$

and replacing u_j^{n-1} by v_j^n in (1.5.22).

In Fourier space (1.5.22) and (1.5.23) become

$$(1+\alpha)\hat{u}^{n+1} = 2\alpha\cos\xi h\,\hat{u}^n + (1-\alpha)\hat{v}^n$$
$$\hat{v}^{n+1} = \hat{u}^n$$

(1.5.24)

where $\alpha = 2k/h^2$.

The amplification, or solution matrix of (1.5.24) is

$$\begin{pmatrix} \dfrac{2\alpha}{1+\alpha}\cos\xi h & \dfrac{1-\alpha}{1+\alpha} \\ 1 & 0 \end{pmatrix}$$

(1.5.25)

It is shown in RICHTMYER and MORTON (Sect. 4.11, Condition 2) that this matrix satisfies the stability criterion provided that its elements are bounded and all its eigenvalues, λ_i (with the possible exception of one), lie inside the unit circle.

The eigenvalues of (1.5.25) are

$$\lambda = \frac{\alpha\cos\xi h \pm \sqrt{1-\alpha^2\sin^2\xi h}}{1+\alpha} = \lambda_1, \lambda_2 \text{ say}$$

Then $|\lambda_1| \leq 1$ and $|\lambda_2| < 1$ so that the scheme is always stable.

RICHTMYER and MORTON show that the truncation error of this scheme is

$$\left(\frac{k}{h}\right)^2 \left(\frac{\partial^2 u}{\partial t^2}\right)^n_j + O(h^2) + O(k^2) + O(k^4/h^2)$$

Thus the Dufort-Frankel scheme is *consistent* with the heat conduction equation iff $k \to 0$ faster than $h \to 0$ (for example, if $k/h^2 = m$, a constant). If, on the other hand, $k/h = l$, a constant, then the scheme is consistent, not with the heat conduction equation, but with the hyperbolic equation

$$\frac{\partial u}{\partial t} - \frac{\partial^2 u}{\partial x^2} + l^2\frac{\partial^2 u}{\partial t^2} = 0.$$

References

Chorin, A.J.: Course Notes, Mathematics 228 B: Numerical Analysis. Berkeley: University of California.

Miller, K.: Course Notes, Mathematics 228 B: Numerical Analysis. Berkeley: University of California.

Richtmyer, R. D., Morton, K. W.: *Difference Methods for Initial Value Problems*, 2nd Ed. New York: Wiley 1967.

CHAPTER 2

The Godunov Schemes

2.1 The Origins of Godunov's First Scheme

Monotonic Difference Schemes: In one of his earliest papers concerned with numerical schemes for solving equations of Gas Dynamics GODUNOV (1959) seeks an alternative to the Method of Characteristics. He proposes three main requirements for such schemes. Firstly, they should retain the simplicity of Characteristics Methods while overcoming the inconveniences introduced by shearing and distortion of characteristics networks. Secondly, they should be able to include consideration of surfaces of discontinuity such as shock waves and fluid interfaces. Thirdly, when applied to linearized equations they should predict a solution for physical variables which is in qualitative agreement with analytical solutions.

To satisfy the third requirement GODUNOV seeks difference schemes for simple, model, wave like equations which are monotonic in character. This means that if such a scheme is applied to an initial value problem in which an unknown $u=u(x,t)$ is monotonic initially then the computed distributions of u at all later times are also monotonic.

Consider the one directional wave equation

$$\frac{\partial u}{\partial t} = \frac{\partial u}{\partial x} \tag{2.1.1}$$

with the initial conditions $u_0 = u(x,0) = f(x)$, where $f(x)$ is monotonic in x.

The analytic solution of (2.1.1) is that for a wave of constant shape, namely,

$$u = f(x+t) \tag{2.1.2}$$

Clearly this satisfies the monotonic property at all times.

We now solve (2.1.1) by a first order difference scheme and begin with the triangle scheme. In developing this we use the notation

$$u_0 = u(t_0, x_0), \quad u^0 = u(t_0 + \tau, x_0), \quad u_1 = u(t_0, x_0 + h), \quad u_{-1} = u(t_0, x_0 - h)$$

Then

$$\frac{\partial u}{\partial t} = \frac{u^0 - \frac{1}{2}(u_1 + u_{-1})}{\tau}, \quad \frac{\partial u}{\partial x} = \frac{u_1 - u_{-1}}{2h}$$

and (2.1.1) gives the difference equation

$$u^0 = \frac{1}{2}(u_1 + u_{-1}) + \frac{\tau}{2h}(u_1 - u_{-1})$$

or, if $r = \tau/h$

$$u^0 = \frac{1+r}{2}u_1 + \frac{1-r}{2}u_{-1} \tag{2.1.3}$$

Suppose that the initial function $f(x)$ is a step function defined by

$$u_k = 0, \qquad k \leq 0 \quad (u_k = u(0, kh))$$
$$u_k = 1, \qquad k \geq 1.$$

Then at $t = \tau$ we find

$$u^k = \frac{1+r}{2}u_{k+1} + \frac{1-r}{2}u_{k-1}$$

so

$$u^k = 0 \quad \text{if} \quad k \leq -1,$$
$$u^0 = \frac{1+r}{2},$$
$$u^1 = \frac{1+r}{2},$$
$$u^k = 1 \qquad k \geq 2.$$

Since $r \leq 1$ (to ensure stability of the triangle scheme) $1 + r/2 \leq 1$ and u remains monotonic at $t = \tau$. Similarly, u is monotonic at all later times. Since a general monotonic function can be represented by a distribution of step functions and since the difference equations are linear the monotonic property applies for all initially monotonic functions $f(x)$.

It can be shown that the most accurate monotonic first order scheme is identified by the formula

$$u^0 = ru_1 + (1-r)u_0 \tag{2.1.4}$$

with $r \leq 1$. It can also be shown that there are no second order schemes satisfying the monotonic property.

Following GODUNOV, we now extend these ideas to a pair of first order equations

$$\frac{\partial u}{\partial t} = A \frac{\partial v}{\partial x}, \qquad \frac{\partial v}{\partial t} = B \frac{\partial u}{\partial x} \tag{2.1.5}$$

where A and B are constants.

These can be written in the following characteristic form

$$\frac{\partial}{\partial t}\left(u + \sqrt{\frac{A}{B}}\,v\right) = \sqrt{AB}\,\frac{\partial}{\partial x}\left(u + \sqrt{\frac{A}{B}}\,v\right)$$
$$\frac{\partial}{\partial t}\left(u - \sqrt{\frac{A}{B}}\,v\right) = -\sqrt{AB}\,\frac{\partial}{\partial x}\left(u - \sqrt{\frac{A}{B}}\,v\right) \tag{2.1.6}$$

Eqs. (2.1.6) have wave solutions of constant shape

$$u + \sqrt{\frac{A}{B}}\,v = F_{+}(x + \sqrt{AB}\,t)$$
$$u - \sqrt{\frac{A}{B}}\,v = F_{-}(x - \sqrt{AB}\,t) \tag{2.1.7}$$

which clearly satisfy the monotonic property.

In solving (2.1.5) (or (2.1.6)) by finite differences a short calculation establishes that the first order scheme satisfying the monotonic property yields the formulae

$$u^0 = u_0 - \frac{\tau A}{2h}(v_1 - v_{-1}) + \frac{\tau\sqrt{AB}}{2h}(u_1 - 2u_0 + u_{-1})$$
$$v^0 = v_0 + \frac{\tau B}{2h}(u_1 - u_{-1}) + \frac{\tau\sqrt{AB}}{2h}(v_1 - 2v_0 + v_{-1}) \tag{2.1.8}$$

Linearized Lagrangian Equations: The results for (2.1.6) can be interpreted physically in connection with the one dimensional unsteady equations of motion, simplified by the use of a linear relation between pressure and specific volume.

If the pressure, velocity and specific volume are denoted by p, u and v the Lagrangian equations may be written

$$\frac{\partial u}{\partial t} + B \frac{\partial p(v)}{\partial x} = 0$$
$$\frac{\partial v}{\partial t} - B \frac{\partial u}{\partial x} = 0 \tag{2.1.9}$$

Introducing the linear relation (equivalent to the Karman-Tsien approximation)

$$p(v) = -\frac{A}{B}(v-v_0) + p_0 \qquad (2.1.10)$$

Eqs. (2.1.9) reduce to (2.1.5). Reintroducing p, the monotonic difference solutions of (2.1.9) may then be written

$$u^0 = u_0 - \frac{\tau B}{h}\left[\left(\frac{p_1+p_0}{2} - \sqrt{\frac{A}{B}}\frac{u_1-u_0}{2}\right) - \left(\frac{p_0+p_{-1}}{2} - \sqrt{\frac{A}{B}}\frac{u_0-u_{-1}}{2}\right)\right]$$

$$v^0 = v_0 + \frac{\tau B}{h}\left[\left(\frac{u_1+u_0}{2} - \frac{p_1-p_0}{2\sqrt{\frac{A}{B}}}\right) - \left(\frac{u_0+u_{-1}}{2} - \frac{p_0-p_{-1}}{2\sqrt{\frac{A}{B}}}\right)\right]$$

$$(2.1.11)$$

We now introduce the notation

$$P_{m+1/2} = \frac{p_{m+1}+p_m}{2} - \sqrt{\frac{A}{B}}\frac{u_{m+1}-u_m}{2}$$

$$U_{m+1/2} = \frac{u_{m+1}+u_m}{2} - \frac{p_{m+1}-p_m}{2\sqrt{\frac{A}{B}}}$$

$$(2.1.12)$$

Formulae (2.1.11) then take the simpler form

$$u^0 = u_0 - \frac{\tau B}{h}(P_{1/2} - P_{-1/2})$$

$$v^0 = v_0 + \frac{\tau B}{h}(U_{1/2} - U_{-1/2})$$

$$(2.1.13)$$

The solutions (2.1.13) can be interpreted in terms of an elementary breakdown problem for the Lagrangian equations.

Consider two adjacent cells in the x, t network bounded by $t=t_0, t=t_0+\tau$ and $x=-1/2h, x=3/2h$. Conditions in the left and right cells are distinguished by integers 0 and 1, respectively (see Fig. 2.1).

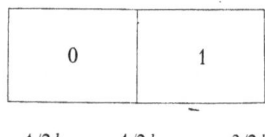

$-1/2h$ $1/2h$ $3/2h$ Fig. 2.1   Basic cells for Lagrangian equations

Up to time t_0 conditions in cells 0 and 1 are defined by p_0, u_0, v_0 and p_1, u_1, v_1. If cell boundaries are instantaneously broken at time t_0 then conditions for $t > t_0$ will be governed by the acoustic wave equations (2.1.5) (or (2.1.6)). From each cell boundary left and right acoustic waves will be propagated with speed $\sqrt{AB}$. These carry infinitesimal discontinuities in pressure, which may be either expansions or compressions. The magnitudes of the discontinuities are such that the pressure in between the two waves is constant. The jump relations across the wave fronts originating at $x = 1/2h$ at time $t = t_0$ are derived from the integral form of the first of (2.1.9) namely,

$$\oint [u\,dx - Bp(v)\,dt] = 0 \tag{2.1.14}$$

taken around a closed contour including the fronts and the boundaries $t = t_0$, $t = t_0 + \tau$. To the first order, (2.1.14) can be written

$$[u]\,dx - B[p]\,dt = 0 \tag{2.1.15}$$

where $[\ ]$ denotes change across a right or left moving front.

Now $dx/dt = \pm\sqrt{AB}$ on the right and left fronts, respectively, so if P and U denote the values of p and u on $x = 1/2h$ just after breakdown, (2.1.15) gives the following two jump relations

$$(U_{1/2} - u_1)\sqrt{\frac{A}{B}} - (P_{1/2} - p_1) = 0$$

$$(U_{1/2} - u_0)\sqrt{\frac{A}{B}} + (P_{1/2} - p_0) = 0$$

with the solutions

$$P_{1/2} = \frac{1}{2}(p_1 + p_0) - \frac{1}{2}\sqrt{\frac{A}{B}}(u_1 - u_0)$$

$$\tag{2.1.16}$$

$$U_{1/2} = \frac{1}{2}(u_1 + u_0) - \frac{1}{2}\sqrt{\frac{B}{A}}(p_1 - p_0)$$

The expressions (2.1.16) are identical with formulae (2.1.12) so that the values of P and U needed in (2.1.13) can be determined directly from acoustic breakdown relations.

It should be noted that (2.1.16) are valid only up to a time $(>t_0)$ at which wave fronts from neighboring boundaries $x = -1/2h$ or $x = 3/2h$ arrive at $x = 1/2h$. This means that we must take

$$\tau < h/\sqrt{AB},$$

a condition already satisfied to ensure stability of the monotonic scheme identified by (2.1.11).

We now extend these concepts to the one dimensional unsteady equations of motion in Eulerian form.

2.2 Godunov's First Scheme. One-Dimensional Eulerian Equations

The monotonic scheme developed by GODUNOV for the Lagrangian equations in one dimension is developed for the Eulerian equations in one and more dimensions by GODUNOV, ZABRODIN and PROKOPOV (1961):

Eulerian Equations in One Dimension: When the integral forms of the conditions of conservation of mass, momentum and energy are written out for one dimensional unsteady motion (see ROZHDESTVENSKII and YANENKO, 1968) we obtain, in the Eulerian formulation,

$$\oint \{ \varrho \, dx - \varrho u \, dt \} = 0$$

$$\oint \{ \varrho u \, dx - (p + \varrho u^2) dt \} = 0 \tag{2.2.1}$$

$$\oint \left\{ \varrho \left(e + \frac{1}{2} u^2 \right) dx - \varrho u \left(e + \frac{p}{\varrho} + \frac{1}{2} u^2 \right) dt \right\} = 0$$

The integrals are taken around a closed contour in the x, t plane. The contour may include surfaces of discontinuity such as shocks and interfaces.

We consider a column of gas moving parallel to the x axis at time $t = t_0$ and divide this into equally spaced segments of length h. We denote segment boundaries by $x_m, x_{m+1}, \ldots$ and identify the segment between x_m and x_{m+1} by suffix $m + 1/2$. We wish to evaluate values of unknowns in each segment at a later time $t_0 + \tau$ and identify these by an index $m + 1/2$.

We suppose that the distributions of unknowns with x are given at time $t = t_0$ and approximate these by staircase functions such that $\varrho_{m+1/2}, u_{m+1/2}, p_{m+1/2}, e_{m+1/2}$ are constant in each cell.

To calculate $\varrho^{m+1/2}$, etc., we allow the segment boundaries $x = x_m, x_{m+1} \ldots$ to break at time $t = t_0$ and apply relations (2.2.1) around each basic cell bounded by $x = x_m$, $x = x_{m+1}$, $t = t_0$, $t = t_0 + \tau$, assuming constant values on each boundary. As in Sect. 2.1, values on cell boundaries after breakdown are denoted by capitals. The cell formulae of Sect. 1 are unaffected by the nonlinearity of the Eulerian equations and yield the following relations

$$\varrho^{m+1/2} = \varrho_{m+1/2} - \frac{\tau}{h}\left\{(RU)_{m+1} - (RU)_m\right\}$$

$$(\varrho u)^{m+1/2} = (\varrho u)_{m+1/2} - \frac{\tau}{h}\left\{(P+RU^2)_{m+1} - (P+RU^2)_m\right\} \qquad (2.2.2)$$

$$\left[\varrho\left(e+\frac{u^2}{2}\right)\right]^{m+1/2} = \left[\varrho\left(e+\frac{u^2}{2}\right)\right]_{m+1/2} - \frac{\tau}{h}\left\{\left[RU\left(E+\frac{P}{R}+\frac{U^2}{2}\right)\right]_{m+1}\right.$$
$$\left. - \left[RU\left(E+\frac{P}{R}+\frac{U^2}{2}\right)\right]_m\right\}$$

These are supplemented at all points by the equation of state

$$p = p(\varrho, e) \qquad (2.2.3)$$

The quantities P, R, U, E are determined from breakdown formulae, which in turn are solutions of basic interaction problems at cell boundaries. Although these formulae are complicated it should be noted that P, R, U, E are constant on each boundary up to $t = t_0 + \tau$ provided that no interactions of waves from adjoining cell boundaries take place. For sufficiently small τ shocks and contact discontinuities originating at $x = x_m$ have constant speeds and expansion waves are governed by similarity solutions of the type $U = U\{(x-x_m)/(t-t_m)\}$ (see (1.3.12), (1.3.13), Introduction) which give constant U on $x = x_m$ for all $t < t_0 + \tau$.

Breakdown Formulae: These are given in detail in GODUNOV et al. (1961) and we will simply explain their derivation briefly. A change from the linearized Lagrangian equations of Sect. 2.1 must be noted. In the former case the velocity and pressure on $x = x_m$ after breakdown always agree with the corresponding values on the contact discontinuities. This may not be true for the full Eulerian equations since breakdown disturbances move with the acoustic speed relative to the local fluid, rather than with absolute acoustic speed. We therefore distinguish between values on $x = x_m$ (denoted by capitals) and values on the contact discontinuity (denoted by suffix c.d.).

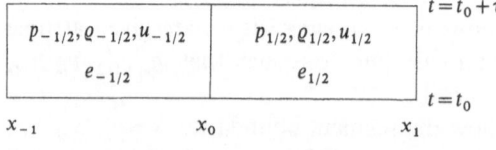

$t = t_0 + \tau$

| $p_{-1/2}, \varrho_{-1/2}, u_{-1/2}$ | $p_{1/2}, \varrho_{1/2}, u_{1/2}$ |
| $e_{-1/2}$ | $e_{1/2}$ |

$t = t_0$

x_{-1} x_0 x_1

Fig. 2.2 Cell breakdown for Eulerian equations

Consider two adjacent cells bounded by x_{-1}, x_0, x_1 and $t = t_0$, $t = t_0 + \tau$. Without loss of generality we consider a perfect gas for which

$$p = (\gamma - 1)\varrho e \qquad (2.2.3\,a)$$

where γ is the constant specific heat ratio. At time $t=t_0$ conditions in the left cell are denoted by suffix $-1/2$ and in the right cell by suffix $1/2$. The nature of the breakdown across $x=x_0$ at $t=t_0$ depends on the relative values of these quantities. After breakdown one of the boundaries is always a contact discontinuity and the waves originating at $x=x_0$ are either both expansion waves, both shock waves, or one of each, depending on whether $p_{c.d.}$ is smaller than both $p_{1/2}$ and $p_{-1/2}$, greater than both values or has an intermediate value.

Let us consider an intermediate case in which $p_{-1/2}>p_{c.d.}>p_{1/2}$. Then to the right we have three shock relations which, after elimination of the shock speed, give two equations for $p_{c.d.}$, $u_{c.d.}$ and v_R (v_R is the specific volume to the right of the contact discontinuity). To the left the simple wave relation

$$u_{-1/2} - \frac{2}{\gamma-1}(\gamma p_{-1/2} v_{-1/2})^{1/2} = u_{c.d.} - \frac{2}{\gamma-1}(\gamma p_{c.d.} v_L)^{1/2}$$

together with the isentropic conditions

$$p_{-1/2} v_{-1/2}^{\gamma} = p_{c.d.} v_L^{\gamma}$$

are satisfied. We thus have four relations to determine $p_{c.d.}$, $u_{c.d.}$, v_R, v_L.

GODUNOV derives the following compact expressions for $p_{c.d.}$ and $u_{c.d.}$, after breakdown at a general cell boundary $x=x_m$. These cover all possibilities simultaneously

$$p_{c.d.} = \frac{b_m p_{m-1/2} + a_m p_{m+1/2} + a_m b_m (u_{m-1/2} - u_{m+1/2})}{a_m + b_m}$$

$$u_{c.d.} = \frac{a_m u_{m-1/2} + b_m u_{m+1/2} + p_{m-1/2} - p_{m+1/2}}{a_m + b_m}$$

(2.2.4)

where

$$a_m = \left\{ \tfrac{1}{2}\left[(\gamma+1)p_{c.d.} + (\gamma-1)p_{m-1/2}\right] \varrho_{m-1/2} \right\}^{1/2} \quad \text{if} \quad p_{c.d.} \geq p_{m-1/2}$$

$$a_m - \frac{\gamma-1}{2\gamma} \{\gamma p_{m-1/2} \varrho_{m-1/2}\}^{1/2} \frac{1-p_{c.d.}/p_{m-1/2}}{1-(p_{c.d.}/p_{m-1/2})^{(\gamma-1)/2\gamma}} \quad \text{if} \quad p_{c.d.} < p_{m-1/2}$$

(2.2.5)

The expressions for b_m (2.2.6) are obtained by replacing the suffixes $m-1/2$ by $m+1/2$ in (2.2.5).

Eqs. (2.2.4) and (2.2.5) are exact and, if $x=x_m$ is a finite discontinuity, must be solved as they stand, using the following iterative process. First estimate the value of $p_{c.d.}$. Then solve the appropriate relations (2.2.5) and (2.2.6) for a_m and b_m and substitute in (2.2.4) to recalculate $p_{c.d.}$. If the latter value disagrees with the estimate repeat the process with the most recently found $p_{c.d.}$ value and iterate until a converged value of $p_{c.d.}$ is calculated.

This full iteration scheme need only be applied when a cell boundary coincides with a shock wave. In general, the changes in values of dependent variables across a cell boundary are small (of order of the cell length) and (2.2.4)–(2.2.6) can be simplified by the acoustic approximation. We then get the following explicit formulae

$$a_m = b_m = \left\{\frac{\gamma}{4}(p_{m-1/2} + p_{m+1/2})(\varrho_{m-1/2} + \varrho_{m+1/2})\right\}^{1/2}$$

$$p_{\text{c.d.}} = \frac{p_{m+1/2} + p_{m-1/2}}{2} + a_m \frac{u_{m-1/2} - u_{m+1/2}}{2} \qquad (2.2.7)$$

$$u_{\text{c.d.}} = \frac{u_{m+1/2} + u_{m-1/2}}{2} + \frac{p_{m-1/2} - p_{m+1/2}}{2 a_m}$$

To determine U, P, E and R, the motion of the three key boundaries following breakdown must be examined. Left and right acoustic waves move with speeds

$$D_{\text{L}} = u_{m-1/2} - \frac{a_m}{\varrho_{m-1/2}}, \qquad D_{\text{R}} = u_{m+1/2} + \frac{a_m}{\varrho_{m+1/2}}$$

respectively, while the contact discontinuity has velocity $u_{\text{c.d.}}$. For times $t > t_0$ these boundaries divide the region near $x = x_m$ into four zones, shown in Fig. 2.3.

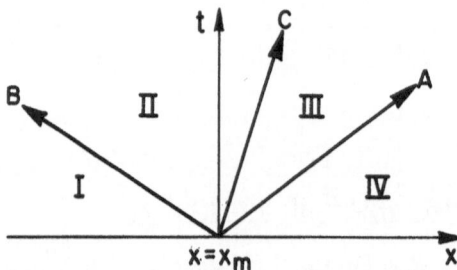

Fig. 2.3 Key boundaries in breakdown process

Values of the physical variables in Zones I and IV are the same as before breakdown and are therefore defined by suffices $m-1/2$ and $m+1/2$, respectively. In Zones II and III the values of $u_{\text{c.d.}}$ and $p_{\text{c.d.}}$ are found from (2.5.7). The density is determined from

$$\frac{\varrho_{\text{L}}}{\varrho_{m-1/2}} = \frac{(\gamma+1)p_{\text{c.d.}} + (\gamma-1)p_{m-1/2}}{(\gamma-1)p_{\text{c.d.}} + (\gamma+1)p_{m-1/2}} \qquad (2.2.8)$$

and a corresponding equation for ϱ_{R} with $m-1/2$ replaced by $m+1/2$.

Four cases need to be considered. If D_L and D_R are both positive all the boundaries $0B$, $0A$ and $0C$ are to the right of $x = x_m$ and the post breakdown values of U, P, R are the same as those before breakdown (Zone I values). If D_L and D_R are both negative U, P, R take Zone IV values. If D_L is negative and D_R positive we use (2.2.7) to calculate P and U; in addition, we take $R = \varrho_L$ (2.2.8) if $u_{c.d.}$ is positive and $R = \varrho_R$ if $u_{c.d.}$ is negative.

Application to One-Dimensional Shock Problems: To clarify Godunov's first scheme we apply it to the problem of a piston moving with given constant speed into a uniform gas at rest (assumed to obey a perfect gas law). The analytical solution to this problem is well known and reduces to an expression for the shock speed in terms of the piston speed.

To apply the scheme in its original form we estimate the length of gas column which will be disturbed by the piston in a standard time interval (say 1 sec.) and divide this into n equal intervals of length h. We next choose a time step τ to satisfy the stability criterion $\tau < h/\text{max}$ propagation speed. The space grid is shown in Fig. 2.4. Cell boundaries are denoted by suffixes m

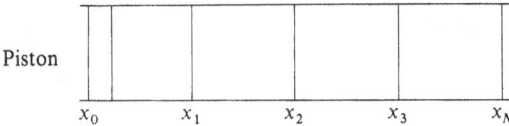

Piston

$x_0 \qquad x_1 \qquad x_2 \qquad x_3 \qquad x_N$ Fig. 2.4   Cells in piston problem

and cells are distinguished by 1/2 integers $m + 1/2$, suffix 0 corresponding to the piston. Denote the undisturbed values of the physical variables in the gas column by $p_\infty, \varrho_\infty, e_\infty$. Then, at time $t = 0$,

$$p = p_\infty, \qquad \varrho = \varrho_\infty, \qquad e = e_\infty, \qquad u = 0 \text{ in all cells except } \tfrac{1}{2}.$$

In cell 1/2, $p = p_\infty$, $\varrho = \varrho_\infty$, $e = e_\infty$, $u = \tfrac{1}{2}V$, where $V = $ pistin speed.

To calculate values at $t = \tau$ we apply breakdown formulae (2.2.4) and (2.2.5) at the boundary $x = x_1$ only. Then

$$u^{1/2} = \tfrac{1}{2}(V + U_1), \qquad p^{1/2} = \tfrac{1}{2}(p_\infty + P_1), \dots$$
$$u^{3/2} = \tfrac{1}{2}(U_1), \qquad p^{3/2} = \tfrac{1}{2}(p_\infty + P_1), \dots$$

so that, at time $t = \tau$, two cells are disturbed. At time $t = 2\tau$ breakdown formulae must be applied on $x = x_1$ and $x = x_2$. The process is continued for successively higher time steps until a steady state is reached, identified by uniform conditions in all cells disturbed and a constant propagation speed on the right hand boundary of the disturbance. This calculation is carried out in GODUNOV (1959).

An alternative way to treat this problem is to use, instead of a fixed network, a moving network which expands with the length of gas column disturbed. In this the far left boundary coincides with the piston and the far right boundary with the shock wave. The region between the two boundaries is divided into a suitable fixed number of equal intervals. The cells are all disturbed in this approach and the cell formulae (2.2.2) are modified to take account of the motion of the network (moving network cell formulae are discussed in more detail in Sect. 2.4). The calculation is started at a time $t=\tau$, rather than $t=0$, to allow a disturbed region to be formed. At each time step the non-linear breakdown formulae are applied on the far right boundary $x=x_N$ while the acoustic formulae are used on all other boundaries.

A sample calculation using the moving network was carried out by two Berkeley students, Michael GROSS and Robert HAGERTY, with the following conditions

Piston speed $= 0.428 \cdot 10^5$ cm/sec

$$\varrho_\infty = 1.225 \cdot 10^{-3} \text{ g/cm}^3, \qquad p_\infty = 1.013 \cdot 10^6 \text{ dyne/cm}^2.$$

100 space steps were used and the modified Godunov scheme was applied for 400 cycles. The evolution of the pressure distribution is shown in Fig. 2.5 and it is seen that the approach to uniform state is very satisfactory. The same scheme was also applied to the problem of normal shock reflection from a wall with equal success.

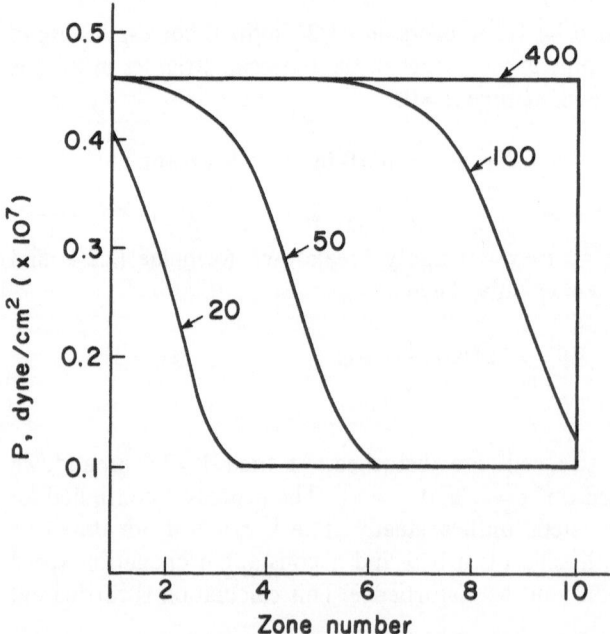

Fig. 2.5   Growth of shock in piston problem

2.3 Godunov's First Scheme in Two and More Dimensions

To generalize Godunov's first scheme to unsteady motion in two dimensions we start with the Eulerian equations in integral form, referred to Cartesians,

$$\oiint \{\varrho\,dx\,dy + \varrho u\,dy\,dt + \varrho v\,dx\,dt\} = 0$$

$$\oiint \{\varrho u\,dx\,dy + (p + \varrho u^2)\,dy\,dt + \varrho uv\,dx\,dt\} = 0$$

$$\oiint \{\varrho v\,dx\,dy + \varrho uv\,dy\,dt + (p + \varrho v^2)\,dx\,dt\} = 0 \tag{2.3.1}$$

$$\oiint \left\{ \varrho\left(e + \frac{u^2 + v^2}{2}\right)dx\,dy + \varrho u\left(e + \frac{p}{\varrho} + \frac{u^2 + v^2}{2}\right)dy\,dt \right.$$

$$\left. + \varrho v\left(e + \frac{p}{\varrho} + \frac{u^2 + v^2}{2}\right)dx\,dt \right\} = 0$$

(see ROZHDESTVENSKII and YANENKO, 1968).

These surface integrals are taken over a closed surface in the x, y, t plane. To obtain generalized cell formulae GODUNOV et al. (1961) divide each plane $t = $ constant (called the xy plane) into a rectangular grid with mesh lengths h_x and h_y, respectively. A typical x, y cell is shown in Fig. 2.6.

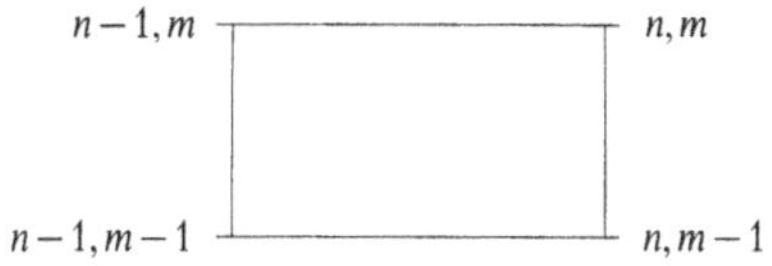

Fig. 2.6   Cell in two dimensional flow

Suffix n refers to the x mesh point, suffix m to the y mesh point. The cell is identified by the means of the bounding x and y suffixes, in the case shown by $n - 1/2$, $m - 1/2$.

We apply (2.3.1) to a box bounded by the planes $x = x_{n-1}$, $x = x_n$, $y = y_m$, $y = y_{m-1}$, $t = t_0$, $t = t_0 + \tau$. Values on the x, y faces are assumed constant and denoted by suffixes $n - 1/2$, $m - 1/2$ $(t = t_0)$ and indices $n - 1/2$, $m - 1/2$ $(t = t_0 + \tau)$. On the other faces one dimensional breakdown formulae are applied, on the assumption that waves propagated in the x and y directions are independent and do not interact. The breakdown value of P on the box face bounded by the plane $x = x_{n-1}$ is denoted by suffixes $n - 1$, $m - 1/2$.

In this way four cell formulae are obtained from (2.3.1); the first of these (mass conservation) gives

$$\varrho^{n-1/2,\,m-1/2} = \varrho_{n-1/2,\,m-1/2} - \frac{\tau}{h_x}\left[(RU)_{n,\,m-1/2} - (RU)_{n-1,\,m-1/2}\right]$$

$$- \frac{\tau}{h_y}\left[(RV)_{n-1/2,\,m} - (RV)_{n-1/2,\,m-1}\right] \tag{2.3.2}$$

This, with the remaining three cell relations and the equation of state determine p, ϱ, e, u, v, on the x, y face $n-1/2$, $m-1/2$ at the new time $t = t_0 + \tau$.

At general cell boundaries the breakdown values P, R, E, U, V are determined from the acoustic formulae described in Sect. 2.2. In evaluating $(U)_{n, m-1/2}$, for example, we keep the 1/2 integer y suffix constant and consider breakdown across the face $x = x_n$. We first evaluate $p_{\text{c.d.} m-1/2}$ and $u_{\text{c.d.} m-1/2}$ to identify which of the four breakdown cases applies. We then use the appropriate one dimensional formula for U as described in Sect. 2.2.

Applications: GODUNOV et al. (1970) reformulated their two dimensional scheme in spherical polars and then applied it to calculate uniform flow of a perfect gas (air) past a sphere at $M_\infty = 4.0$. They used a fixed length network based on equally spaced sections bounded by sphere radii. Their results compared well with those found for the same problem by BELOTSERKOVSKII, using the Method of Integral Relations. The shock shape and sonic line are shown in Fig. 2.7. In this application acoustic formulae can be used throughout in the θ direction. In the radial direction acoustic formulae are used at all interior points and the full shock formulae are only applied at the outermost cell boundaries.

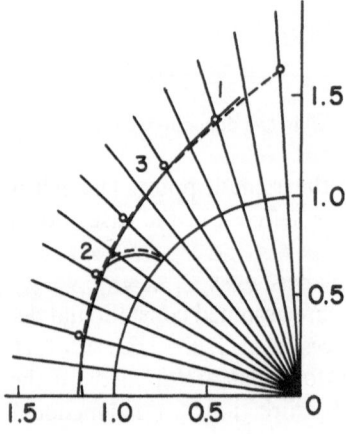

Fig. 2.7 Shock wave and sonic line in blunt body problem

The method has been extended to three dimensions and applied in modified form to calculate supersonic and hypersonic flow past a number of reentry body configurations ranging from cones at angles of attack to bell shaped bodies used for exploration of outer planets such as Mars and Jupiter. These are described in MASSON et al. (1969), MASSON and TAYLOR (1971), TAYLOR and MASSON (1970), and MORETTI (1967). A complete list of references is given in TAYLOR (1974).

2.4 Godunov's Second Scheme

The principal deficiency in the original Godunov scheme described in the preceding sections is that it is only of first order accuracy and therefore requires a large number of time steps in application to practical problems, particularly when an approach to steady state is required. In an attempt to overcome this objection GODUNOV and his collaborators (ALALYKIN et al., 1970) proposed a more elaborate predictor-corrector scheme of second order accuracy. In applying this to one dimensional Gas Dynamics the calculation of change in flow conditions at each time step is in two parts. In the first half time step, the predictor, determining conditions on an intermediate layer, a three point implicit formula is used in each of three equations of motion, written in characteristic form. This leads to a second order matrix difference equation which is solved by a double sweep method in between the two extreme boundaries of the motion. For the second half step, the corrector, two formulations are proposed. In the first the equations of motion are written in integral form and integrated around each basic network cell lying between the initial (known) layer and the final layer reached at the end of a whole time interval. On each side of the cell conditions are assumed to be constant. The integrations yield explicit formulae for the values of dependent variables at network points on the final layer. In the other formulation of the corrector step, the characteristic forms of the equations are again used to derive explicit difference formulae for the unknowns in the final layer.

The network is a movable grid formed by equally spaced constant time lines and curves tied to two extreme boundaries in the problem, for example, a piston curve on the left and a characteristic or shock line on the right. These curves are usually equally spaced. The number of nodal points on a constant time line can be increased or decreased if expansions or additional shocks should develop in the interior of the flow.

The second scheme has been applied to three problems, the growth of a shock formed by a piston accelerating to uniform speed from rest, wave formation of electrodes in plasma, and open channel flow with non-linear surface waves. These are all hyperbolic problems, although the governing equations change in detail from problem to problem. We shall describe the scheme in terms of the equations of one dimensional unsteady Gas Dynamics.

Basic Difference Scheme: The scheme is first applied to the model equation

$$\frac{\partial u}{\partial t} + A\frac{\partial u}{\partial r} = 0 \tag{2.4.1}$$

where A is a constant, t is the time, and r distance measured from a plane, line or point of symmetry. This represents acoustic propagation of u in one direction. The rectangular network shown in Fig. 2.8 is used in the finite difference scheme.

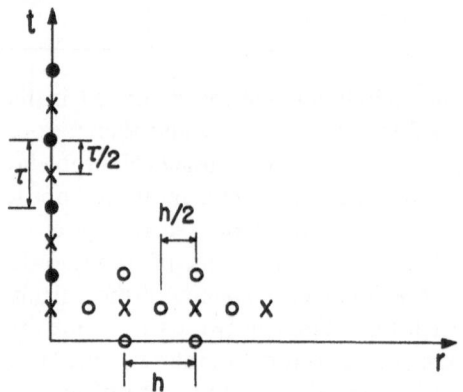

Fig. 2.8   Network in the second Godunov scheme

At network points we write $u(t,r)=u(m\tau,nh)=u_n^m$, where τ is the constant time step and h the constant spacing.

We suppose that $u(m\tau,nh)$ is known for a certain m at all n. We wish to calculate $u((m+1)\tau,nh)$. To do this we first find $u_n^{m+1/2}$ at $t=(m+1/2)\tau$ from the implicit formula

$$\frac{u_n^{m+1/2}-u_n^m}{\tau/2}+A\,\frac{u_{n+1}^{m+1/2}-u_{n-1}^{m+1/2}}{2h}=0 \tag{2.4.2}$$

Eq. (2.4.2) is a second order difference equation which is solved by a double sweep method (described below) between the left boundary $n=1$ and the right boundary $n=N$. (One boundary condition can be given at one end, say $n=1$. To ensure uniqueness (2.4.2) at $n=N$ is replaced by a simple difference relation between u_N and u_{N-1}.)

When $u_n^{m+1/2}$ has been calculated for all integral n, values at half integral points are determined from the interpolation formula

$$u_{n+1/2}^{m+1/2}=(1-\alpha)\,\frac{u_{n+1}^{m+1/2}+u_n^{m+1/2}}{2}+\alpha\,\frac{u_{n+2}^{m+1/2}+u_{n-1}^{m+1/2}}{2} \tag{2.4.3}$$

In general we take $\alpha=0.25$.

In the second half step we calculate u_n^{m+1} from the explicit formula

$$\frac{u_n^{m+1}-u_n^m}{\tau}+A\,\frac{u_{n+1/2}^{m+1/2}-u_{n-1/2}^{m+1/2}}{h}=0 \tag{2.4.4}$$

The combined scheme is accurate to order τ^2,h^2 and is stable, provided that $0<\alpha\le0.25$, for all τ/h.

We now describe the general scheme to be applied to the movable network shown in Fig. 2.9 used for solving the equations of unsteady flow in one space variable. The grid is formed of equally spaced lines $t=$ constant (correspond-

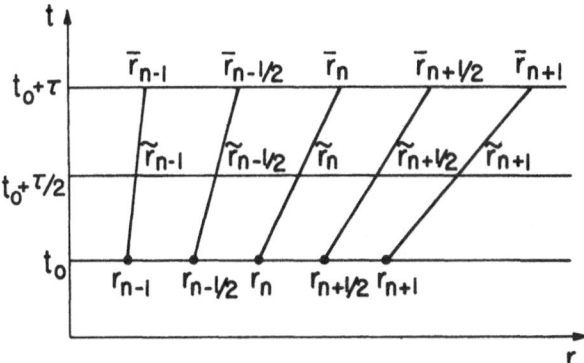

Fig. 2.9   Moving network in the second Godunov scheme

ing to whole time steps τ and half steps $\tau/2$) and equally spaced curves bounded by definite boundaries at the left and right ends of the region of calculation.

We suppose that conditions are known on a basic layer $t=t_0$ at network points $r=r_1,\ldots,r_n,\ldots,r_N$. As with the model equation we calculate conditions at the next time step $t=t_0+\tau$ in two steps. In the first half step we apply an implicit scheme to find conditions on the intermediate layer $t=t_0+\tau/2$, $r=\tilde{r}_1,\ldots,\tilde{r}_n,\ldots,\tilde{r}_N$. We then integrate the equations of motion in divergence form around a basic cell to give formulae for the unknowns on the final layer $t=t_0+\tau$, $r=\bar{r}_1,\ldots,\bar{r}_n,\ldots,\bar{r}_N$.

The dependent variables are taken in generalized form

$$\mu = \mu(u), \qquad \pi = \pi(p), \qquad \sigma = \sigma(S).$$

For the first half step the equations of motion are written in characteristic form

$$\left(\frac{\partial \mu}{\partial t} + \lambda \frac{\partial \pi}{\partial t}\right) + (u+c)\left(\frac{\partial \mu}{\partial r} + \lambda \frac{\partial \pi}{\partial r}\right) = -\mu'(u)uc\frac{Q'(r)}{Q(r)} \tag{2.4.5}$$

$$\left(\frac{\partial \mu}{\partial t} - \lambda \frac{\partial \pi}{\partial t}\right) + (u-c)\left(\frac{\partial \mu}{\partial r} - \lambda \frac{\partial \pi}{\partial r}\right) = \mu'(u)uc\frac{Q'(r)}{Q(r)} \tag{2.4.6}$$

$$\frac{\partial \sigma}{\partial t} + u(\mu)\frac{\partial \sigma}{\partial r} = 0 \tag{2.4.7}$$

where $\lambda = \mu'(u)/\{\varrho c \pi'(p)\}$, c is the speed of sound and $Q(r)$ is an area function

$$Q(r) = r^j$$

$\quad j = 0 \quad$ plane symmetry

$\quad j = 1 \quad$ cylindrical symmetry

$\quad j = 2 \quad$ spherical symmetry.

If, in (2.4.5), (2.4.6) and (2.4.7) we replace time derivatives along coordinate lines $r=$ constant by time derivatives along the moving network lines we find

$$\left(\frac{\partial \mu}{\partial t} + \lambda \frac{\partial \pi}{\partial t}\right) + (u+c-V)\left(\frac{\partial \mu}{\partial r} + \lambda \frac{\partial \pi}{\partial r}\right) = -\mu'(u)uc\frac{Q'(r)}{Q(r)} \tag{2.4.8}$$

and two similar equations, where V is the speed of the network, i. e.,

$$V = \frac{\tilde{r}_n - r_n}{\tau/2} \tag{2.4.9}$$

We now apply the implicit formula (2.4.2) to (2.4.5), (2.4.6), (2.4.7), taking

$$A = (u_n + c_n - V_n), \quad (u_n - c_n - V_n), \quad (u_n - V_n),$$

respectively, and so obtain

$$a_n \tilde{\mu}_{n+1} + \lambda_n a_n \tilde{\pi}_{n+1} + \tilde{\mu}_n + \lambda_n \tilde{\pi}_n - a_n \tilde{\mu}_{n-1} - \lambda_n a_n \tilde{\pi}_{n-1} = f_n \tag{2.4.10}$$

$$b_n \tilde{\mu}_{n+1} - \lambda_n b_n \tilde{\pi}_{n+1} + \tilde{\mu}_n - \lambda_n \tilde{\pi}_n - b_n \tilde{\mu}_{n-1} + \lambda_n b_n \tilde{\pi}_{n-1} = g_n \tag{2.4.11}$$

$$k_n \tilde{\sigma}_{n+1} + \tilde{\sigma}_n - k_n \tilde{\sigma}_{n-1} = l_n \tag{2.4.12}$$

where

$$a_n, b_n = \frac{\frac{1}{2}\tau(u_n \pm c_n) + r_n - \tilde{r}_n}{(\tilde{r}_{n+1} - \tilde{r}_{n-1})\kappa} \quad *$$

$$f_n = -\mu'_n u_n c_n \frac{Q'_n}{Q_n}\frac{\tau}{2} + \mu_n + \lambda_n \pi_n$$

$$g_n = \mu'_n u_n c_n \frac{Q'_n}{Q_n}\frac{\tau}{2} + \mu_n - \lambda_n c_n$$

$$l_n = \sigma_n$$

$$k_n = \frac{\frac{1}{2}\tau(u_n) + r_n - \tilde{r}_n}{(\tilde{r}_{n+1} - \tilde{r}_{n-1})\kappa}$$

---

* The parameter κ is introduced, partly so that difference formulae can be reduced to first order form near boundaries and partly to allow for a predictor step different from $\tau/2$. We must statisfy $1/2 < \kappa \leq 1$ for stability and in general take $\kappa = 1$.

Eqs. (2.4.10)–(2.4.12) can be written in matrix form

$$A_n \tilde{U}_{n+1} + B_n \tilde{U}_n + C_n \tilde{U}_{n-1} = G_n \tag{2.4.13}$$

where $\tilde{U}_n$ is a vector $(\tilde{\mu}_n, \tilde{\pi}_n, \tilde{\sigma}_n)$ and A_n, B_n, C_n, G_n are known matrix coefficients.

Eq. (2.4.13) is supplemented by boundary conditions at the left and right ends and solved by the method of double sweep.

Once $\tilde{U}_n$ has been determined for all integral n, interpolation formulae of type (2.4.3) are used to find $\tilde{U}$ at half integral points $\ldots n-1/2, n+1/2, \ldots$.

For the second half step we use the equations of motion in integral form

$$\oint (\varrho Q \, dr - \varrho u Q \, dt) = 0$$

$$\oint \{ \varrho u Q \, dr - (p + \varrho u^2) Q \, dt \} = \int\int p Q' \, dr \, dt \tag{2.4.14}$$

$$\oint \left[\varrho \left(\frac{u^2}{2} + E \right) Q \, dr - \varrho u \left(E + \frac{p}{\varrho} + \frac{u^2}{2} \right) Q \, dt \right] = 0$$

and apply these to the full line circuit shown in Fig. 2.6. On each side of this cell we represent the integrand by its value at the mid-point.

We then arrive at explicit formulae of the type

$$\begin{aligned} \bar{\varrho}_n = \frac{1}{(\bar{r}_{n+1/2} - \bar{r}_{n-1/2}) \bar{Q}_n} & \{ Q_n \varrho_n (r_{n+1/2} - r_{n-1/2}) \\ & + \tilde{\varrho}_{n+1/2} \tilde{Q}_{n+1/2} (\bar{r}_{n+1/2} - r_{n+1/2} - \tau \tilde{u}_{n+1/2}) \\ & - \tilde{\varrho}_{n-1/2} \tilde{Q}_{n-1/2} (\bar{r}_{n-1/2} - r_{n-1/2} - \tau \tilde{u}_{n-1/2}) \} \end{aligned} \tag{2.4.15}$$

Adjacent to boundaries we replace formulae (2.4.15) by explicit difference formulae derived directly from (2.4.5)–(2.4.7) (equivalent to (2.4.4) for the model equation).

Boundary Conditions: Two types of boundary must be considered, external and internal. There are two external boundaries at the left and right ends of the region to be calculated.

For illustration, suppose that the left boundary is a given piston path. There u_1 and σ_1 are known (entropy is conserved along the piston path). The relation (2.4.13) corresponding to the $u-c$ direction can be used but must be written as a first order relation between points 1 and 2. At point 2 we have three equations (2.4.13) which can be solved to give $\tilde{U}_3$ in terms of $\tilde{U}_2$. Proceeding to the right we obtain relations between $\tilde{U}_{n+1}$ and $\tilde{U}_n$ for successive n. The last of these corresponds to $n = N-1$. At $n = N$ we have one difference relation (corresponding to the $u+c$ direction) but write this in first order form (since $n = N+1$ does not exist). We also have two shock relations, after eliminating the shock velocity. Thus we obtain 6 relations between $\tilde{U}_{N-1}$ and

$\tilde{U}_N$ and can solve for $\tilde{U}$ at $r = r_N$. The inverse sweep then gives values of $\tilde{U}$ for $r = r_{N-1}, r_{N-2}, \ldots, r_1$.

At an internal boundary, separating a region M on the left from a region $M+1$ on the right, say, we must be able to transfer (2.4.13) across the boundary.

Suppose the internal boundary is a shock wave, as shown in Fig. 2.10.

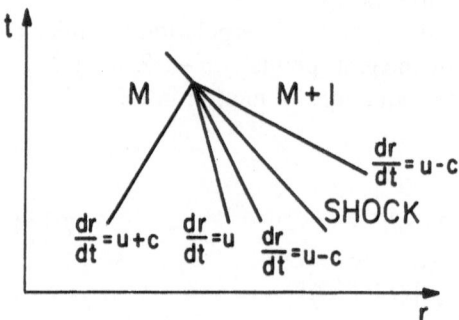

Fig. 2.10 Transition across an interior boundary

Denote the last two points in M by $N-1$, N.

Denote the first two points in $M+1$ by 1, 2.

Then from direct sweep from the left we have three relations between $\tilde{U}_{N-1}$ and $\tilde{U}_N$. In region $M+1$ we have one difference relation (in the $u-c$ direction) written as a first order relation between $\tilde{U}_1$ and $\tilde{U}_2$. We have two shock relations between $\tilde{U}_N$ and $\tilde{U}_1$. When we combine these with (2.3.3) for $\tilde{U}_1, \tilde{U}_2, \tilde{U}_3$ we can solve for $\tilde{U}_3$ in terms of $\tilde{U}_2$ and so continue direct sweep to the external boundary on the right.

On the final layer we use more conventional techniques for calculating unknowns on boundaries. At interior points near boundaries we can find the final layer values of unknowns from cell or other explicit formulae and values at boundaries are found by solving boundary conditions locally.

2.5 The Double Sweep Method

In Sect. 2.4 we referred to the solution of the difference equation on the intermediate layer by a double sweep method. This is a technique for solving second order difference equations with two point boundary conditions. It was originally proposed by GEL'FAND and LOKUTSIEVSKII and is described in detail in GODUNOV and RYABENKII (1964). We now outline this important method which is used, not only in the second Godunov scheme, but also in the BVLR scheme described in Chapt. 3.

Following GODUNOV and RYABENKII (1964), we consider the solution of the linear second order difference equation

$$a_n u_{n-1} - 2 b_n u_n + c_n u_{n+1} = g_n \quad (1 \leq n \leq N-1) \tag{2.5.1}$$

with end conditions $u_0 = \phi$, $u_N = \psi$, where a_n, b_n, c_n are given coefficients dependent on the mesh point $x = nh$ and ϕ and ψ are given values.

We begin with the difference equations at the left end

$$a_1 u_0 - 2 b_1 u_1 + c_1 u_2 = g_1 \tag{2.5.2}$$

Since $u_0 = \phi$ (2.5.2) can be written

$$u_1 = L_{3/2} u_2 + K_{3/2} \tag{2.5.3}$$

where

$$L_{3/2} = \frac{c_1}{2 b_1}, \qquad K_{3/2} = \frac{a_1 \phi - g_1}{2 b_1} \tag{2.5.4}$$

We now eliminate u_1 between (2.5.3) and the next difference equation

$$a_2 u_1 - 2 b_2 u_2 + c_2 u_3 = g_2 \tag{2.5.5}$$

to obtain

$$u_2 = L_{5/2} u_3 + K_{5/2} \tag{2.5.6}$$

where

$$L_{5/2} = \frac{c_2}{2 b_2 - a_2 L_{3/2}}, \qquad K_{5/2} = \frac{a_3 K_{3/2} - g_2}{2 b_2 - a_2 L_{3/2}} \tag{2.5.7}$$

We continue this process of elimination step by step, moving to the right. At a general point we obtain the formula

$$u_n = L_{n+1/2} u_{n+1} + K_{n+1/2} \tag{2.5.8}$$

where

$$L_{n+1/2} = \frac{c_n}{2 b_n - a_n L_{n-1/2}}, \qquad K_{n+1/2} = \frac{a_n K_{n-1/2} - g_n}{2 b_n - a_n L_{n-1/2}} \tag{2.5.9}$$

The last expression of type (2.5.8) is

$$u_{N-1} = L_{N-1/2} u_N + K_{N-1/2} \tag{2.5.10}$$

Now at the right hand boundary $u_N = \psi$ and (2.5.10) therefore determines u_{N-1}. Moving one step to the left we then obtain u_{N-2} from the formula

$$u_{N-2} = L_{N-3/2} u_{N-1} + K_{N-3/2} \tag{2.5.11}$$

and subsequent shifts to the left yield the values $u_{N-3}, \ldots, u_2, u_1$.

The double sweep method of solving the problem in (2.5.1) is therefore in two parts. In the first, called the direct sweep, we move successively to the right, calculating the coefficients $L_{n+1/2}$, $K_{n+1/2}$ from (2.5.9) at each step starting with $n=1$, and ending with $n=N-1$. We then carry out the reverse sweep in which we successively calculate the unknowns u_n from (2.5.8), starting with $n=N-1$ and finishing with $n=2$.

Double sweep enables us to transfer the left hand boundary condition to the right end so that we can evaluate both u_N and u_{N-1}. We then have enough conditions to solve our second order difference equation (2.5.1) starting at the right end.

It is shown in GODUNOV and RYABENKII (1964) that the double sweep process is stable and leads to a unique solution of problem (2.5.1) provided that $a_n > \delta$, $c_n > \delta$, $b_n > 1/2(a_n + c_n) + \delta$ where δ is any small positive quantity. It is also shown that the error in u_n is of the same order as any error in the boundary values ϕ and ψ.

The double sweep method can be generalized to a matrix equation rather than the equation for the single unknown u_n. This generalized form is used in Godunov's second scheme when solving on the intermediate layer.

2.6 Execution of the Second Scheme on the Intermediate Layer

In applying the second Godunov scheme to the equations of unsteady motion in one space variable we need to solve the second order matrix difference equation (2.4.13)

$$A_n \tilde{U}_{n+1} + B_n \tilde{U}_n + C_n \tilde{U}_{n-1} = G_n \tag{2.6.1}$$

where

$$A_n = \begin{pmatrix} a_n & \lambda_n a_n & 0 \\ b_n & -\lambda_n b_n & 0 \\ 0 & 0 & k_n \end{pmatrix} \qquad B_n = \begin{pmatrix} 1 & \lambda_n & 0 \\ 1 & -\lambda_n & 0 \\ 0 & 0 & 1 \end{pmatrix}$$

$$C_n = \begin{pmatrix} -a_n & -\lambda_n a_n & 0 \\ -b_n & +\lambda_n b_n & 0 \\ 0 & 0 & -k_n \end{pmatrix} \qquad G_n = \begin{pmatrix} f_n \\ g_n \\ l_n \end{pmatrix}$$

a_n, b_n, f_n, g_n, k_n are all defined in Sect. 2.5

$$\tilde{U}_n = \{\tilde{\mu}_n(u), \tilde{\pi}_n(p), \tilde{\sigma}_n(S)\}\ .$$

To begin with we suppose that we solve (2.6.1) in conjunction with boundary conditions given at the extreme left and extreme right ends. We then need six scalar boundary conditions which can either be in terms of the unknowns $\tilde{U}_n$ themselves $(n=1, n=N)$ or in terms of linear relations between $\tilde{U}_n$ and $\tilde{U}_{n+1}$ $(n=1, n=N-1)$. In either case the left end condition gives a matrix relation

$$\tilde{U}_1 = P_{3/2}\tilde{U}_2 + Q_{3/2} \tag{2.6.2}$$

where $P_{3/2}$ and $Q_{3/2}$ are given; $P_{3/2}$ is a square matrix of rank 3 and $Q_{3/2}$ is a column vector.

If we eliminate $\tilde{U}_1$ between (2.6.1) and (2.6.2) we will derive a relation similar in form to (2.6.2) between $\tilde{U}_2$ and $\tilde{U}_3$. We then move one step to the right to relate $\tilde{U}_3$ and $\tilde{U}_4$. At a general point in this direct sweep we arrive at the matrix relation

$$\tilde{U}_{n-1} = P_{n-1/2}\tilde{U}_n + Q_{n-1/2} \tag{2.6.3}$$

where

$$P_{n+1/2} = -(B_n + C_n P_{n-1/2})^{-1} A_n \tag{2.6.4}$$

$$Q_{n+1/2} = (B_n + C_n P_{n-1/2})^{-1}(G_n - C_n Q_{n-1/2})$$

We evaluate the matrices $P_{n+1/2}, Q_{n+1/2}$ step by step for successive values of $n=3,4,\ldots,N-1$.

At $n=N$ we have (2.6.3)

$$\tilde{U}_{N-1} = P_{N-1/2}\tilde{U}_N + Q_{N-1/2} \tag{2.6.5}$$

together with a linear matrix boundary condition between $\tilde{U}_N$ and $\tilde{U}_{N-1}$. These can be solved for $\tilde{U}_N$ and $\tilde{U}_{N-1}$. We then carry out the reverse sweep, solving (2.6.3) for $n=N-1,\ldots,2$ successively moving to the left to determine all remaining values of $\tilde{U}_n$. The scheme is valid provided that $\|P_{n+1/2}\| \le 1$.

If an internal boundary arises in the problem a means must be found· to carry the double sweep process across it. The details of this together with other changes required near boundaries are considered in the next section.

2.7 Boundary Conditions on the Intermediate Layer

The difference equations to be solved on the intermediate layer in Godunov's second scheme are of second order although the differential equations they approximate are of first order. In physical problems this can lead to an undetermined system of algebraic equations for values of unknowns at network points. To overcome this defect and to guarantee uniqueness of solutions GODUNOV supplements the intermediate layer equations at certain boundary points with first order difference equations.

The situation is illustrated by the model problem governed by the equation

$$\frac{\partial u}{\partial t} + A \frac{\partial u}{\partial r} = 0 \tag{2.7.1}$$

in the interval $0 \le r \le 1$, with $u = u_0(t)$, $r = 0$, and $u = f(r)$, $t = 0$. The solution of this problem does not require a boundary condition at the right end $r = 1$.

When we apply the second scheme to this we divide $[0,1]$ into $N-1$ equal intervals at nodes $r_1 = 0$, $r_2 = h, \ldots, r_{N-1} = (N-1)h$, $r_N = Nh = 1$. On the intermediate layer we have the second order difference equations

$$\frac{u_n^{m+1/2} - u_n^m}{\tau/2H} + A \frac{u_{n+1}^{m+1/2} - u_{n-1}^{m+1/2}}{2h} = 0 \tag{2.7.2}$$

$(1/2 < H \le 1$, for a general intermediate sub-time step). Eq. (2.7.2) is satisfied for $n = 2, \ldots, N-1$ and so yields $N-2$ algebraic equations. With the left boundary condition at $n = 1$ we therefore have $N-1$ equations to determine N unknowns. To close the system and to achieve uniqueness consistent with the analytical solution GODUNOV adds the following first order difference equation at the right end

$$\frac{u_N^{m+1/2} - u_N^m}{\tau/2H} + A \frac{u_N^{m+1/2} - u_{N-1}^{m+1/2}}{h} = 0 \tag{2.7.3}$$

This can be written as a difference boundary condition at the right end

$$-2a u_{N-1}^{m+1/2} + (1 + 2a) u_N^{m+1/2} = f_N \tag{2.7.4}$$

where $a = A\tau/4Hh$, $f_N = u_N^m$.

Similar difference boundary conditions are applied at open ends in Gas Dynamics problems. For example, in the problem of a piston moving to the right with prescribed speed into a column of gas at rest, the right boundary is initially a $+$ characteristic which later develops into a shock wave. Both before and after shock formation the $+$ characteristic equation (2.3.8) applies at right end nodal points. At the two right end nodes $n = N-1$, $n = 1$ we apply the first order difference equation

$$-2a_N(\tilde{\mu}_{N-1}+\lambda_N\tilde{\pi}_{N-1})+(1+2a_N)(\tilde{\mu}_N+\lambda_N\tilde{\pi}_N)=f_N \tag{2.7.5}$$

where

$$a_N=\frac{\frac{1}{2}\tau(u_N+c_N)+r_N-\tilde{r}_N}{2(\tilde{r}_N-\tilde{r}_{N-1})H}.$$

In general, when a boundary is reached one or more first order difference relations of type (2.7.5) are used to connect the network point on the boundary and its immediately adjacent nodal point.

To explain the procedure we consider a Gas Dynamics problem in which a left running shock wave is an internal boundary as shown in Fig. 2.11.

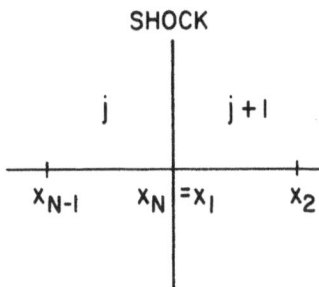

Fig. 2.11   Grid points near an interior shock point

Denote the regions to the left and right of the shock by indices j and $j+1$, respectively. Moving from left to right the boundary point in j is denoted by x_N and that in $j+1$ by x_1.

Now, at a point on the j side of the shock all three characteristic directions can be drawn, while on the $j+1$ side only one direction $(u-c)$ exists. Therefore we may write three first order relations of type (2.7.4) between the last two points of j, namely

$$B_N\tilde{U}_N+C_N\tilde{U}_{N-1}=G_N \tag{2.7.6}$$

where B_N, C_N are 2×3 matrices so that (2.7.6) represents three scalar equations.

Across the shock wave we have two relations between $\tilde{U}_N$ and $\tilde{U}_1$ after eliminating the shock velocity (one mechanical shock condition and one Hugoniot relation). We may linearize these equations in terms of conditions on the basic layer and write them in the form

$$R\tilde{U}_N+\bar{R}=L\tilde{U}_1+\bar{L} \tag{2.7.7}$$

where $R, \bar{R}, L, \bar{L}$ are known and (2.7.7) represents two scalar equations.

Finally, in region $j+1$ the shock cuts off two characteristic directions but we can still use the $u-c$ equation to derive a first order relation between values at x_1 and x_2

$$A_2 \tilde{U}_2 + B_1 \tilde{U}_1 = G_1 \tag{2.7.8}$$

where A_2, B_1 are row vectors, giving one scalar equation.

We regard (2.7.6), (2.7.7), (2.7.8) as six algebraic equations to determine six unknowns $\tilde{U}_N(\tilde{\mu}_N, \tilde{\pi}_N, \tilde{\sigma}_N)$ and $\tilde{U}_1(\tilde{\mu}_1, \tilde{\pi}_1, \tilde{\sigma}_1)$. If we solve these for $\tilde{U}_N$ in terms of $\tilde{U}_{N-1}$ and $\tilde{U}_2$ (eliminating $\tilde{U}_1$) the direct sweep relations (2.6.5) become relations between $\tilde{U}_{N-1}$ and $\tilde{U}_2$ and we have therefore succeeded in jumping over the internal boundary in the sweep process. We then continue this to the extreme right boundary, solve for all unknowns there and perform the reverse sweep ((2.7.6)–(2.7.8) now determine $\tilde{U}_N$ and $\tilde{U}_{N-1}$, since $\tilde{U}_2$ and $\tilde{U}_1$ are known from the last two sweep relations in region $j+1$).

2.8 Procedure on the Final Layer

The first step in the final layer calculations is to interpolate for values of unknowns at half integer points on the intermediate layer. At a general internal point away from boundaries we use the four point formula (2.4.3) which has the effect of smoothing the intermediate layer values and also of stabilizing the predictor step process. Adjacent to a boundary point, points to the right or left are cut off and we must use the simple mean value

$$u_{n+1/2}^{m+1/2} = \frac{u_{n+1}^{m+1/2} + u_n^{m+1/2}}{2} \tag{2.8.1}$$

Next, at an interior point on the final layer we use cell formulae of type (2.4.15) (see Fig. 2.6) to calculate $U_n(\mu, \pi, \sigma)$.

At certain points we may not be able to construct a cell as shown in Fig. 2.6 because unknowns are discontinuous or many valued at $r=r_n$ or $r=\bar{r}_n$. In this case we use formulae of type (2.7.4), viz.,

$$\bar{\varrho}_n = \frac{2}{(\bar{r}_{n+1} - \bar{r}_{n-1})\tilde{Q}_r} \left\{ Q_n \varrho_n \frac{\tilde{r}_{n+1} - \tilde{r}_{n-1}}{2} + \tilde{\varrho}_{n+1/2} \tilde{Q}_{n+1/2} (\bar{r}_n - r_n - \tau \tilde{u}_{n+1/2}) \right.$$

$$\left. - \tilde{\varrho}_{n-1/2} \tilde{Q}_{n-1/2} (\bar{r}_n - r_n - \tau u_{n-1/2}) \right\} \tag{2.8.2}$$

(For example, we could use relations of type (2.8.2) to the left of the shock discussed in Sect. 2.7.)

At an internal left running shock wave (see Fig. 2.11) separating regions j and $j+1$ we proceed as follows. At the last point x_N in region j $\bar{u}_N, \bar{p}_N, \bar{S}_N$ are determined from three characteristic relations derived directly from (2.7.5)–(2.7.7) applied over the whole time interval τ:

$$\bar{u}_N^j \left(1 + \frac{\bar{a}_N^{(j)}}{2}\right) + \frac{\bar{p}_N^j}{\varrho_N^j c_N^j}\left(1 + \frac{\bar{a}_N^j}{2}\right) = \bar{A}_N^j \tag{2.8.3}$$

and two similar equations where $\bar{a}_N$ and $\bar{A}_N$ depend on $\bar{u}_{N-1}, \bar{p}_{N-1}$ and values on the initial layer. At the first point in region $j+1$ we have a similar relation

$$\bar{u}_1^{j+1}\left(1 - \frac{\bar{b}_1^{j+1}}{2}\right) - \frac{\bar{p}_1^{j+1}}{\varrho_1^{j+1} c_1^{j+1}}\left(1 - \frac{\bar{b}_1^{j+1}}{2}\right) = B_1^{j+1} \tag{2.8.4}$$

Across the boundary we have two shock relations, including the shock velocity, between $(u_N^j, p_N^j, \varrho_N^j)$ and $(u_1^{j+1}, p_1^{j+1}, \varrho_1^{j+1})$ (one Hugoniot relation and one mechanical shock condition)

$$\bar{E}_N^j - \bar{E}_1^{j+1} + \frac{\bar{p}_N^j + p_1^{j+1}}{2}(v_N^j - v_1^{j+1}) = 0 \tag{2.8.5}$$

$$\bar{u}_N^j - \bar{u}_1^{j+1} = \{-(\bar{p}_N^j - \bar{p}_1^{j+1})(\bar{v}_N^j - \bar{v}_1^{j+1})\}^{1/2}$$

We solve (2.8.3) for $\bar{u}_N, \bar{p}_N, \bar{S}_N$ in j and then (2.8.4), (2.8.5) become three equations to determine $\bar{u}_1, \bar{p}_1$ and $\bar{S}_1$ in $j+1$.

At other internal boundaries (say contact discontinuities or right running shocks) the detailed procedure is changed but in all cases we solve six equations for six unknowns.

2.9 Applications of the Second Godunov Scheme

Three applications of this scheme are discussed in ALALYKHIN et al. (1970). The first is to the model problem of shock formation in a column of gas compressed by a piston accelerated to uniform velocity from rest. The scheme is started just after the piston has reached constant speed where the pressure distribution is as shown in Fig. 2.12a. At this time the left boundary is the piston and the right boundary is a known plus characteristic, which is treated as a shock wave at subsequent times. The calculated distribution at later times is shown in Figs. 2.12b,c,d and the approach to constant state shock-piston relation is seen to be quite rapid. Also, the oscillations frequently found behind the shock when applying artificial viscosity schemes to this problem are quickly eliminated in the second scheme.

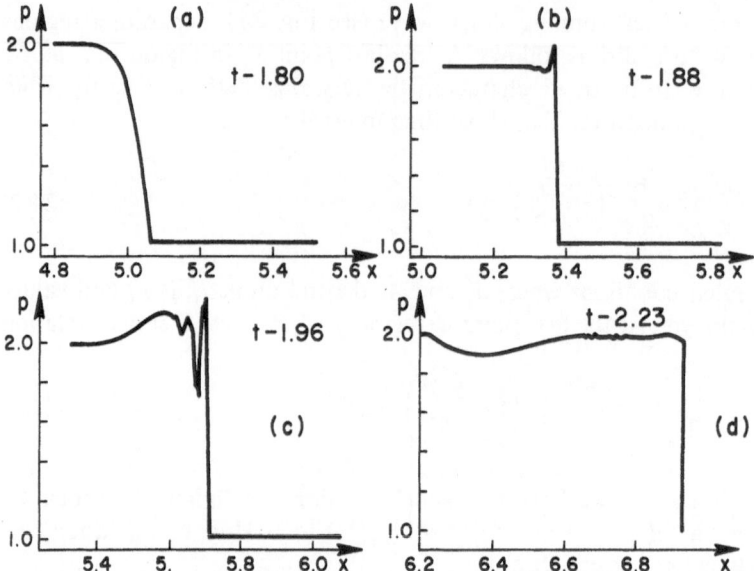

Fig. 2.12   Shock development using Godunov's second scheme

The second application, to calculate wave formation on the sides of electrodes in a dense plasma, was presented by GODUNOV at the first International
Conference on Numerical Methods in Fluid Mechanics in Novosibirsk in
1969 and is published in English in GODUNOV et al. (1970).

The third application is to the dam break problem. Approximate analytical
solutions of this problem are discussed in WHITHAM (1974). The application
of the second scheme to the shallow water formulation of the problem is
described in English by VASILIEV (1971). The initial conditions and boundaries
arising in the problem are shown in Fig. 2.13. The right boundary (2) is the

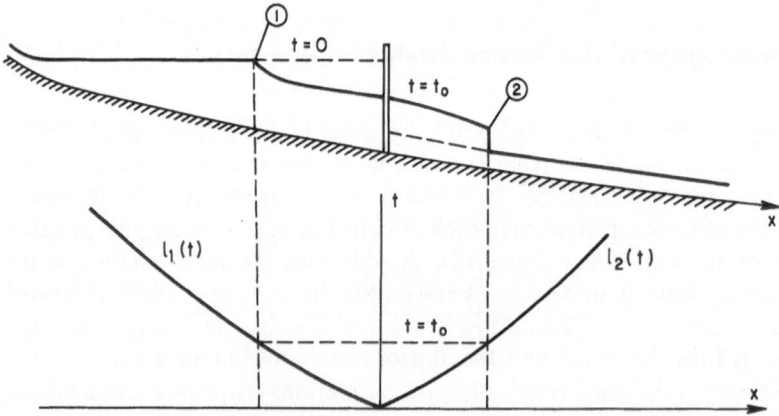

Fig. 2.13   The dam break problem

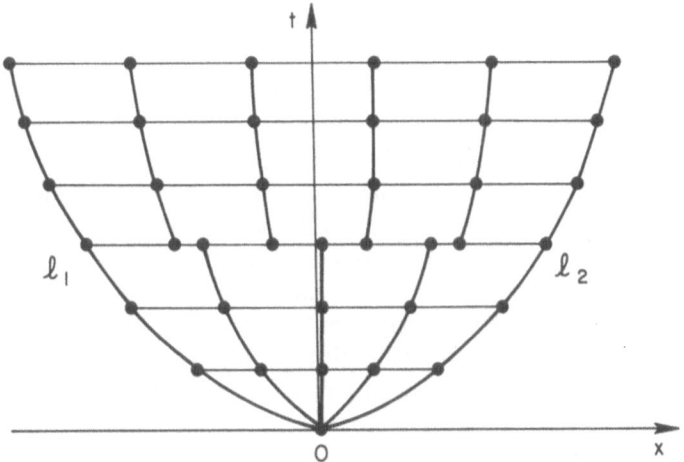

Fig. 2.14   The moving grid in the dam break problem

bore equivalent to a right moving shock and the left boundary (1), the depression wave, is a left moving characteristic. The shallow water equations lead to two equations in characteristic form for the surface elevation and volume discharge per cross section. The predictor-corrector scheme of Sect. 2.4 is applied essentially unchanged with two external boundaries (1) and (2) and no internal boundaries. The moving grid is shown in Fig. 2.14 and it should be noted that the number of grid points is changed after a short time to give better resolution. Figure 2.15 shows the development of the bore from the dam break in a non-prismatic bed during the first thirty minutes.

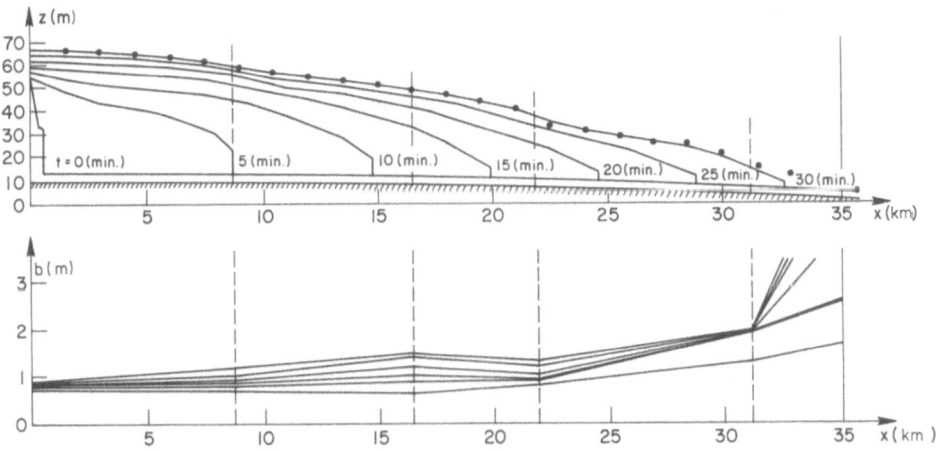

Fig. 2.15   Propagation of a bore in a non-prismatic channel

2.10 Glimm's Method

As noted in 2.1–2.3, Godunov's original scheme is of first order accuracy. In an attempt to improve on this situation, a variant on Godunov's first scheme was proposed in recent years, mainly in the United States. The variant originated with a paper by GLIMM (1955), concerned with the proof of existence of solutions to systems of non-linear hyperbolic partial differential equations. The ideas presented in this paper have been explained, developed and improved, and converted into a numerical method in a series of papers by CHORIN and his colleagues. The most important of these are CHORIN (1976, 1977), CONCUS and PROSKUROWSKI (1979), SOD (1977, 1978) and COLELLA (1982). We shall refer to all these papers and also to an excellent term paper by WIGTON (1978).

We begin by outlining Glimm's method for solving hyperbolic partial differential equations (unsteady in one space coordinate) in conservation form. Essentially the method reduces to the solution of a series of Riemann problems by a Random Choice Method. We then clarify the procedure in the case of a simple model equation, the linearized wave equation in one direction, explaining the need for use of a random variable. Returning to the general equations of one-dimensional Gas Dynamics we describe various random choice possibilities, including the van der Corput scheme, and make accuracy estimates of the Glimm schemes. The following section is devoted to the solution of Riemann problems, particularly in relation to Glimm's method. The extension of the method to problems with cylindrical or spherical symmetry is then described (for such problems the governing equations are no longer in conservation form). Finally, some remarks are made on the applications of Glimm's method to multi-dimensional problems.

2.11 Outline of Solution for Gas Dynamic Equations

The equations of motion for unsteady flow of a gas in one dimension, in Eulerian form, may be written

$$\frac{\partial U}{\partial t} + \frac{\partial}{\partial x} F(U) = 0, \tag{2.11.1}$$

where

$$U = \begin{pmatrix} \varrho \\ \varrho u \\ e \end{pmatrix}, \quad F(U) = \begin{pmatrix} \varrho u \\ p + \varrho u^2 \\ (p+e)u \end{pmatrix}$$

and ϱ denotes the density, u the velocity, p the pressure, e the specific internal energy. Eq. (2.1) represents a system of first order partial differential equations of hyperbolic type, written in conservation form. We seek a solution of the system satisfying the initial condition

$$U(x, 0) = \text{given function of } x. \tag{2.11.2}$$

To solve the initial value problem (2.11.1), (2.11.2) by Glimm's method we proceed as follows. We introduce a network in the (x, t) plane with step size h in the x direction and k in the t direction and calculate $U(x, t)$ approximately at network points (ih, nk), writing

$$U_i^n = U(ih, nk).$$

As in Godunov's first scheme, we represent the initial distribution of $U(x, 0)$ by a step function

$$U(x, 0) = U(ih, 0), \quad (i - \tfrac{1}{2})h < x \leq (i + \tfrac{1}{2})h$$
$$i = 0, 1, 2, \ldots \tag{2.11.3}$$

In each interval $ih < x \leq (i+1)h$ we solve the following simplified initial value problem.
Solve

$$U_t + [F(U)]_x = 0 \tag{2.11.1}$$

with initial conditions

$$U(x, 0) = U(ih, 0), \qquad x \leq (i + \tfrac{1}{2})h$$
$$U(x, 0) = U((i+1)h, 0), \quad x > (i + \tfrac{1}{2})h \tag{2.11.4}$$

The solution is used in the time interval $0 \leq t \leq \tfrac{1}{2}k$ in order to calculate U at time $t = 1/2k$. The IVP (2.11.1), (2.11.4) is a Riemann problem and, for the equation of gas dynamics (with perfect gas behavior) can be solved analytically, as discussed in Sect. 2.2.

The solution determines the distribution of U at time $t = 1/2k$. We represent this distribution by a new step function

$$U(x, \tfrac{1}{2}k) = U(ih, \tfrac{1}{2}k), \quad (i - \tfrac{1}{2})h \leq x \leq (i + \tfrac{1}{2})h$$

and now repeat the procedure used at $t = 0$ for solving a series of Riemann problems. This is continued at successive half time intervals $t = nk, n = 1/2,$ 1, 3/2, ...

Both the first Godunov scheme and the present Glimm scheme agree up to this point, in using successive solutions of Riemann problem. Neither method uses the whole information derived from such solutions; instead each uses the data at a selected point in the relevant interval. In the Godunov scheme the Riemann solution is always evaluated at the boundary between two steps. In the Glimm scheme the Riemann solution is calculated at a point chosen at random in the interval $(-1/2h, 1/2h)$ on either side of the step function boundary.

To clarify the Glimm procedure, and to emphasize the difference between the Godunov and Glimm schemes we examine the solutions of a linearized model of Eq. (2.11.1), namely, the equation representing propagation of acoustic waves in one direction. This discussion is based on a term paper prepared by WIGTON (1978).

2.12 The Glimm Scheme for Simple Acoustic Waves

The differential equation

$$\frac{\partial U}{\partial t} + a \frac{\partial U}{\partial x} = 0 \quad (a > 0), \tag{2.12.1}$$

where a is constant, represents propagation of linear acoustic waves in the positive x direction only. If U has the dimensions of velocity, a represents the constant speed of sound. The general solution of Eq. (2.12.1)

$$U(x, t) = f(x - a t) \tag{2.12.2}$$

represents propagation of waves to the right only with no change of form. Such a solution is often called a *simple wave*. The function f is determined by initial conditions, in particular, the wave form at time $t = 0$.

The curves (straight lines) $x - a t = \text{const}$ are characteristics and any member of the family can carry a discontinuity in the initial wave form. In particular the solution of the special initial value problem:

Solve Eq. (2.12.1) with

$$
\begin{aligned}
U(x, 0) &= 0, \quad x \le 0, \\
U(x, 0) &= 1, \quad x > 0
\end{aligned}
\tag{2.12.3}
$$

is simply

$$
\begin{aligned}
U(x, t) &= 0, \quad x \le a t, \\
U(x, t) &= 1, \quad x > a t.
\end{aligned}
\tag{2.12.4}
$$

In other words, the initial step function 0–1 is propagated without change along $x = at$ (see Fig. 2.16).

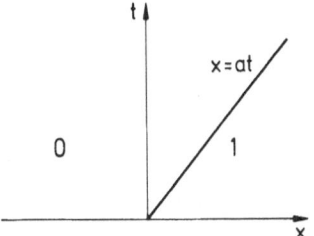

Fig. 2.16

To see how the Glimm or Godunov methods apply to this IVP solve the Riemann problem for Eq. (2.12.1)

$$U = 0, \quad ih < x \leq (i + \tfrac{1}{2})h, \quad t = nk,$$
$$U = 1, \quad (i + \tfrac{1}{2})h < x < (i + 1)h, \quad t = nk.$$

If we transfer the origin $\{(i + 1/2)h, nk\}$ to $(0, 0)$ the solution to this problem is

$$U(x, t) = 0, \quad x \leq at,$$
$$U(x, t) = 1, \quad x > at. \tag{2.12.4}$$

We wish to calculate the solution at $\{(i + 1/2)h, (n + 1/2)k\}$. In the Godunov scheme we use the Riemann solution at $X = 0$ (X is the transformed coordinate) so that

$$u_{n+1/2}^{i+1/2} = 0.$$

To ensure stability of the finite difference scheme used to solve Eq. (2.12.1) we require that

$$ah > k.$$

It follows that $\{(i + 1)h, (n + 1/2)h\}$ is to the right of $X - aT = 0$ and hence

$$u_{n+1/2}^{i+1} = 1.$$

In other words, the Godunov scheme always moves the discontinuity 0–1 through a distance $h/2$ to the right. After N whole time steps, or $(2N \cdot 1/2)$

time steps, the discontinuity 0-1, as determined by the Godunov scheme would be at the position

$$2N \cdot \frac{h}{2} = Nh.$$

If T is now the total time elapsed $T = Nk$ and shift of discontinuity is

Th/k.

According to the exact solution the shift in the discontinuity is

aT.

Since the Courant-Friedrichs-Lewy condition

$h/k > a$

is satisfied, the Godunov scheme overestimates the shift in the discontinuity by

$\{h/k - a\} T$.   (2.12.5)

When applying the Glimm scheme to this problem we evaluate the Riemann solution (2.12.4), not uniquely at the point $X = 0$, but at a randomly chosen point on $t = (n + 1/2)k$ within the interval $(-h/2, h/2)$ on either side of the node, $[(i + 1/2)h, (n + 1/2)k]$. In view of the CFL condition the characteristic $X = aT$ will always intersect the interior of the interval $(-h/2, h/2)$. Therefore, each time the randomly chosen point is taken to the left of $X = aT$, the discontinuity 0-1 will be shifted $h/2$ to the right. In view of result (2.12.5), the Glimm scheme will also overestimate the shift in the discontinuity 0-1 to the right unless a certain number of sampled points in $(-h/2, h/2)$ are taken to the right of $X = aT$. A simple calculation (due to WIGTON) shows that the number of samples to the left of the discontinuity must be proportional to A/h where $A = (h/2) + (ah/2)$. The balance of samples (those to the right of 0-1) is then proportional to $2N - (A/h)$.

Finally, using Elementary Statistics, it can be shown that the numbers of samples to the left and to the right of $X = aT$ will only be in the correct proportion if the samples are randomly distributed *uniformly* over the interval $(-h/2, h/2)$.

In summary, this elementary example shows that, in general, the Godunov scheme (at least in the form given here) introduces an error in the location of a propagated discontinuity. This results from the fact that the Riemann solution is always calculated at the actual discontinuity boundary. In the Glimm scheme, where the Riemann solution is sampled at uniformly distributed random points within $(-h/2, h/2)$ on either side of the discontinuity, this error can be eliminated, or at least reduced to a very small order quantity. For both methods the accuracy in locating the discontinuity increases as the CFL condition is more closely satisfied.

2.13 Random Choice for the Gas Dynamic Equations

As explained in Sect. 2.11, the Glimm scheme, as applied to the equations of one-dimensional unsteady flow reduced to the following: Suppose the solution of

$$U_t + [F(U)]_x = 0 \qquad (2.13.1)$$

(fixed by initial conditions at time $t=0$) is known at network points $x=jh$, $j+0, \pm 1, \pm 2, \ldots$ up to time $t=nk$. This solution is given by step functions

$$
\begin{aligned}
U(x, t) &= U(ih, nk), & x &\leq (i+\tfrac{1}{2})h \\
U(x, t) &= U((i+1)h, nk), & x &> (i+\tfrac{1}{2})h.
\end{aligned}
\qquad (2.13.2)
$$

For each i we solve the Riemann problem (2.13.1), (2.13.2) with $x=(i+1/2)h$ as origin in the half time interval $nk \leq t \leq (n+1/2)k$.

We evaluate this solution at a randomly chosen point in the interval $(-1/2h, 1/2h)$ on either side of $x=(i+1/2)h$ at time $t=(n+1/2)k$ and take this as the solution at $x=(i+1/2)h$, $t=(n+1/2)k$.

Thus, in the Glimm scheme

$$U((i+\tfrac{1}{2})h, (n+\tfrac{1}{2})h) = V\{(i+\tfrac{1}{2}+\theta)h, (n+\tfrac{1}{2})k\}, \qquad (2.13.3)$$

where V is the solution of the Riemann problem (2.13.1), (2.13.2) and θ is a randomly chosen number in the range $(-1/2, 1/2)$.

In the period since Glimm's method has been proposed as a useful numerical technique for solving practical problems governed by equations of type (2.13.1) various suggestions have been made about the best procedure for choosing θ.

It can be shown that the accuracy of Glimm's scheme improves when the numbers θ are close to equipartition on $(-1/2, 1/2)$ (i.e., when the number of θ's in any subinterval of $(-1/2, 1/2)$ becomes roughly proportional to the length of the subinterval). The original Glimm proposal was to chose the θ's at random (which leads to rough equipartition after a large number of steps). However, one can do better. If it is known in advance how many time steps will be required in a problem with no boundaries, and if intermediate answers are not sought, we can simply divide $(-1/2, 1/2)$ into $2n$ subintervals of equal length, $2n$ = number of half steps, and then put θ_j in the middle of the j-th subinterval (this proposal is due to Lax). An intermediate strategy, which allows for intermediate answers and for boundaries, and uses a pair of mutually prime integers, was proposed by CHORIN (1976), with error estimates in CHORIN (1977). The original Glimm proposal is rather inaccurate, and the Chorin sampling leads to an error which is roughly $O(h)$ (but includes no diffusive error). Problems in which the method performs particularly well are discussed in CHORIN (1977).

The sampling process was further investigated by COLELLA (1982) in his Ph. D. research. COLELLA concluded that the best scheme, in terms of rapid approach to uniform distributions of numbers on $(-1/2, 1/2)$ and the accuracy of the Glimm solution, uses the van der Corput sequence [see HAMMERSLEY and HANDSCOMB (1965)]. This is explained as follows:

Let

$$\sum_{k=0}^{m} i_k 2^k = n, \quad i_k = 0, 1$$

be the binary expansion of the positive integer $n = 1, 2, \ldots$.

Let

$$a^n = \sum_{k=0}^{m} i_k 2^{-(k+1)}.$$

Then, in the range $(0, 1)$ a^n defines van der Corput sampling for successive n. It is simple to convert a^n to corresponding sampling in $(-1/2, 1/2)$.

It is instructive to calculate the first few values of a^n:

$$n = 1, \quad 1 = 1.2^0 + 0.2^1, \qquad a^1 = (1/2) = 0.5,$$
$$n = 2, \quad 2 = 0.2^0 + 1.2^1, \qquad a^2 = 1/4 = 0.25,$$
$$n = 3, \quad 3 = 1.2^0 + 1.2^1, \qquad a^3 = 0.75,$$
$$n = 4, \quad 4 = 0.2^0 + 0.2^1 + 1.2^2, \quad a^4 = (1/8) = 0.125.$$

We note that if

n is odd $a^n \geq 0.5$

. n is even $a^n < 0.5$.

If we apply Glimm's method with van der Corput sampling we divide the total time of the calculation, T, into $2N$ half steps. For each successive half time step we choose (in formula (2.13.3))

$$\theta_j = a^j - \tfrac{1}{2}, \quad j = 1, \ldots, 2N.$$

In his investigation, COLELLA (1982) studies this sampling in detail and gives rigorous error bounds for calculation of both shock points and non-discontinuity points. He shows that the error in the shock location is at most of order $h^{1/2}|\log h|$ while the error in the solution away from discontinuity points is at most of order $h|\log h|$. When COLELLA applies van der Corput sampling to the one-dimensional shock propagation problem, he obtains excellent results, much superior to those based on earlier samplings.

HOLT and LI (1982) apply this version of Glimm's method to the shallow water equations, solving the dam break problem as an example. The numerical solution is practically identical with the analytical solution.

2.14 Solution of the Riemann Problem

The solution of the Riemann problem for one-dimensional unsteady flow is the basis for the Breakdown Formulae used by Godunov in his first scheme (GODUNOV et al. 1961). These formulae are discussed in Sect. 2.2 and in greater detail in the original Godunov paper. It is not necessary here to repeat all the earlier discussion. We simply need to modify the procedure previously proposed to permit the calculation of the Riemann solution at a general point *near* a breakdown boundary rather than just at the boundary itself.

In the notation adopted in Sect. 2.2, suffix m refers to the breakdown boundary, suffix $m-1/2$ refers to the cell to the left, suffix $m+1/2$ refers to the cell to the right and suffix c.d. refers to the contact discontinuity (which moves always after cell breakdown).

As before, we solve equation (2.2.4) for $p_{c.d.}$ and $u_{c.d.}$ (by an iterative process). We can then regard the contact discontinuity as a piston, with known speed $u_{c.d.}$ compressing the gas to the right and expanding the gas to the left. The solution to the right of the contact discontinuity is then represented by uniform shock equations while that to the left is given by a simple wave. Thus the whole solution in $(-h/2, h/2)$ on either side of the breakdown boundary can be represented by algebraic formulae and can be sampled at the randomly selected point used in the Glimm scheme. Good illustrations of possible Riemann solutions are given in CONCUS and PROSKU-ROWSKI (1979).

2.15 Extension to Unsteady Flow with Cylindrical or Spherical Symmetry

The equations of motion for unsteady flow of gas with cylindrical or spherical symmetry differ from Eqs. (2.11.1) in the addition of a nonhomogeneous term containing the radial coordinate. As a consequence, the equations are no longer in conservation form and are not strictly satisfied by solutions of Riemann problems discussed earlier.

To overcome this difficulty SOD (1977) solves the governing equation by operator splitting. Firstly, he drops the nonhomogeneous term, and applies Glimm's method over two half time steps, exactly as described previously, to obtain estimates of the vector U at the end of a whole time step. He then substitutes this estimate in the nonhomogeneous term and solves a simple equation (containing only the time derivative) to find the correction to U. This predictor-corrector method was applied by SOD to calculate the propagation of both diverging and converging cylindrical shocks.

The equations of motion are

$$\frac{\partial U}{\partial t} + \frac{\partial}{\partial r}\{F(U)\} = -W(U), \tag{2.15.1}$$

where

$$U = \begin{pmatrix} \varrho \\ \varrho u \\ e \end{pmatrix}, \quad F(U) = \begin{pmatrix} \varrho u \\ \varrho u^2 + p \\ (p+e)u \end{pmatrix}, \quad W(U) = (v-1)\begin{pmatrix} \varrho u/r \\ \varrho u^2/r \\ u(p+e)/r \end{pmatrix},$$

$v=0$, plane symmetry, $v=1$, cylindrical symmetry, $v=1$, spherical symmetry.

To solve these equations SOD proceeds as follows:
(i) Solve the IVP

$$\frac{\partial U}{\partial t} + \frac{\partial}{\partial r}\{F(U)\} = 0 \tag{2.15.2}$$

over the network $(ih, (n+1)k)$ given

$U(ih, nk)$.

Glimm's method is used here, exactly as described in Sect. 2.13. The method is applied over two half time steps $nk \le t \le (n+1/2)k$ and $(n+1/2)k \le t \le (n+1)k$, using a randomly chosen variable in $(-1/2h, 1/2h)$ on either side of a nodal point $r=(i+1/2)h$ at which to evaluate the solution of the appropriate Riemann problem.
(ii) Substitute the solution of (2.15.1) at time $t=(n+1)k$ in $-W(U)$ and then solve the equation

$$\frac{\partial U}{\partial t} = -W(U) \tag{2.15.3}$$

by a simple explicit finite difference scheme. Specifically, if $\tilde{U}_i^{n+1}$ is the solution of the IVP (2.15.2) at $t=(n+1)k$, $r=ih$ we solve

$$\frac{U_i^{n+1}-U_i^n}{\Delta t} = W(\tilde{U}_i^{n+1}).$$
(2.15.4)

We then repeat steps (i) and (ii) at time $t=(n+1)k$ using initial data U_i^{n+1} determined from Eq. (2.15.4).

HOLT and LI (1980) applied this method to the problem of propagation of finite amplitude waves (plane or cylindrical) on shallow water. The governing equations are similar in form to those treated by SOD. In the water wave application it was found that accuracy of the predictor-corrector method could be improved considerably if the corrector was applied after each *half* time step instead of after only *whole* time steps. This improvement applies also to the gas dynamic equations considered by SOD.

More recently the modified splitting scheme has been applied by FLORES and HOLT (1981) to determine the characteristics of underwater explosions.

2.16 Remarks on Multi-Dimensional Problems

CHORIN (1976) considered the extension of Glimm's method to unsteady flow in two dimensions. The governing equations, which, for plane flow, can be written in conservation form, are

$$\varrho_t + (\varrho u)_x + (\varrho v)_y = 0,$$
$$(\varrho u)_t + (p + \varrho u^2)_x + (\varrho uv)_y = 0,$$
$$(\varrho v)_t + (\varrho uv)_x + (p + \varrho v^2)_y = 0,$$
$$e_t + \{(e+p)n\}_x + \{(e+p)v\}_y = 0.$$
(2.16.1)

We form a network with nodes $x=ih$, $y=jl$, $t=nk$ and apply Glimm's method in fractional steps at each half time interval. Firstly, we drop the y derivatives in (2.16.1) and solve Riemann problems for the x, t equations in the time interval, evaluating the solution at randomly chosen points in $(-1/2h, 1/2h)$ about each node $\{(i+1/2)h, jl, (n+1/2)k\}$. A similar procedure is applied in the y direction to yield values at $\{(i+1/2)h, (j+1/2)l, (n+1/2)k\}$. Independent sets of van der Corput random numbers are used in the two directions. The same procedure is used in the next half time step.

COLELLA (1982) made an extensive study of multi-dimensional problems and concluded that Glimm's method is only partially satisfactory in this case. The method follows contact discontinuities with reasonable accuracy but can only follow strong shocks by introducing artifical viscosity into the governing equations. Much further work is needed to make Glimm's method effective in the multi-dimensional case.

References

Alalykin, G. B., Godunov, S. K., Kireeva, I. L., Pliner, L. H.: *Solutions of One Dimensional Problems in Gas Dynamics in Moving Networks*. Moscow: NAUKA 1970.

Chorin, A. J.: J. Comp. Phys. **22**, 517-533 (1976).

Chorin, A. J.: J. Comp. Phys. **25**, 253-272 (1977).

Colella, P.: SIAM J. Sci. Comput. **3**, 76-110 (1982).

Concus, P., Proskurowski, W.: J. Comp. Phys. **30**, 153-166 (1979).

Flores, J., Holt, M.: J. Comp. Phys. **44**, 377-387 (1981).

Glimm, J.: Comm. Pure Appl. Math. **18**, 697-715 (1955).

Godunov, S. K.: Mat. Sborn. **47**, 271 (1959).

Godunov, S. K., Zabrodin, A. V., Prokopov, G. P.: USSR Comp. Math. Math. Phys. **6**, 1020-1050 (1961).

Godunov, S. K., Deribas, A. A., Zabrodin, A. V., Kozin, N. S.: J. Comp. Phys. **5**, 517-539 (1970).

Godunov, S. K., Ryabenkii, V. W.: *The Theory of Difference Schemes*. New York: Wiley 1964.

Holt, M., Li, K.-M.: Phys. Fluids **24**, 816 (1981).

Masson, B. S., Taylor, T. D., Foster, R. M.: AIAA J, **7**, 694 (1969).

Masson, B. S., Taylor, T. D.: Polish Fluid Dynamic Transactions **5** (1971).

Moretti, G.: AIAA J, **5** (1967).

Rozhdestvenskii, B. L., Yanenko, N. N.: *Theory of Quasilinear Hyperbolic Partial Differential Equations*. Moscow: NAUKA 1970.

Sod, G. A.: J. Fluid Mech. **83**, 785-794 (1977).

Sod, G. A.: J. Comp. Phys. **27**, 1-31 (1978).

Taylor, T. D., Masson, B. S.: J. Comp. Phys. **5** (1970).

Taylor, T. D.: AGARDograph No. 187, 1974.

Vasiliev, O. F.: *Lecture Notes in Physics No. 8* (Ed. M. Holt), p. 410. Berlin-Heidelberg-New York: Springer 1971.

Whitham, G. B.: *Linear and Non-linear Waves*. New York: Wiley 1974.

Wigton, L.: Term paper, Course ME 266, University of California, Berkeley, 1978.

CHAPTER 3

The BVLR Method

3.1 Description of Method for Supersonic Flow

An alternative finite difference method for calculating steady high speed flow past bodies of general shape was proposed by BABENKO and VOSKRESENSKII (1961) and applied to circular conical flows by BABENKO, VOSKRESENSKII, LYUBIMOV and RUSANOV (1963). GONIDOU (1967), in generalizing this application to elliptic cones, uses the shorter description, BVLR method, and this has now been generally adopted. The method was first developed for purely supersonic flows but was later extended by RUSANOV (1968) and by LYUBIMOV and RUSANOV (1970) to mixed subsonic-supersonic flows and applied to calculate a series of complex blunt body flow fields. This extension is discussed in Sect. 3.2.

In supersonic steady flow the equations of motion are hyperbolic and one of the coordinates, x say, can always be chosen to be time like; we only require that this coordinate direction be inclined at a sufficiently small angle to the velocity vector. The BVLR method is a marching technique for advancing the integration of the equations of motion from a plane $x = $ constant to an adjacent plane downstream.

Consider the problem of uniform supersonic flow past a body of revolution in general, at angle of attack, shown in Fig. 3.1. If the nose of the body is smooth, as shown, the disturbed flow field is in two parts, (i) mixed subsonic-

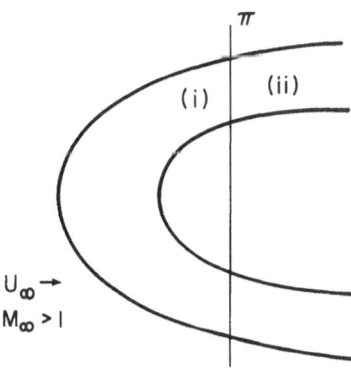

Fig. 3.1   Flow regions near blunt body

supersonic near the body stagnation point, and (ii) purely supersonic. If the nose is pointed the flow in region (i) grows from a conical flow. We suppose that flow in region (i) is known (its calculation for blunt bodies is discussed in Sect. 3.2), and that values of the dependent variables, including the position and slope of the shock wave, are prescribed on the whole plane π.

The equations of motion for steady flow can be written, in cylindrical polars based on the body,

$$A' \frac{\partial X}{\partial z} + B' \frac{\partial X}{\partial r} + C' \frac{\partial X}{\partial \phi} + \Gamma' = 0 \tag{3.1.1}$$

where

$$A' = \begin{pmatrix} u & 0 & 0 & 1/\varrho & 0 \\ 0 & u & 0 & 0 & 0 \\ 0 & 0 & u & 0 & 0 \\ \varrho a^2 & 0 & 0 & u & 0 \\ \varrho & 0 & 0 & 0 & u \end{pmatrix} \qquad B' = \begin{pmatrix} v & 0 & 0 & 0 & 0 \\ 0 & v & 0 & 1/\varrho & 0 \\ 0 & 0 & v & 0 & 0 \\ 0 & \varrho a^2 & 0 & v & 0 \\ 0 & \varrho & 0 & 0 & v \end{pmatrix}$$

$$C' = \frac{1}{r} \begin{pmatrix} w & 0 & 0 & 0 & 0 \\ 0 & w & 0 & 0 & 0 \\ 0 & 0 & w & 1/\varrho & 0 \\ 0 & 0 & \varrho a^2 & w & 0 \\ 0 & 0 & \varrho & 0 & w \end{pmatrix} \tag{3.1.2}$$

and X, Γ' are the column vectors

$$\Gamma' = \frac{1}{r} \begin{pmatrix} 0 \\ -w^2 \\ wv \\ \varrho a^2 v \\ \varrho v \end{pmatrix} \qquad X = \begin{pmatrix} u \\ v \\ w \\ p \\ \varrho \end{pmatrix} \tag{3.1.3}$$

Let the equations of the body and shock wave be

$$r = G(z, \phi) \tag{3.1.4}$$

$$r = F(z, \phi) \tag{3.1.5}$$

respectively. If suffix v denotes the component of the velocity normal to the shock the five shock equations are

$$\varrho q_v = \varrho_\infty q_{v\infty}$$

$$p + \varrho_\infty q_{v\infty} q_v = p + \varrho q_{v\infty}^2$$

$$h + \tfrac{1}{2} q_v^2 = h_\infty + \tfrac{1}{2} q_{v\infty}^2 \qquad (3.1.6)$$

$$u + v F_z = u_\infty + v_\infty F_z$$

$$v \frac{F_\varphi}{F} + w = v_\infty \frac{F_\varphi}{F} + w_\infty$$

where

$$q_v = (u F_z - v + w F_\varphi/F)/(1 + F_z^2 + F_\varphi^2/F^2)^{1/2}\ .$$

The solid wall condition is

$$u G_z - v + w \frac{G_\varphi}{G} = 0 \qquad (3.1.7)$$

In (3.1.7) G is given and for a body of revolution $G_\varphi = 0$. Eqs. (3.1.6) contain six unknowns $(u, v, w)_{\text{shock}}$, p, ϱ and F.

We consider the following problem. Given values of X on a plane π we wish to calculate a solution in planes $\pi_1, \pi_2, \ldots$ downstream of π which satisfies (3.1.1) and the boundary conditions (3.1.6), (3.1.7).

If the z component of velocity is everywhere supersonic we can choose the planes $\pi_1, \pi_2, \ldots$ as planes $z = \text{constant}$.

We change our coordinate system so that the coordinate crossing the flow field is tied to the body and shock geometry; thus we take

$$x = z, \qquad \xi = \xi(z, r, \phi), \qquad \theta = \phi$$

so that $\xi = 0$ on the wall, $\xi = 1$ on the shock. A convenient, and normally acceptable form for ξ is

$$\xi = \{r - G(z, \phi)\}/\{F(z, \phi) - G(z, \phi)\} \qquad (3.1.8)$$

The choice of ξ is restricted by the conditions that the BVLR difference scheme for solving (3.1.1) with boundary conditions (3.1.6), (3.1.7) should be stable and convergent. This is discussed at length in BABENKO et al. (1963) and leads to two inequalities to be satisfied by ξ. Their significance is explained in terms of Fig. 3.2, showing the projections of stream surface, plus and minus characteristic surfaces on the plane $\phi=$ constant and the $\xi=$ constant direction. For a well conditioned BVLR solution the $\xi=$ constant direction must lie above the streamline and $+$ Mach line direction while it lies below the $-$ Mach line direction. Eq. (3.1.8) will normally meet this requirement.

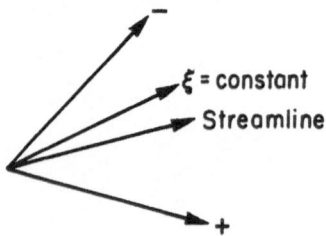

Fig. 3.2 Characteristic directions in three-dimensional flow

Following the change of variables the region in which a solution is sought is

$$x > x_0, \qquad 0 \le \theta \le 2\pi, \qquad 0 \le \xi \le 1$$

while (3.1.1) are replaced by

$$A\frac{\partial X}{\partial x} + B\frac{\partial X}{\partial \xi} + C\frac{\partial X}{\partial \theta} + \Gamma = 0 \tag{3.1.9}$$

where $A = A'$, $B = \xi_z A' + \xi_r B' + \xi_\phi C'$, $C = C'$, $\Gamma = \Gamma'$.

Difference Scheme

We construct a rectangular network with steps

$$\Delta x = \tau, \qquad \Delta \xi = \frac{1}{M} = h_1, \qquad \Delta \theta = \frac{2\pi}{L} = h_2$$

where M and L are integers.
Network nodes have coordinates

$$x^n = n\tau + x^0, \qquad \xi_m = mh_1, \qquad \theta_l = lh_2$$

while network functions are written

$$f(x^n, \xi_m, \theta_l) = f^n_{m,l}$$

n, m, l all being integers.
We call the set of points with the same n a layer. The set with the same n and l is called a ray.

In replacing (3.1.9) by a finite difference equation we use centered difference formulae for the derivatives, modified by the addition of an artificial viscosity term in the x direction. These generally are of second order accuracy. The finite difference approximation to (3.1.9) can be written in symbolic form

$$a^n_{m+1/2,l} X^{n+1}_{m+1,l} + b^n_{m+1/2,l} X^{n+1}_{m,l} = f_{m+1/2,l} \tag{3.1.10}$$

We solve (3.1.10) by an iterative process. At the beginning of each cycle of iteration we use the latest estimates of $X^{n+1}_{m,l}$ to calculate the coefficients $a^n_{m+1/2,l}, b^n_{m+1/2,l}, f^n_{m+1/2,l}$. In the following cycle we can then regard (3.1.10) as linear difference equations for X with known coefficients. The iteration is terminated when the initial and final values of X differ by a sufficiently small quantity.

Eq. (3.1.10) represents $5ML$ scalar difference equations. In addition, for each l we have a wall condition and five shock conditions, giving a total of $5ML+6L$ equations. Since the unknowns in these equations correspond only to the l ray, we can divide them into L groups of $5M+6$ equations along each ray $\begin{pmatrix} n,l \\ \text{constant} \end{pmatrix}$.

For each ray we therefore must solve the equations

$$a_{m+1/2} X_{m+1} + b_{m+1/2} X_m = f_{m+1/2}, \quad m=0,1,\ldots,M-1. \tag{3.1.11}$$

To carry this out the double sweep method is used. The body surface condition can be written

$$\mu_0 X_0 = g_0 \tag{3.1.12}$$

where

$$\mu_0 = \omega_0 \{G_x, -1, G_\theta/G, 0, 0\}$$
$$\omega_0 = \{G_x^2 + 1 + G_\theta^2/G^2\}^{1/2}.$$

We use direct sweep to transfer condition (3.1.12) to the shock point. At a general point the formula

$$\mu_m X_m = g_m \tag{3.1.13}$$

is established, with the recurrence formulae

$$\mu_{m+1} = \omega_{m+1} \mu_m (b^{-1} a)_{m+1/2}$$
$$g_{m+1} = \omega_{M+1} [\mu_m (b^{-1} f)_{m+1/2} - g_m]$$

ω_m is a constant chosen to normalize μ_m [$\omega_0 = \{G_x^2 + 1 + G_\theta^2/G^2\}^{-1/2}$, for example]. At each step in the direct sweep μ_m and g_m are calculated. At the shock we have

$$\mu_M X_M = g_M \tag{3.1.14}$$

together with the shock equations (3.1.6). These form six equations for the six unknowns X_M and F. To perform the inverse sweep linear combinations of (3.1.11) and (3.1.13) are formed to yield a relation of the form

$$X_m = c_m X_{m+1} + d_m \tag{3.1.15}$$

such that $\|c_m\| < 1$, required for stability. Simple inversion of (3.1.11) is not stable since $\|(b^{-1}a)_{m+1/2}\| > 1$. The details of the derivation of (3.1.15) are complicated and are discussed in LYUBIMOV and RUSANOV (1970).

BABENKO et al. applied their method to the problem of supersonic flow past a circular cone. The correct solution of this problem is independent of length scale and is determined by functions of the angular variables (r/z and ϕ in cylindrical polars). To apply the BVLR method we estimate initial values of the unknowns and the shock position in a plane π. We then apply the above process step by step downstream up to a plane π_n where the condition

$$\|X_n - X_{n-1}\| < \delta$$

is satisfied, δ being an assigned small quantity defining the accuracy required.

In the cone calculations the symmetrical zero incidence flow was first determined (ϕ variation absent). The angle of incidence was then increased successively by increments of $1°$ so that each new flow field found was a small departure from one just previously found. In this way BABENKO et al. (1963)

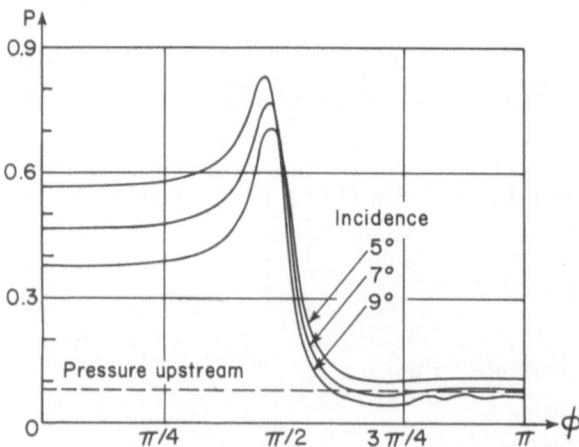

Fig. 3.3   Pressure on an elliptic cone at incidence (angles of attack $5°$, $7°$, $9°$, $M_\infty = 7$)

were able to provide a complete set of cone tables for a range of Mach numbers, cone angles and angles of attack.

GONIDOU (1967) applied the method to circular cones at moderate angles of attack, including examples where the angle of incidence slightly exceeds the cone angle. He shed some light on the behavior of constant entropy surfaces on the leeward side of the cone, confirming the phenomenon of lift off of vortical singularity at high angles of attack. GONIDOU also extended the method to apply to cones of elliptic cross section. Figure 3.3 shows the pressure distribution for a moderately flat elliptic cone of axis ratio 3:1, while Fig. 3.4 shows the isobars and shock shape for this cone at angle of incidence 9°. It is noteworthy that extensive regions of almost constant pressure exist on both the windward and leeward faces.

We shall return to the supersonic yawed cone problem in Chapt. 5.

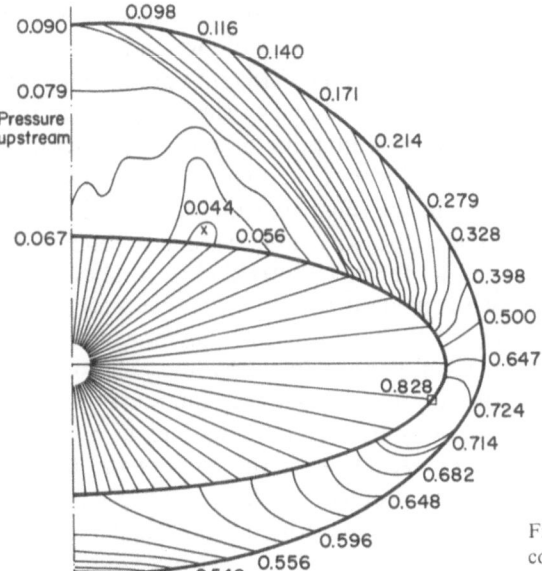

Fig. 3.4 Isobars around an elliptic cone (meridian plane half angle = 20°, axis ratio = 1/3. $M_\infty = 7$)

3.2 Extensions to Mixed Subsonic-Supersonic Flow. The Blunt Body Problem

The BVLR method as presented in Sect. 3.1 is applicable strictly to supersonic flows, in particular, flows past pointed bodies or the supersonic parts of flows past blunt nosed bodies. To start the method we need data on some initial plane π shown in Fig. 3.1. To determine these data in the case of flow past smooth bodies we must solve the so-called blunt body problem.

In the general blunt body problem, illustrated in Fig. 3.1, we consider uniform supersonic flow past a three dimensional body of smooth and continuous shape. Since there must be a stagnation point at the nose of the body it is surrounded by a detached shock wave. At the head of the body, near the stagnation point, there is a subsonic region which is bounded by the body, a sonic surface and a shock wave.

In the case of steady flow the flow field can be divided into three regions, a uniform supersonic flow upstream of the shock, a non-uniform subsonic region behind the shock, and non-uniform supersonic region downstream of the surface. If we know the values of the unknowns along a surface π extending completely over the latter region we can use the BVLR method to calculate the flow further downstream. But conditions on this surface depend on the subsonic flow region. The problem of calculating the flow is very difficult and several methods have been proposed to achieve this.

These methods are of two types, the first applicable to the steady form of the problem, as presented here, and the second starting with an unsteady formulation. In the steady formulation the equations of motion are of mixed elliptic-hyperbolic type and a non-linear version of Tricomi's problem must be solved. If the boundaries of the flow were fixed and of simple form (say rectangular or circular) we could solve these equations by replacing derivatives by finite differences and applying a relaxation method. Unfortunately, the boundary conditions are non-linear in the sense that the shape of the shock wave, sonic line and limiting characteristics are unknown and depend on the solution inside the region bounded by them. UCHIDA and YASUHARA (1956), among others, tried to apply such a method using an iterative process, but found poor convergence. Experience has shown that the only practical way to solve the steady problem is to represent the unknowns by interpolation series in all but one of the independent variables so that the equations of motion can be reduced to a system of ordinary differential equations to determine the coefficient in the series. This representation is the basis of the Method of Integral Relations, applied to this problem by BELOTSERKOVSKII (1960) and of Telenin's method (1964). These will be discussed in Chapt. 5 and 6, respectively.

In the second type of method for solving the blunt body problem the flow is regarded as unsteady. If we introduce time dependence into the problem the equations of motion are of hyperbolic type and can be solved by a finite difference method (marching method). Godunov's method is the first of this type and is a generalization of the basic method used to calculate one dimensional problems of Gas Dynamics. The second unsteady method is of fairly recent origin and was developed by LYUBIMOV and RUSANOV (1970). It is a generalization of the BVLR method and based on the same principles.

The BVLR method, when applied to conical flow, is started in a plane π and calculates flow conditions in successive planes $\pi_1, \pi_2, \ldots, \pi_n, \pi_{n+1}, \ldots$ downstream. The calculation is terminated when values in π_n and π_{n+1} differ by sufficiently small amounts.

In the Lyubimov-Rusanov generalization of BVLR there are three space coordinates x_1, x_2, x_3 and the time t. Initially, at time $t=0$ we give the values of the unknowns in a hyperplane $\phi(x_1, x_2, x_3, 0)=0$ and calculate the flow in successive hyperplanes $\phi(x_1, x_2, x_3, t)=0$, $t=\tau$, $t=2\tau, \ldots, n\tau$. We terminate the calculation when the condition $\|X^{n+1} - X^n\| < \varepsilon$ is satisfied for sufficiently small $\varepsilon > 0$.

The governing equations are generalizations of (3.1.1).

$$\frac{\partial X}{\partial t} + A' \frac{\partial X}{\partial q_1} + B' \frac{\partial X}{\partial q_2} + C' \frac{\partial X}{\partial q_3} = \Gamma \tag{3.2.1}$$

where X and F are five element column vectors and A', B', C' are 5×5 matrices similar to (3.1.2) and q_1, q_2, q_3 are generalized space coordinates. We transform to coordinates ξ, η, ζ tied to the geometry of the body and enclosing shock wave. In the case of a body of revolution at angle of attack we use the cylindrical polars fixed in the body, z, r, θ. To choose ξ, η, ζ in this case refer to Fig. 3.5 showing a section of the body and shock wave in a meridian plane.

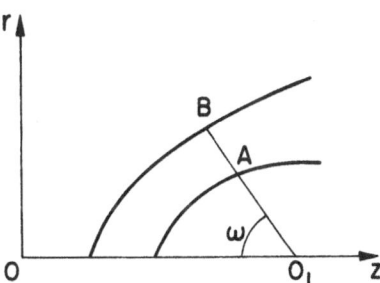

Fig. 3.5   Coordinate system for blunt body problem

We take a new origin 0_1 on the axis of symmetry within the body. The general radius vector through 0_1 meets the body at A and the shock at B.

We write

$$0_1 A = G(\eta, \theta, t)$$

$$0_1 B = F(\eta, \theta, t)$$

$$R = G + \xi(F - G)$$

taking $\eta = \omega$, $\xi = (R - G)/(F - G)$, $\zeta = \theta$, so that $\xi = 0$ on the body, $\xi = 1$ on the shock.

Eqs. (3.2.1), after transformation to new coordinates ξ, η, ζ, t are written in the following finite difference form, analogous to (3.1.10),

$$a^n_{m+1/2, k, l} X^{n+1}_{m+1, k, l} + b^n_{m+1/2, k, l} X^{n+1}_{m, k, l} = \pi^n_{m+1/2, k, l} \tag{3.2.2}$$

where a, b, c are matrix coefficients containing transfer operators in all directions and n, m, k, l are integers corresponding to network values of t, ξ, η and θ, respectively.

We define ξ and θ as in steady flow with $\xi_m = m h_1$ $(m = 0, 1, ..., M)$, $M h_1 = 1$, $\theta_l = l h_3$ $(l = 0, ..., L)$, $L h_3 = 2\pi$. Suffix k corresponds to the angular variable η—the angle between a ray and the central axis of a spherical polar coordinate system (frequently the body axis of symmetry). Then $\eta_k = k h_2$, $k = 0, 1, ..., K$. Suffix n corresponds to the time $t = n\tau$ where τ is the time step. Rays for which k has values different from $0, 1, K$ and l has values other then $0, L$ are called interior rays. All other rays are called boundary rays.

Eqs. (3.2.2) can be decomposed into independent equations along rays of the type

$$a_{m+1/2} X_{m+1} + b_{m+1/2} X_m = \pi_{m+1/2} \tag{3.2.3}$$

which are solved by double sweep, transferring the solid wall boundary conditions to the shock and returning to the body. The shock equations are implicit and must be solved by an iterative process. The sweep process is discussed in Sect. 3.3.

LYUBIMOV and RUSANOV have carried out an impressive series of applications of their method to a wide variety of three dimensional blunt body configurations. These include a family of elliptic paraboloids and Fig. 3.6 shows

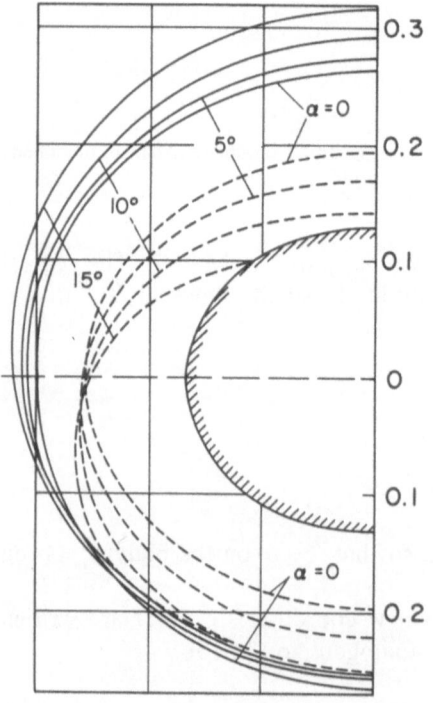

Fig. 3.6 Sections of shock waves and sonic lines on the elliptic paraboloid $z = 0.5 \, (5 \cos^2\theta + 3 \sin^2\theta) r^2$

end views (looking downstream) of the flow field for a paraboloid with a 5:3 elliptic cross section at angles of attack ranging from $0°$ to $15°$. The solid lines are sections of the shock surface, while the dashed lines are sonic line sections.

3.3 The Double Sweep Method for Unsteady Three-Dimensional Flow

The double sweep method applied along rays between the body and shock is presented more clearly by LYUBIMOV and RUSANOV (1970) than in the earlier, steady BVLR account (BABENKO et al., 1963). In particular, the difficulties found with stability in the inverse sweep for the steady formulation are now overcome by separating one of the equations of motion from the sweep process and solving this after each step in the sweep has been completed.

A distinction is made between general interior rays and boundary rays in the sweep process. It will be sufficient here to describe the procedure for interior rays.

To simplify the sweep process it is necessary to take advantage of the specific structure of the coefficient matrices $a_{m+1/2}$ and $b_{m+1/2}$.

Now the object of the sweep process is to establish a succession of relations of the type

$$\mu_m X_m = g_m \tag{3.3.1}$$

starting with the surface boundary condition and using (3.2.2) for $m = 0, 1, \dots$.

LYUBIMOV and RUSANOV observe that this process is simpler (and also stable in the inverse sweep) if one equation is at first omitted from (3.2.3).

The matrix $b_{m+1/2}$, written in full, is

$$b_{m+1/2} = \begin{pmatrix} b_0 & 0 & 0 & b_1 & 0 \\ 0 & b_0 & 0 & b_2 & 0 \\ 0 & 0 & b_0 & b_3 & 0 \\ b_{41} & b_{42} & b_{43} & b_0 & 0 \\ b_{51} & b_{52} & b_{53} & 0 & b_0 \end{pmatrix}_{m+1/2}.$$

The $a_{m+1/2}$ matrix has the same structure with b replaced everywhere by a. The elements are given by

$$b_{4s} = \varrho^2 c^2 b_s \qquad b_{5s} = \varrho^2 b_s \qquad s = 1,2,3$$
$$a_0 = 2 - b_0 \qquad a_{ij} = -b_{ij} \qquad i = 4,5$$
$$a_j = -b_j \qquad j = 1,2,3.$$

In the surface boundary condition (3.1.12) (which still applies) $\mu_{5,0}=0$. Accordingly, as can easily be verified, when $\mu_{5,1}$ is calculated by eliminating X_0 from (3.1.12) and (3.2.3), because of the structure of the matrices $\{a_{1/2}\}$ and $\{b_{1/2}\}$, we find that $\mu_{5,1}=0$. Continuing the calculations for higher m we find that $\mu_{5,m}=0$ for all m.

We can therefore simplify the sweep process by omitting from it the last of (3.2.3) (the equation giving the density ϱ).

For the reduced system the notation

$$
\tilde{a}_{m+1/2} = \begin{pmatrix} a_0 & 0 & 0 & a_1 \\ 0 & a_0 & 0 & a_2 \\ 0 & 0 & a_0 & a_3 \\ a_{41} & a_{42} & a_{43} & a_0 \end{pmatrix}_{m+1/2}
\qquad
\tilde{b}_{m+1/2} = \begin{pmatrix} b_0 & 0 & 0 & b_1 \\ 0 & b_0 & 0 & b_2 \\ 0 & 0 & b_0 & b_3 \\ b_{41} & b_{42} & b_{43} & b_0 \end{pmatrix}_{m+1/2}
$$

$$
\tilde{X}_m = \begin{pmatrix} u \\ v \\ w \\ p \end{pmatrix}
\qquad
\tilde{\pi} = \begin{pmatrix} \pi_1 \\ \pi_2 \\ \pi_3 \\ \pi_4 \end{pmatrix}
$$

$$
\tilde{\mu}_m = \{\mu_{1,m}, \mu_{2,m}, \mu_{3,m}, \mu_{4,m}\}
$$

is used and (3.2.3), (3.3.1) are replaced by

$$
\tilde{b}_{m+1/2} \tilde{X}_m + \tilde{a}_{m+1/2} \tilde{X}_{m+1} = \tilde{\pi}_{m+1/2} \tag{3.3.2}
$$

$$
\tilde{\mu}_m \tilde{X}_m = g_m \tag{3.3.3}
$$

LYUBIMOV and RUSANOV show that

$$
\tilde{\mu}_{m+1} = \frac{\tilde{\mu}_m(\tilde{\beta}\tilde{a})_{m+1/2}}{\|\tilde{\mu}_m(\tilde{\beta}\tilde{a})_{m+1/2}\|}, \qquad
g_{m+1} = \frac{\tilde{\mu}_m(\tilde{\beta}\tilde{a})_{m+1/2} - g_m \Delta_{m+1/2}}{\|\tilde{\mu}_m(\tilde{\beta}\tilde{a})_{m+1/2}\|}
$$

where $\tilde{\beta}_{m+1/2}$ is the transpose of the matrix of signed minors of the elements $\tilde{b}_{m+1/2}$ and $\Delta_{m+1/2}$ is the determinant of $\tilde{b}_{m+1/2}$. In normalizing $\tilde{\mu}_{m+1}$ any norm of the finite dimensional space can be used, for example, $\|\tilde{\mu}\| = \max |\tilde{\mu}_i|$.

The full expression for the elements of $a_{m+1/2}$, $b_{m+1/2}$ and their minors are given in LYUBIMOV and RUSANOV (1970).

Eq. (3.3.3) is solved step-by-step up to $m=M$. The relation is then combined with the shock relation (the unsteady version of (3.1.6) with absolute velocities replaced by velocities relative to the shock) and solved for X_M and D (the shock speed). The reverse sweep is then performed for successive $m=M-1, M-2,\dots$. First we solve (3.3.2) for $\tilde{X}_m$ in terms of $\tilde{X}_{m+1}$. Then we solve the missing equation of (3.2.3) to give ϱ_m in terms of ϱ_{m+1} and X_{m+1}. LYUBIMOV and RUSANOV show that the reverse sweep for the reduced system (3.3.2) is stable, in contrast to the situation with the full system (3.2.3).

3.4 Worked Problem. Application to Circular Arc Airfoil

The application of the BVLR and Lyubimov-Rusanov methods to the full problem of supersonic flow past a general blunt nosed body requires a great deal of programming, well beyond the capacity of a graduate student working part time. Nevertheless, all the principal features of the method are brought out if its application is restricted to two dimensional steady flow. This simplified calculation was performed as the term assignment for the author's class, Mechanical Engineering 266, in 1975 by two graduate students in Mechanical Engineering at the University of California, Berkeley, D. R. Shanti Gunewardana and K. S. Chang.

The reader is invited to attempt the problem and then compare his results with the solution given below.

Problem

Calculate supersonic flow past a circular cylindrical airfoil at zero angle of attack. The free stream Mach number is 2.5 and $h/c = 8$.

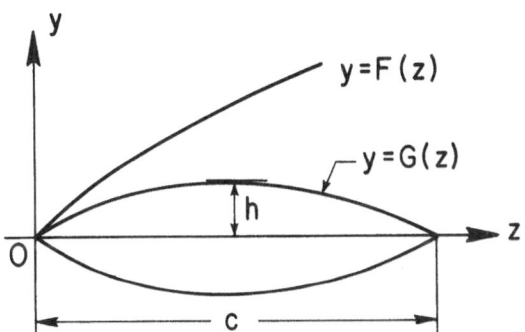

Fig. 3.7   Circular arc airfoil

In Cartesian coordinates, the equation of the body is

$$y = G(z) = -\tfrac{15}{16} + ((\tfrac{17}{16})^2 - (z - \tfrac{1}{2})^2)^{1/2}$$

$$G_z = -\frac{(z - \tfrac{1}{2})}{((\tfrac{17}{16})^2 - (z - \tfrac{1}{2})^2)^{1/2}} \, .$$

Formulation and Solution (by Gunewardana and Chang)

Initial data: Flow conditions up to a small distance downstream from the leading edge are determined using a two dimensional double wedge approximation. Let this small distance be x_0. In 2−D Cartesian coordinates, the equations of motion (3.1.1) reduce to

$$A' \frac{\partial X}{\partial z} + B' \frac{\partial X}{\partial y} = 0 \tag{3.4.1}$$

where

$$A' = \begin{pmatrix} u & 0 & \dfrac{1}{\varrho} & 0 \\ 0 & u & 0 & 0 \\ \varrho a^2 & 0 & u & 0 \\ \varrho & 0 & 0 & u \end{pmatrix}, \quad B' = \begin{pmatrix} v & 0 & 0 & 0 \\ 0 & v & \dfrac{1}{\varrho} & 0 \\ 0 & \varrho a^2 & v & 0 \\ 0 & \varrho & 0 & v \end{pmatrix}, \quad X = \begin{pmatrix} u \\ v \\ p \\ \varrho \end{pmatrix}$$

Eqs. (3.1.9) become

$$A \frac{\partial X}{\partial x} + B \frac{\partial X}{\partial \xi} = 0 \tag{3.4.2}$$

where

$$A = A'$$
$$B = \xi_z A' + \xi_y B'$$
$$\xi = \frac{y - G(z)}{F(z) - G(z)}$$
$$\xi_z = -\frac{G_z + \xi(F_z - G_z)}{(F - G)} \quad \text{and} \quad \xi_y = \frac{1}{(F - G)}.$$

Eqs. (3.1.6) and (3.1.7) become

$$\varrho V_v = \varrho_\infty V_{v\infty} \quad \text{where} \quad V_v = \frac{u F_x - v}{(1 + F_x^2)^{1/2}}$$

$$p + \varrho_\infty V_{v\infty} V_v = p_\infty + \varrho_\infty V_{v\infty}^2$$

$$h + \frac{1}{2} V_v^2 = h_\infty + \frac{1}{2} V_{v\infty}^2 \quad \text{where} \quad h = \frac{\gamma}{\gamma - 1} \frac{p}{\varrho} \tag{3.4.3}$$

$$u + v F_x = u_\infty + v_\infty F_x$$

$$u G_x - v = 0$$

The system of difference equations (see LYUBIMOV and RUSANOV (5.1), (5.2) with $\sigma = 0$) will now become

$$A_{m+1/2}^{n+1/2} \left[X_{m+1}^{n+1} - X_{m+1}^n + X_m^{n+1} - X_m^n \right]$$
$$+ B_{m+1/2}^{n+1/2} 2\kappa_1 \left[\alpha(X_{m+1}^{n+1} - X_m^{n+1}) + \beta(X_{m+1}^n - X_m^n) \right] = 0 \tag{3.4.4}$$

with the following boundary conditions,

$$(\varrho V_{\nu})_M^{n+1} = (\varrho V_{\nu})_{\infty}^{n+1}$$

where

$$(V_{\nu})_M^{n+1} = \frac{u_M^{n+1}(F_x)^{n+1} - v_M^{n+1}}{(1 + ((F_x)^{n+1})^2)^{1/2}}$$

$$[p + (\varrho_{\infty} V_{\nu\infty}) V_{\nu}]_M^{n+1} = [p + \varrho V_{\nu}^2]_{\infty}^{n+1}$$

$$h(p_M^{n+1}, \varrho_M^{n+1}) + \tfrac{1}{2}(V_{\nu M}^{n+1})^2 = h_{\infty} + \tfrac{1}{2}(V_{\nu\infty}^{n+1})^2$$

$$u_M^{n+1} + (F_x)^{n+1} v_M^{n+1} = u_{\infty} + (F_x)^{n+1} v_{\infty}$$

$$G_x^{n+1} u_0^{n+1} - v_0^{n+1} = 0.$$

The location of the shock F^{n+1} is determined by

$$F^{n+1} = F^n + \frac{\tau}{2}(F_x^{n+1} + F_x^n).$$

We now have $4M+5$ equations with $4M+5$ unknowns.
By definition,

$$f_{m+1/2}^{n+(q/2)} = \tfrac{1}{4}(f_{m+1}^n + f_{m+1}^{n+(q)} + f_m^n + f_m^{n+(q)})$$

where q is the iteration with known flow conditions.
Hence we get

$$A_{m+1/2}^{n+(q/2)}\left[X_{m+1}^{n+(q+1)} - X_{m+1}^n + X_m^{n+(q+1)} - X_m^n\right]$$
$$+ B_{m+1/2}^{n+(q/2)} 2\kappa_1\left[\alpha(X_{m+1}^{n+(q+1)} - X_m^{n+(q+1)}) + \beta(X_{m+1}^n - X_m^n)\right] = 0 \tag{3.4.5}$$

It can be expressed in the form

$$a_{m+1/2}^{n+(q/2)} X_{m+1}^{n+(q+1)} + b_{m+1/2}^{n+(q/2)} X_m^{n+(q+1)} = f_{m+1/2} \tag{3.4.6}$$

where

$$a_{m+1/2}^{n+(q/2)} = A_{m+1/2}^{n+(q/2)} + 2\alpha\kappa_1 B_{m+1/2}^{n+(q/2)}$$
$$b_{m+1/2}^{n+(q/2)} = A_{m+1/2}^{n+(q/2)} - 2\alpha\kappa_1 B_{m+1/2}^{n+(q/2)}$$
$$f_{m+1/2} = A_{m+1/2}^{n+(q/2)}(X_{m+1}^n + X_m^n) - B_{m+1/2}^{n+(q/2)} 2\beta\kappa_1(X_{m+1}^n - X_m^n)$$
$$= X_{m+1}^n\left[A_{m+1/2}^{n+(q/2)} - 2\beta\kappa_1 B_{m+1/2}^{n+(q/2)}\right] + X_m^n\left[A_{m+1/2}^{n+(q/2)} + 2\beta\kappa_1 B_{m+1/2}^{n+(q/2)}\right] \tag{3.4.7}$$

or simply

$$a_{m+1/2} X_{m+1} + b_{m+1/2} X_m = f_{m+1/2} \qquad (3.4.8)$$

with

$$a_{m+1/2} = \begin{pmatrix} a_0 & 0 & a_1 & 0 \\ 0 & a_0 & a_2 & 0 \\ a_{31} & a_{32} & a_0 & 0 \\ a_{41} & a_{42} & 0 & a_0 \end{pmatrix} \qquad b_{m+1/2} = \begin{pmatrix} b_0 & 0 & b_1 & 0 \\ 0 & b_0 & b_2 & 0 \\ b_{31} & b_{32} & b_0 & 0 \\ b_{41} & b_{42} & 0 & b_0 \end{pmatrix} \qquad (3.4.9)$$

where

$$a_0 = u + 2\alpha\kappa_1(\xi_z u + \xi_y v)$$

$$b_0 = u - 2\alpha\kappa_1(\xi_z u + \xi_y v)$$

$$a_1 = \frac{1}{\varrho} + 2\alpha\kappa_1\left(\xi_z \frac{1}{\varrho}\right)$$

$$b_1 = \frac{1}{\varrho} - 2\alpha\kappa_1\left(\xi_z \frac{1}{\varrho}\right)$$

$$a_2 = 2\alpha\kappa_1\left(\xi_y \frac{1}{\varrho}\right)$$

$$b_2 = -a_2$$

and

$$a_{3i} = \varrho^2 a^2 a_i \qquad i = 1, 2$$

$$a_{4i} = \varrho^2 a_i \qquad i = 1, 2$$

$$b_{3i} = \varrho^2 a^2 b_i \qquad i = 1, 2$$

$$b_{4i} = \varrho^2 b_i \qquad i = 1, 2 .$$

Forward and Reverse Sweeps: The boundary value problem associated with boundary conditions at $m = 0$, i. e., (3.2.2), and (3.3.1) with $m = M$, and (3.1.6) is now solved using the double sweep method.

Before the double sweep can be performed, it is helpful to make certain observations pertaining to the structure of (3.2.3).

It is clear from the structure of the matrices (3.4.9) that the first three equations of (3.4.8) do not contain the unknown ϱ. Then, in the double sweep method, let us write the boundary condition at the body in the form

$$\mu_0 X_0 = g_0$$

where

$$\mu_0 = \omega_0(G_z, -1, 0, 0), \qquad g_0 = 0$$

and

$$\omega_0^{-1} = (G_z^2 + 1)^{1/2}, \qquad G_z = \frac{dG(z)}{dz} \quad [\text{note } z = x].$$

Also let us assume a relationship of the form

$$\mu_m X_m = g_m.$$

Re-writing the recurrence formulae below (3.1.13),

$$\mu_{m+1} = \omega_{m+1} \mu_m (b^{-1} a)_{m+1/2}$$

and

$$g_{m+1} = \omega_{m+1} \left(\mu_m (b^{-1} f)_{m+1/2} - g_m \right)$$

where ω_{m+1} is a normalizing factor that makes $\|\mu_{m+1}\| = 1$. Now we can determine the μ_m and g_m specified in (3.3.1).

However, in view of the observation made earlier pertaining to the matrices $a_{m+1/2}$ and $b_{m+1/2}$, it can be verified that $\mu_{m,4} = 0$ for all m. Hence, we can simplify the forward sweep by omitting the 4th equation in (3.2.3).

For this purpose let us introduce the following notation:

$$\tilde{a}_{m+1/2} = \begin{pmatrix} a_0 & 0 & a_1 \\ a & a_0 & a_2 \\ a_{31} & 0 & a_0 \end{pmatrix} \qquad \tilde{\mu}_m = (\mu_{m,1}, \mu_{m,2}, \mu_{m,3})$$

$$\tilde{b}_{m+1/2} = \begin{pmatrix} b_0 & 0 & b_1 \\ 0 & b_0 & b_2 \\ b_{31} & 0 & b_0 \end{pmatrix} \qquad \tilde{f}_{m+1/2} = \begin{pmatrix} f_1 \\ f_2 \\ f_3 \end{pmatrix}$$

and

$$\tilde{X}_m = \begin{pmatrix} u \\ r \\ p \end{pmatrix}$$

Now we have the following equations in place of (3.2.3) and (3.3.1):

$$\tilde{b}_{m+1/2} \tilde{X}_m + \tilde{a}_{m+1/2} \tilde{X}_m = \tilde{f}_{m+1/2} \tag{3.4.10}$$

$$\tilde{\mu}_m \tilde{X}_m = \tilde{g}_m \tag{3.4.11}$$

where, of course,

$$\tilde{\mu}_{m+1} = \omega_{m+1} \tilde{\mu}_m (\tilde{b}^{-1} \tilde{a})_{m+1/2}$$

and

$$\tilde{g}_{m+1} = \omega_{m+1} (\tilde{\mu}_m (\tilde{b}^{-1} \tilde{f})_{m+1/2} - \tilde{g}_m).$$

From the above relations $\tilde{\mu}_m$ and $\tilde{g}_m$ can be found for $m=0,1,2,...,M$. The final expression for the forward sweep will be,

$$\tilde{\mu}_M \tilde{X}_M = \tilde{g}_m.$$

This relation is now solved simultaneously with the boundary conditions at the shock.

Solution of the Shock Equations: The scheme given below is described in detail in NASA TT F-380.

Let us introduce the auxiliary variables $\tilde{u}$, $\tilde{v}$, $\tilde{g}$, y_1, and y_2 which will be determined from the following equations:

$$u_M = u_\infty + \tilde{u} \qquad\qquad\qquad v_M = v_\infty + \tilde{v}$$

$$y_1 = V_{v\infty}^2 \qquad\qquad\qquad y_2 = V_{v\infty} (\tilde{u}^2 + \tilde{v}^2)^{1/2}$$

$$\tilde{g}_M = g_M - (\mu_{M,1} u_\infty + \mu_{M,2} v_\infty + \mu_{M,3} P_\infty).$$

The shock equations (3.1.6), after some transformations, may be written in the following form. From here on μ refers to the components of vector $\tilde{\mu}$. Dropping the subscript M,

$$\mu_1 \tilde{u} + \mu_2 \tilde{v} = \tilde{g} - \mu_3 \varrho_\infty y_2$$

$$u_\infty \tilde{u} + v_\infty v = -y_2 \qquad\qquad\qquad\qquad (3.4.12)$$

$$\tilde{u}^2 + \tilde{v}^2 = y_2^2/y_1$$

$$p = p_\infty + \varrho_\infty y_2$$

$$\varrho = \varrho_\infty y_1/(y_1 - y_2) \qquad\qquad\qquad\qquad (3.4.13)$$

$$h(p,\varrho) = h(p_\infty, \varrho_\infty) + y_2 - \tfrac{1}{2} y_2^2/y_1$$

$$F_x = -\frac{\tilde{u}}{\tilde{v}} \qquad\qquad\qquad\qquad\qquad (3.4.14)$$

Solving for $\tilde{u}$ and $\tilde{v}$ from the first two equations of (3.4.12) in terms of y_2 and then substituting them into the third equation, we have

$$y_1 (c_0 y_2^2 + c_1 y_2 + c_2) - c_3 y_2^2 = 0 \qquad\qquad\qquad (3.4.15)$$

where

$$c_0 = (\mu_2 - \mu_3 \varrho_\infty v_\infty)^2 + (\mu_3 \varrho_\infty u_\infty - \mu_1)^2$$

$$c_1 = 2[(\mu_2 - \mu_3 \varrho_\infty v_\infty) v_\infty - (\mu_3 \varrho_\infty u_\infty - \mu_1) u_\infty] \tilde{g}$$

$$c_2 = (v_\infty^2 + u_\infty^2) \tilde{g}^2$$

$$c_3 = (\mu_1 v_\infty - \mu_2 u_\infty)^2 .$$

Now, substituting the values of p and ϱ from (3.4.13) into the third equation of (3.4.13) and noting that $h = \{k/(k-1)\} (p/\varrho)$, we get the following expression:

$$y_1 = \left(\frac{k+1}{2} y_2 + k \frac{p_\infty}{\varrho_\infty}\right) \tag{3.4.16}$$

Substituting (3.4.16) in (3.4.15), we get the following cubic equation for y_2:

$$(c_0 y_2^2 + c_1 y_2 + c_2)\left(\frac{k+1}{2} y_2 + k \frac{p_\infty}{\varrho_\infty}\right) - c_3 y_2^2 = 0 \tag{3.4.17}$$

where $k = 1.4$ for our problem.

Only one value of y_2 obtained from the above cubic equation has physical meaning. From the first equation of (3.4.13), it is clear that $y_2 > 0$, or else a rarefaction shock would result. After some analysis, it is shown in NASA TT F-380 that the cubic equation in y_2 has exactly two positive roots and that we can attribute physical meaning to only the smaller positive root. Having found y_2 it is a simple matter to determine u_M, v_M, p_M, ϱ_M, and F_x.

Now we are ready to perform the reverse sweep.

Reverse Sweep: The reverse sweep is performed by making successive calculations for $\tilde{X}_{M-1}, \tilde{X}_{M-2}, \dots, \tilde{X}_0$. The omitted fourth equation of (3.4.8) is used to solve for ϱ_{m+1} in terms of ϱ_m and $\tilde{X}_m$. The following reverse sweep relations are obtained as a result.

$$f_1^* = f_1 - a_0 u_{m+1} + b_1 p_{m+1}$$

$$f_2^* = f_2 - a_0 v_{m+1} + b_2 p_{m+1}$$

$$p_m = \frac{b_0 g_m - (\mu_{m,1} f_1^* + \mu_{m,2} f_2^*)}{b_0 \mu_{m,3} - (b_1 \mu_{m,1} + b_2 \mu_{m,2})}$$

$$u_m = b_0^{-1}(f_1^* - b_1 p_m)$$

$$v_m = b_0^{-1}(f_2^* - b_2 p_m)$$

and finally,

$$\varrho_m = b_0^{-1}\{f_4 - a_0 \varrho_{m+1} - \varrho_{m+1}^2 [b_1(u_m - u_{m+1}) + b_2(v_m - v_{m+1})]\} .$$

It is important to note that for each m during the forward and reverse sweeps we have to store only the following quantities: $\mu_{m,1}$, $\mu_{m,2}$, $\mu_{m,3}$, g_m, a_0, a_1, a_2, b_0, b_1, ϱ^2, $\varrho^2 a^2$, f_1, f_2, f_3, and f_4. The above observation results in considerable storage economy.

3.5 Results and Discussion

The chord was divided into 100 equal parts, making $\tau = 0.01$, and the normalized distance from the body to the shock was divided into 20 equal parts, making $h_1 = 0.05$. The computer output was compared with the results obtained from an approximate method using shock expansion theory (see Appendix) (HAYES and PROBSTEIN, 1959). Figure 3.8 gives a Calcomp plot of the shock curve. Figure 3.9 shows a comparison between this curve and that obtained by shock expansion theory. A flow chart for the computer program used is attached.

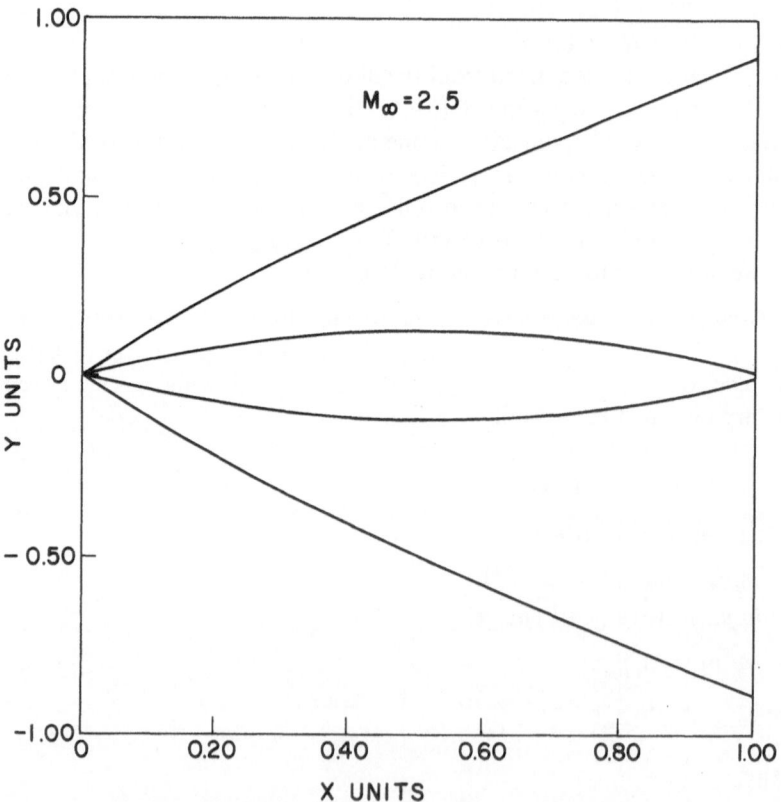

Fig. 3.8   Shock shape calculated on circular arc airfoil by BVLR method

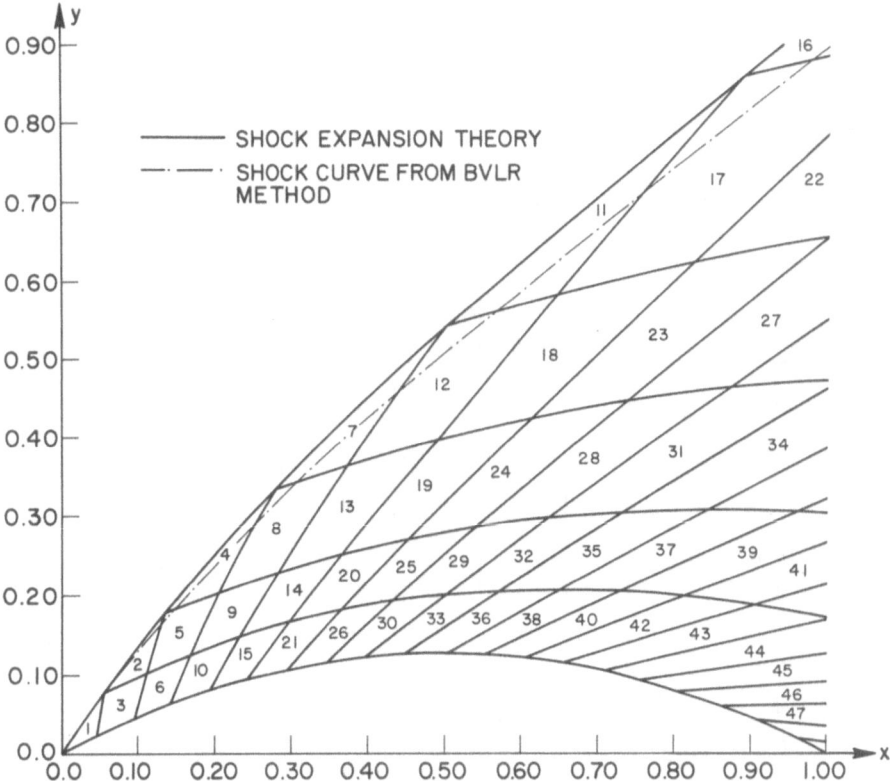

Fig. 3.9   Comparison of BVLR method with shock expansion method

LYUBIMOV and RUSANOV (1970) give a description of the extended BVLR method that can be used for the general problem of unsteady flow past a blunt body. This reference also gives algorithms that could be used to perform a stable double sweep. However, the interested reader is advised to read BABENKO et al. (1963) first, since the contents of LYUBIMOV and RUSANOV are very advanced.

Appendix—Shock Expansion Theory

The result of the BVLR method may be examined by another exact numerical scheme like the Method of Characteristics, but the latter is quite lengthy and onerous. Instead, we choose the shock-expansion theory which gives a fairly accurate approximation of the solution of the hypersonic flow over two dimensional sharp-nosed airfoils for which the shock is attached at the leading edge and the flow behind the shock is supersonic.

FLOW CHART

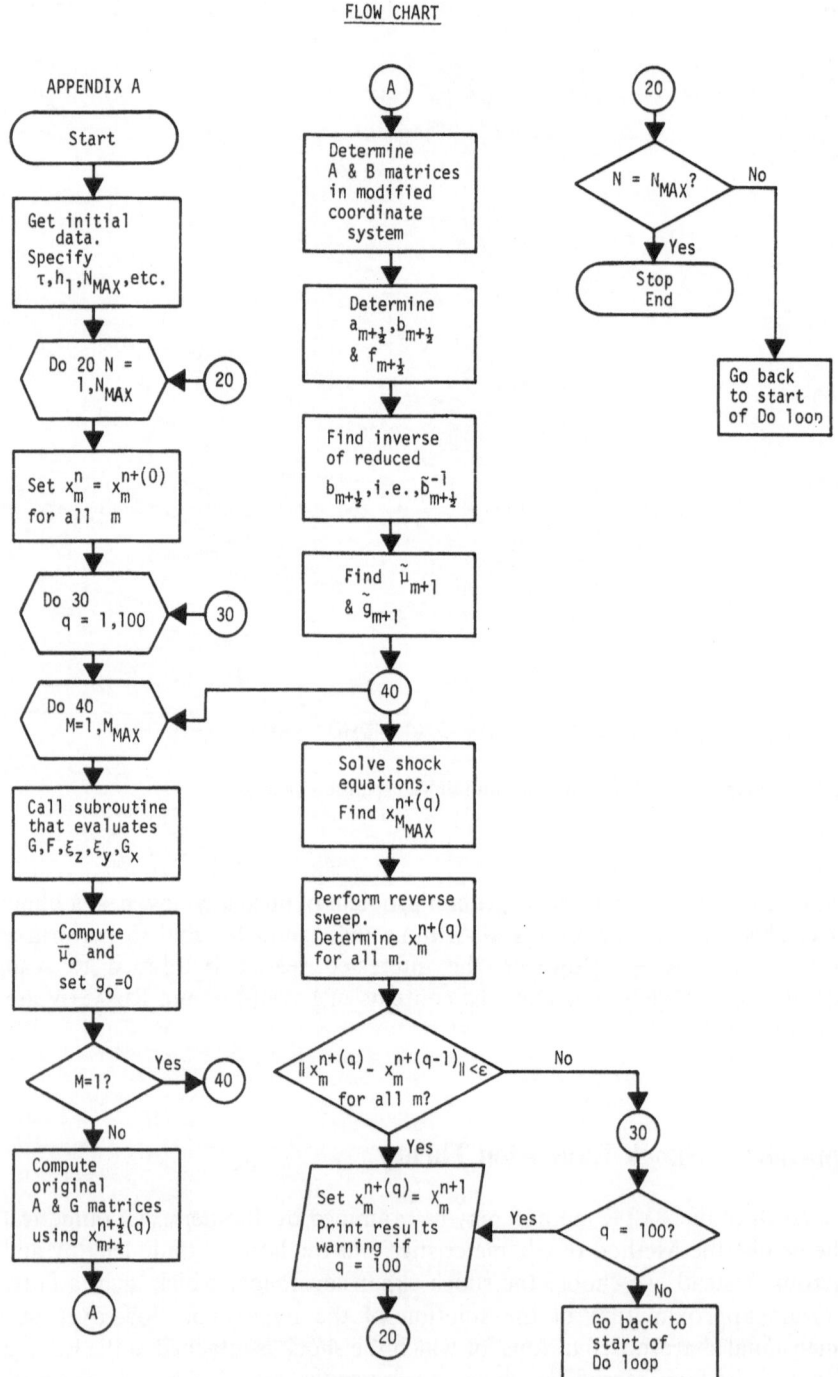

APPENDIX A

Start

Get initial data. Specify τ, h_1, N_{MAX}, etc.

Do 20 N = 1, N_{MAX} 20

Set $x_m^n = x_m^{n+(0)}$ for all m

Do 30 q = 1,100   30

Do 40 M=1, M_{MAX}

Call subroutine that evaluates G, F, ξ_z, ξ_y, G_x

Compute $\bar{\mu}_0$ and set $g_0 = 0$

M=1?   Yes   40

No

Compute original A & G matrices using $x_{m+\frac{1}{2}}^{n+\frac{1}{2}(q)}$

A

A

Determine A & B matrices in modified coordinate system

Determine $a_{m+\frac{1}{2}}, b_{m+\frac{1}{2}}$ & $f_{m+\frac{1}{2}}$

Find inverse of reduced $b_{m+\frac{1}{2}}$, i.e., $\tilde{b}_{m+\frac{1}{2}}^{-1}$

Find $\tilde{\mu}_{m+1}$ & $\tilde{g}_{m+1}$

40

Solve shock equations. Find $x_{M_{MAX}}^{n+(q)}$

Perform reverse sweep. Determine $x_m^{n+(q)}$ for all m.

$\| x_m^{n+(q)} - x_m^{n+(q-1)} \| < \epsilon$ for all m?   No

Yes

Set $x_m^{n+(q)} = x_m^{n+1}$ Print results warning if q = 100

20

20

N = N_{MAX}?   No

Yes

Stop End

Go back to start of Do loop

q = 100?   Yes

No

30

Go back to start of Do loop

The airfoil surface pressure and shock position can be well approximated by this method if the curvature of the airfoil is small. In the shock-expansion method, the airfoil characteristics are computed by assuming that the flow behind the leading edge shock is the same as an isentropic Prandtl-Meyer expansion, with only a single family of principal characteristics taken into account. Reflections of the characteristics from the shock wave are neglected since their strength is negligibly small for a weak shock. The field is not taken to be homentropic, however, and the Mach lines are not taken to be straight as in a Prandtl-Meyer flow. This allows the development of a procedure similar to the Method of Characteristics but much simpler than it. The scheme is illustrated in figure 3.9. For further details, see HAYES and PROBSTEIN (1959).

References

Babenko, K. I., Voskresenskii, G. P.: USSR Comp. Math. Math. Phys. 1 (1961).

Babenko, K. I., Voskresenskii, G. P., Lyubimov, A. N., Rusanov, V. V.: *Three Dimensional Flow of Ideal Gases around Smooth Bodies.* NASA TT F-380, 1968 (Russian original published by „NAUKA" Moscow 1964).

Gonidou, R.: La Recherche Aérospatiale No. 120, 1967.

Hayes, W. D., Probstein, R. F.: *Hypersonic Flow Theory.* New York: Academic Press 1959.

Lyubimov, A. N., Rusanov, V. V.: *Gas Flows Past Blunt Bodies,* Part I. NASA TT F-714, 1973 (Russian original published by „NAUKA" Moscow 1970).

Rusanov, V. V.: USSR Comp. Math. Math. Phys. 8, 1968.

Uchida, S., Yasuhara, M.: J. Aero. Sci. 23, 830–845 (1956).

CHAPTER 4

The Method of Characteristics for Three-Dimensional Problems in Gas Dynamics

4.1 Introduction

Although both the Godunov and BVLR methods for solving compressible flow problems in three dimensions use finite difference formulas in rectangular networks their foundations rest on the properties of characteristics of the governing equations. Both before and during the development of these methods, over the past twenty five years, a series of methods of characteristics, as such, have been proposed for three dimensional problems. Many of these compete with finite difference techniques, especially when applied to such problems as unsteady transonic flow, and it is appropriate to include a survey of such methods here. They may be classified under four headings:

1) Methods based entirely on bicharacteristics (THORNHILL (1952), FOWELL (1961), BUTLER (1960), RANSOM et al. (1970)).

2) Methods based on two bicharacteristic families and one noncharacteristic family (COBURN and DOLPH (1949), HOLT (1956)).

3) Optimal methods of type 2) (BRUHN and HAACK (1958), SCHAETZ (1961)).

4) Methods based on characteristic lines not in the bicharacteristic direction (ALBRECHT and URICH (1961). SAUER (1962)).

Type 1) methods have been applied to practical problems in steady supersonic flow. BUTLER determined the supersonic flow past a pointed body of elliptical cross section. FOWELL used the Yawed Cone Problem as the test case for his method. RANSOM et al., extended the method to apply to internal flows. The basic idea in these methods is to connect conditions at an unknown point with the known conditions on an initial surface upstream of it along the backward Mach conoid through the point. Difference relations are established along generators of this conoid (bicharacteristics). The number of these relations (and hence the number of different bicharacteristic directions used) is equal to the number of unknowns (apart from the entropy) to be found at the new point. At each new point in the characteristic network, at least three directions in space have to be determined. These are not related in any definite way to a fixed coordinate system and each calculation at a new point derives no assistance or guidance from calculations at preceding points. The geometry in these methods is therefore very complicated. These methods also leave a more fundamental question unanswered. In any finite difference method based on characteristic properties, the original equations of motion are referred to a transformed characteristic coordinate system rather than the original Cartesian or other fixed coordinate system. One must be sure that the equations in the transformed system are exactly equivalent to the original equations. It is difficult to see how local coordinate systems based on three or more bicharacteristic directions can satisfy the equivalence condition. In fact, TITT (1939), who examined existence and uniqueness properties of hyperbolic equations involving three independent variables, implies that only two bicharacteristic directions and their associated compatibility conditions are independent. This is supported by work described later in this chapter.

In brief, methods of type 1) are criticized on two grounds, a) they are unnecessarily complicated (three dimensional geometry), b) the procedure of using difference relations along more than *two* bicharacteristic directions is of dubious validity. In spite of these reservations it must be conceded that applications to specific problems have produced results agreeing well qualitatively with available observed data.

It can be fairly claimed that the methods of type 2) have a sounder theoretical foundation than those of type 1). They are closely linked with Titt's analytical studies and adopt a similar characteristic coordinate system, defined as follows. On the initial surface (assumed to be space like) a family of nonintersecting curves are drawn. Through each of these curves two characteristic surfaces will pass and these will touch the local Mach conoid at each point along two bicharacteristic curves. In this way, directions along two bicharacteristics and along one noncharacteristic line are defined at each point in the region influenced by the initial data, and the equations of motion are referred to coordinates based on these directions. Once the family of curves has been selected on the initial surface, the characteristic coordinate system is determined throughout the region of interest. The transformed equations are derived from independent linear combinations of the original equations and are clearly equivalent to the original equations.

The author proposed a difference scheme associated with his type 2) method but no practical application has been carried out. Methods developed later appear to be more suitable for fully three dimensional numerical problems. The author's method has been applied successfully to *linearized* characteristics problems, e. g., supersonic flow past bodies of revolution at small angles of attack (HOLT (1959)) and supersonic flow past slightly three dimensional wings (HOLT and YIM (1962)).

Type 2) methods are also suitable for investigation of geometric properties of three dimensional flows, e. g., searching for generalizations of simple wave flows.

Type 3) methods have much in common with those of type 2) although they were developed quite independently. A local coordinate system is again based on *two* bicharacteristic directions. However, instead of relating these to curves on an initial surface, they are defined in terms of the local streamline direction, principal normal and binormal and the transformed equations have a simpler appearance than those arising in type 2) methods. The first method of this type was proposed by Bruhn and Haack (1958). It was clarified and derived in a different manner by Schaetz (1961). Both Bruhn and Haack and Schaetz applied the method, with success, to the problem of unsteady flow through a Laval nozzle. Schaetz also made a trial application to a steady flow problem (with somewhat idealized initial conditions). The emphasis in type 3) methods is on optimization of the characteristics method. An answer is sought to the following question. Given a system of hyperbolic partial differential equations, how can they be combined and transformed to result in a system with the least number of directions of differentiation (and hence to the simplest system of equivalent difference equations). In the case of steady flow, Schaetz finds that the simplest system employs the following four directions:

1) Two bicharacteristic directions in the plane of the velocity vector and the principal normal of the streamline.

2) The velocity direction.

3) The binormal to the streamline.

Equations of motion written with derivatives in these directions are simple extensions of the corresponding two dimensional equations. The task of solving them in a region is complicated by the fact that these directions change from point to point. They are nevertheless undoubtedly simpler than equations arising in type 1) methods since, at a given point, three of the directions are coplanar and the fourth is normal to this common plane.

Two methods of type 4) have been developed recently. The first, by Albrecht and Urich (1961), introduces directional derivatives defined by the coefficients of derivatives of each dependent variable. If there are m independent variables, there will be m^2 such directions. They are not bicharacteristic directions but may lie close to these and are therefore called quasi-characteristic directions. All the directions lie in coordinate planes and the difference method developed by Albrecht and Urich is easier to understand and apply than other methods employing many different directions in space. However, Sauer has pointed out that, in some cases, the domain of dependence defined by quasi-characteristics frequently is inside the corresponding domain defined by an exact analysis. This means that the method does not always converge nor give unique answers, and will not be considered further.

The second method of type 4) is due to Sauer (1962). It is based on the same general approach to three dimensional problems adopted for type 2) and 3) methods, applying characteristic compatibility conditions in only two directions and taking one of the coordinate directions to be noncharacteristic. However, Sauer points out that uniqueness and convergence conditions for solutions of hyperbolic equations can be satisfied by using characteristic directions other than bicharacteristic provided that, in any difference scheme, one always operates inside the local domain of dependence. Accordingly, Sauer keeps one of the coordinate planes fixed and determines characteristic directions for the partial differential equations in such planes or in parallel planes. These directions are not in general bicharacteristics and are called *near characteristics.* The transformed equations, to be solved as difference equations, are derived strictly from *independent* linear combinations of the original equations and are equivalent to these in every sense. Sauer's approach is essentially to treat a three dimensional problem as a sequence of two dimensional problems so that all the detailed knowledge of how to handle shocks, boundaries, simple waves, and so on, can be carried over, with obvious extensions, to three dimensional problems. It is only fair to point out that FERRI (1954) recommended a network based on traces of Mach conoids on planes in his first three dimensional characteristics method (SAUER himself proposed the general idea also in 1950). However, Sauer's latest scheme is based on a rigorous, mathematical derivation with particular attention paid to questions of uniqueness and equivalence. It is also an optimum scheme in the sense defined by SCHAETZ (1961).

Sauer's method is recent and had previously only been worked out for unsteady flow in two dimensions. Its application to steady three dimensional problems is considered in the present chapter but, of course, only in its general aspects. The author feels that among the many Methods of Characteristics Sauer's method offers the best and most economical means of treating steady supersonic flow past three dimensional wings, bodies of revolution at angles of attack, and other lifting re-entry configurations. It would be well worth while to develop it in detail for such problems.

Sauer's method is quite closely connected with the method of BRUHN and HAACK and with Schaetz's method. We shall therefore be concerned with some details of these methods, including a discussion of the application of Sauer's method to steady flow problems. For completeness we precede this with an outline of Butler's method.

4.2 Bicharacteristics Method (BUTLER)

Consider the following system of n partial differential equations of the first order, with n dependent variables u_μ and three independent variables x_i

$$a_{\mu\nu i}\left(\frac{\partial u_\nu}{\partial x_i}\right) = b_\mu \tag{4.2.1}$$

$(i=1,2,3,\ \mu,\nu=1,\ldots,n)$.

If the independent variables x_i are replaced by a new set x_i' (4.2.1) referred to x_i' transform to the following

$$a_{\mu\nu i}'\left(\frac{\partial u_\nu}{\partial x_i'}\right) = b_\mu \tag{4.2.2}$$

where

$$a'_{\mu v i} = a_{\mu v j}\left(\frac{\partial x'_i}{\partial x_j}\right) \tag{4.2.3}$$

Now suppose that u is given on a surface $x'_3 = \text{constant}$ as a differentiable function of x'_1, x'_2. Then (4.2.2) can be regarded as algebraic equations to determine derivatives of u_μ normal to $x'_3 = \text{constant}$, namely, $\partial u_\mu / \partial x'_3$.

These derivatives are determined uniquely if the leading determinant of the system (4.2.2),

$$\Delta = \det\{a'_{\mu v 3}\} \neq 0 .$$

We suppose that Δ factorizes into a quadratic factor and a repeated linear factor so that

$$\Delta = \left\{l_k \frac{\partial x'_3}{\partial x_k}\right\}^{n-2} A_{ij} \frac{\partial x'_3}{\partial x_i} \frac{\partial x'_3}{\partial x_j} \tag{4.2.4}$$

where A_{ij} is symmetric.

If the condition

$$A_{ij} \frac{\partial x'_3}{\partial x_i} \frac{\partial x'_3}{\partial x_j} = 0 \tag{4.2.5}$$

is satisfied at a general point on the surface $x'_3(x_1, x_2, x_3) = \text{constant}$, then the surface touches a quadratic cone at that point. If this cone is real (4.2.1) are said to be hyperbolic.

A surface which satisfies (4.2.5) is called a *characteristic surface* and the envelope of all such surfaces through (x_1, x_2, x_3) is called the *characteristic conoid* through that point. The curve of contact between a characteristic surface and the corresponding envelope is called a *bicharacteristic*. Then along a bicharacteristic the condition

$$A_{ij}^{-1} dx_i dx_j = 0 \tag{4.2.6}$$

is satisfied.

If $x'_3 = \text{constant}$ is characteristic then a further condition must be satisfied to ensure that the derivatives $\partial u_v / \partial x'_3$ are finite. This is called the compatibility condition and may be written

$$E_\mu \frac{\partial u_\mu}{\partial x'_1} + F_\mu \frac{\partial u_\mu}{\partial x'_2} = D , \tag{4.2.7}$$

where E_μ, F_μ, D are functions of u_μ and x_i. The directions x_1' increasing are now chosen so that the curves $x_2' = \text{constant}$ on the characteristic surface are bicharacteristics. Then along a bicharacteristic satisfying (4.2.6)

$$E_\mu du_\mu = [D - F_\mu(\partial u_\mu/\partial x_2')] dx_1' \tag{4.2.8}$$

BUTLER bases his method of characteristics on a numerical technique for solving (4.2.8) along bicharacteristics connecting a new point with points on a given non-characteristic initial surface.

The equations of the bicharacteristics are written in parametric form with two parameters (τ, ϕ). One of the single infinity of bicharacteristic directions through a point $x_i = \alpha_i$ is expressed in the form

$$dx_i = [\lambda_i + \mu_i \cos\theta + v_i \sin\theta] d\tau \tag{4.2.9}$$

where $0 \le \theta < 2\pi$ and λ_i, μ_i, v_i satisfy the relations (derived from (4.2.6))

$$\left.\begin{array}{l} -A_{ij}^{-1} \lambda_i \lambda_j = A_{ij}^{-1} \mu_i \mu_j = A_{ij}^{-1} v_i v_j \\ A_{ij}^{-1} \mu_i v_j = A_{ij}^{-1} v_i \lambda_j = A_{ij}^{-1} \lambda_i \mu_j = 0 \end{array}\right\} \tag{4.2.10}$$

λ_i may be any vector lying inside the characteristic cone.

Along the characteristic direction (4.2.9) the compatibility conditions

$$A_v du_v = \left[B + C_v \left\{ \cos\theta \, v_i \frac{\partial u_v}{\partial x_i} - \sin\theta \, \mu_i \frac{\partial u_v}{\partial x_i} \right\} \right] d\tau \tag{4.2.11}$$

are satisfied where

$$\begin{aligned} A_v &= A_{1v} + A_{2v} \cos\theta + A_{3v} \sin\theta \\ B &= B_1 + B_2 \cos\theta + B_3 \sin\theta \\ C_v &= C_{1v} + C_{2v} \cos\theta + C_{3v} \sin\theta \end{aligned} \tag{4.2.12}$$

where the coefficients on the right of (4.2.12) are functions only of τ.

To trace out a bicharacteristic starting from $x_i = \alpha_i$, an initial value $\theta = \phi$ is assigned and subsequent values of θ are determined from the relation

$$[\lambda_j + \mu_j \cos\theta + v_j \sin\theta] A_{ij}^{-1}(\partial x_i/\partial \phi) = 0 \tag{4.2.13}$$

The direction λ_i (inside the characteristic cone) is chosen so that $C_{1v} = 0$ everywhere. BUTLER now chooses four bicharacteristics connecting $x_i = \alpha_i$ with a space-like initial surface $f(x_i) = 0$. These correspond to initial values of $\phi = 0$, $\pi/2$, π, $3\pi/2$. Then the four corresponding compatibility relations (4.2.11) are multiplied by two independent weighting factors and integrated all round the characteristic conoid. This gives two relations between u_μ at

$x_i = \alpha_i$ and initial values on $f(x_i) = 0$. The additional relations needed are obtained from difference relations along the curves $dx_i = \lambda_i d\tau$ and, if $n > 3$, along characteristic directions $dx_i = l_i d\tau$ defined by the linear factors of the determinant $\Delta = 0$.

Application to Plane Unsteady Flow: The equations of motion for unsteady plane flow of an inviscid gas may be written (after eliminating density derivatives from the continuity equation),

$$\varrho \left\{ \frac{\partial u}{\partial t} + u \frac{\partial u}{\partial x} + v \frac{\partial u}{\partial y} \right\} + \frac{\partial p}{\partial x} = 0 \tag{4.2.14}$$

$$\varrho \left\{ \frac{\partial v}{\partial t} + u \frac{\partial v}{\partial x} + v \frac{\partial v}{\partial y} \right\} + \frac{\partial p}{\partial y} = 0 \tag{4.2.15}$$

$$\frac{\partial p}{\partial t} + u \frac{\partial p}{\partial x} + v \frac{\partial p}{\partial y} + \varrho a^2 \left(\frac{\partial u}{\partial x} + \frac{\partial v}{\partial y} \right) = 0 \tag{4.2.16}$$

where t is the time, x, y are Cartesian coordinates, (u, v) are velocity components in directions (x, y), p is pressure, ϱ density, and a the speed of sound.

We regard (4.2.14)–(4.2.16) as a system of three first order partial differential equations in p, u and v (ϱ is determined from p by relations satisfied along streamlines) and apply the general bicharacteristics method to them.

The characteristic surfaces of (4.2.14)–(4.2.16) through a point (t, x, y) touch a degenerate cubic surface factorizing into the streamline through the point and the quadric cone

$$(dx - u dt)^2 + (dy - v dt)^2 = a^2 dt^2 . \tag{4.2.17}$$

Suppose that the solution of (4.2.14)–(4.2.16) is known on and near a plane $t = t_0 - h$. To determine the solution at a point (t_0, x_0, y_0) introduce parametric coordinates (τ, ϕ) on the characteristic conoid through (t_0, x_0, y_0), taking

$$t = \tau + t_0 - h \tag{4.2.18}$$

and integrating

$$dx = (u + a \cos \theta) dt \tag{4.2.19}$$

$$dy = (v + a \sin \theta) dt \tag{4.2.20}$$

with initial values $x = x_0$, $y = y_0$, $t = t_0$, $\theta = \phi$.

At subsequent points θ is determined from (4.2.13), which reduces to

$$\tan \theta = - \frac{\partial x}{\partial \phi} \bigg/ \frac{\partial y}{\partial \phi} \tag{4.2.21}$$

Integration is performed along bicharacteristics $\phi = $ constant on which the compatibility conditions

$$dp + \varrho a \cos\theta \, du + \varrho a \sin\theta \, dv = -\varrho a^2 s \, d\tau \tag{4.2.22}$$

are satisfied, where

$$s = \sin^2\theta \frac{\partial u}{\partial x} - \sin\theta\cos\theta\left(\frac{\partial u}{\partial y} + \frac{\partial v}{\partial x}\right) + \cos^2\theta \frac{\partial u}{\partial y}.$$

For a perfect gas with constant specific heats an entropy function ψ is defined by

$$\psi = \frac{2}{\gamma-1}\frac{a}{\sigma} \quad \text{where} \quad \sigma = p^{(\gamma-1)/2\gamma}$$

$[\psi$ is a function of S only$]$.
    Eq. (4.2.22) can then be written

$$\psi \, d\sigma + \cos\theta \, du + \sin\theta \, dv = -a s \, d\tau. \tag{4.2.23}$$

Eq. (4.2.23) is written in simple difference form along four bicharacteristics $\phi = 0$, $\pi/2$, π, $3\pi/2$ connecting the point (t_0, x_0, y_0) with the plane $t = t_0 - h$. Then

$$\begin{aligned}
P_1(\sigma_0 - \sigma_1) + U_1(u_0 - u_1) + V_1(v_0 - v_1) &= -\tfrac{1}{2}h\bar{a}_0 v_{y_0} - \tfrac{1}{2}h a_1 s_1 + O(h^3) \\
P_2(\sigma_0 - \sigma_2) + U_2(u_0 - u_2) + V_2(v_0 - v_2) &= -\tfrac{1}{2}h\bar{a}_0 u_{x_0} - \tfrac{1}{2}h a_2 s_2 + O(h^3) \\
P_3(\sigma_0 - \sigma_3) + U_3(u_0 - u_3) + V_3(v_0 - v_3) &= -\tfrac{1}{2}h\bar{a}_0 v_{y_0} - \tfrac{1}{2}h a_3 s_3 + O(h^3) \\
P_4(\sigma_0 - \sigma_4) + U_4(u_0 - u_4) + V_4(v_0 - v_4) &= -\tfrac{1}{2}h\bar{a}_0 u_{x_0} - \tfrac{1}{2}h a_4 s_4 + O(h^3)
\end{aligned} \tag{4.2.24}$$

where

$$\begin{aligned}
P_i &= \tfrac{1}{2}(\bar{\psi}_0 + \psi_i) \\
U_i &= \tfrac{1}{2}(\cos\phi_i + \cos\theta_i) \\
V_i &= \tfrac{1}{2}(\sin\phi_i + \sin\theta_i) \quad i = 1,2,3,4
\end{aligned}$$

(barred quantities are only evaluated to $O(h)$, while unbarred quantities are found correct to $O(h^2)$).
    Suffixes $1,2,3,4$ identify the intersection of bicharacteristics $\phi = 0$, $\pi/2$, π, $3\pi/4$, respectively, through (t_0, x_0, y_0) with the initial surface $t = t_0 - h$. Difference relations (4.2.24) are supplemented by a difference relation in the streamline direction through (t_0, x_0, y_0) (intersecting $t = t_0 - h$ at point 5).

This is

$$P_5(\sigma_0 - \sigma_5) = -\tfrac{1}{2}h\bar{a}_0(u_x + v_y)_0 - \tfrac{1}{2}ha_5(u_x + v_y)_5 + O(h^3) \tag{4.2.25}$$

BUTLER ingeniously combines (4.2.24) and (4.2.25) to establish the following relations

$$\delta\sigma_0 = \frac{2R_5 - (R_1 + R_2 + R_3 + R_4)}{2\bar{\psi}_0} + O(h^3) \tag{4.2.26}$$

$$\delta u_0 = -\tfrac{1}{2}(R_1 - R_3) + O(h^3) \tag{4.2.27}$$

$$\delta v_0 = -\tfrac{1}{2}(R_2 - R_4) + O(h^3) \tag{4.2.28}$$

where $\delta\sigma_0 = (\sigma_0 - \bar{\sigma}_0)$, etc., and

$$R_i = P_i(\bar{\sigma}_0 - \sigma_i) + U_i(\bar{u}_0 - u_i) + V_i(\bar{v}_0 - v_i) + \tfrac{1}{2}a_i s_i \quad i = 1,2,3,4$$

$$R_5 = P_5(\bar{\sigma}_0 - \sigma_5) + \frac{1}{2}ha_5\left(\frac{\partial u}{\partial x} + \frac{\partial v}{\partial y}\right)_5 .$$

Eqs. (4.2.26)–(4.2.28) determine the unknowns at (t_0, x_0, y_0) after the positions of the four bicharacteristic intersection points and the streamline inter-section point on $t = t_0 - h$ have been determined. For the latter we have, to sufficient order of accuracy,

$$\theta_i = \phi_5 + hD_i + O(h^2) \tag{4.2.29}$$

where

$$D_i = -\sin\phi_i\left(\frac{\partial a}{\partial x}\right)_i + \cos\phi_i\left(\frac{\partial a}{\partial y}\right)_i$$

$$+ \cos^2\phi_i\left(\frac{\partial u}{\partial y}\right)_i - \sin\phi_i\cos\phi_i\left(\frac{\partial u}{\partial x} - \frac{\partial v}{\partial y}\right)_i \tag{4.2.30}$$

$$- \sin^2\phi_i\left(\frac{\partial u}{\partial x}\right)_i .$$

The coordinates of the intersection points are then given by

$$x_i = x_0 = \tfrac{1}{2}h(\bar{u}_0 + \bar{a}_0\cos\phi_i + u_i + a_i\cos\phi_i) + O(h^3)$$
$$y_i = y_0 - \tfrac{1}{2}h(\bar{v}_0 + \bar{a}_0\sin\phi_i + v_i + a_i\sin\phi_i) + O(h^3) \quad i = 1,2,3,4 \tag{4.2.31}$$

and

$$x_5 = x_0 - \tfrac{1}{2}h(\bar{u}_0 + u_5) + O(h^3)$$

$$y_5 = y_0 - \tfrac{1}{2}h(\bar{v}_0 + v_5) + O(h^3)$$

(4.2.32)

Eqs. (4.2.31), (4.2.32) and (4.2.26)–(4.2.28) are solved iteratively starting with crude estimates of u_0, v_0, σ_0 and a_0, which need not even be correct to $O(h)$. In each iteration x_i, y_i, θ_i, u_i, v_i, σ_i are first determined $(i = 1, \ldots, 5)$. Eqs. (4.2.26)–(4.2.28) are then solved for σ_0, u_0, v_0, while an improved value of ψ_0 is given by $\psi_0 = \psi_5$. In the next iteration the latest values of σ_0, u_0, v_0 are taken as new values of $\bar{\sigma}_0$, $\bar{u}_0$, $\bar{v}_0$ and the cycle is repeated. BUTLER states that the process is rapidly convergent, noting that if initial estimates are correct to $O(1)$ not more than two iterative cycles are needed to determine u_0, v_0, σ_0, ψ_0 to the required accuracy.

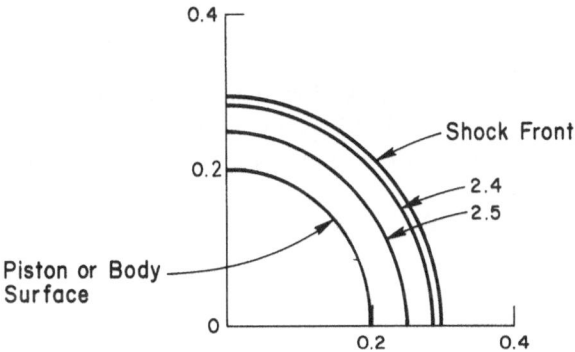

Fig. 4.1 Butler's method applied to a delta wing of variable cross section

BUTLER applies the unsteady formulation to calculate steady hypersonic flow past a body with triangular plan form and spanwise cross section changing gradually from circular near the wing apex to a thin ellipse near the trailing edge. Using hypersonic small disturbance theory (VAN DYKE, 1954) the flow is governed by plane unsteady flow equations, with the coordinate in the undisturbed stream direction acting effectively as the time while the x, y coordinates are in the cross flow plane. The unsteady flow problem solved is that generated by a two dimensional piston, the shape of which is an ellipse with axes increasing with time according to the relations

$$a = t \qquad\qquad 0 \le t \le 1$$

$$b = t \qquad\qquad 0 \le t \le 0.2$$

$$b = t\left[1 - \left\{\frac{t - 0.2}{0.8\,t}\right\}^4\right] \qquad 0.2 \le t \le 1.$$

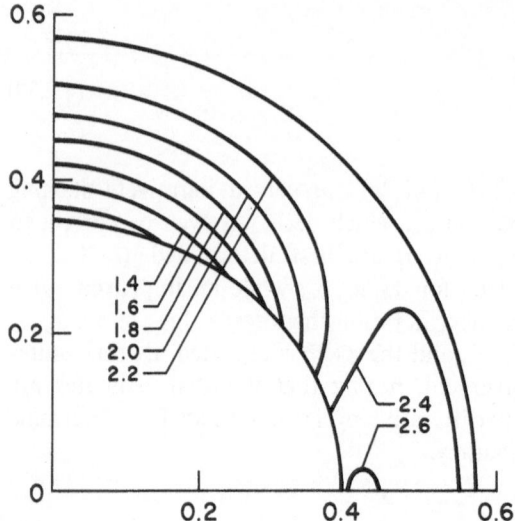

Fig. 4.2   The isobars in the flow past a delta wing of variable cross section

Figures 4.1 and 4.2 show the isobars in a cross flow quadrant at two stations defined by $t=0.2$ and $t=0.4$. The calculated spanwise surface pressure distributions are compared with measured distributions and reasonable agreement is obtained.

RANSOM et al. (1970) have applied Butler's method to a number of internal flow problems.

4.3 Optimal Characteristics Methods
(BRUHN and HAACK, SCHAETZ)

Consider the system of quasilinear first order hyperbolic partial differential equations

$$a^i_{jk} \frac{\partial u^j}{\partial x_i} = b_k \quad i=1,2,\dots,m;\; j,k=1,2,\dots,n \tag{4.3.1}$$

There are therefore m independent and n dependent variables. The a's and b's are given functions of u, x^i. We seek a solution of (4.3.1) satisfying initial data on some given non-characteristic surface. To obtain this in the simplest manner, we wish to derive n linear combinations of (4.3.1).

$$\mu^k_l a^i_{jk} \frac{\partial u^j}{\partial x_i} = \mu^k_l b_k \quad l=1,2,\dots,n \tag{4.3.2}$$

which will contain the least number of independent directional derivatives. As we might expect, some of these directions are characteristic. Others are simply chosen judiciously to reduce the number of non-characteristic directions as much as possible. One requirement to determine the multipliers μ_i^k is that the rank of the matrix

$$M_l = (\mu_l^k a_{jk}^i) \quad \text{should be a minimum.}$$

The technique of optimization is best illustrated by its application to gas dynamics; in particular, to the equations of unsteady axially symmetric flow.

For a perfect gas, with constant specific heats, the governing equations are (in the usual notation)

$$(q \cdot \nabla)p + \frac{\partial p}{\partial t} + a^2 \varrho(\nabla \cdot q) = 0 \tag{4.3.3}$$

$$\nabla p + \varrho(q \cdot \nabla)q + \varrho \frac{\partial q}{\partial t} = 0 \tag{4.3.4}$$

$$(q \cdot \nabla)S + \frac{\partial S}{\partial t} = 0 \tag{4.3.5}$$

Here q is the velocity vector, p the pressure, ϱ the density, and S the entropy.

We introduce unit vectors e_1 and $e_2 = q/q$, where e_1 is normal to e_2 in the meridian plane of q. We define *directional derivatives*

$$\left(\frac{d}{d\sigma}\right)_i = (e_i \cdot \nabla) \quad i = 1, 2 \tag{4.3.6}$$

Then (4.3.3)–(4.3.5) can be replaced by four scalar equations,

$$q\left(\frac{d}{d\sigma}\right)_2 p + \frac{\partial p}{\partial t} + a^2 \varrho\left(\frac{d}{d\sigma}\right)_2 q + a^2 q \varrho\left(\frac{d}{d\sigma}\right)_1 \phi = -\frac{a^2 q \varrho}{r} \sin\phi \tag{4.3.7}$$

$$\left(\frac{d}{d\sigma}\right)_1 p + q^2 \varrho\left(\frac{d}{d\sigma}\right)_2 \phi + q\varrho \frac{\partial \phi}{\partial t} = 0 \tag{4.3.8}$$

$$\left(\frac{d}{d\sigma}\right)_2 p + q\varrho\left(\frac{d}{d\sigma}\right)_2 q + \varrho \frac{\partial q}{\partial t} = 0 \tag{4.3.9}$$

$$q\left(\frac{d}{d\sigma}\right)_2 S + \frac{\partial S}{\partial t} = 0 \tag{4.3.10}$$

Eq. (4.3.10) states that the directional derivative of S (w.r.t. time) along a streamline is zero and can be separated from the other equations. We now consider linear combinations of (4.3.7), (4.3.8), and (4.3.9). The matrix M_l in this case is

$$\begin{pmatrix} \mu^2 & \mu^1 q + \mu^3 & \mu^1 \\ 0 & \mu^1 a^2 \varrho + \mu^3 q \varrho & \mu^3 \varrho \\ \mu^1 a^2 q \varrho & \mu^2 q^2 \varrho & \mu^1 q \varrho \end{pmatrix}$$

The minimum rank of this matrix is 2. To achieve this, the determinant

$$D = \mu^1 a^2 q \varrho^2 [(\mu^3)^2 + (\mu^2)^2 - (\mu^1)^2 a^2] \tag{4.3.11}$$

must vanish.
Hence,

$$either \quad \mu^1 = 0 \tag{4.3.12}$$

$$or \quad a\mu^1 = \pm\sqrt{(\mu^2)^2 + (\mu^3)^2} \tag{4.3.13}$$

To optimize the system, we combine (4.3.13) with $\mu^2 = 0$ (or with $\mu^3 = 0$), not given here. We then obtain the system

$$\left(\frac{d}{dt}\right)_{1,2} p \pm a\varrho \left(\frac{d}{dt}\right)_{1,2} q = -a^2 q \varrho \left[\left(\frac{d}{d\sigma}\right)_1 \phi + \frac{\sin\phi}{r}\right] \tag{4.3.14}$$

$$q \varrho \left(\frac{d}{dt}\right)_3 \phi = -\left(\frac{d}{d\sigma}\right)_1 p \tag{4.3.15}$$

$$\left(\frac{d}{dt}\right)_3 S = 0 \tag{4.3.16}$$

where

$$\left(\frac{d}{dt}\right)_{1,2} = (q \pm a)\left(\frac{d}{d\sigma}\right)_2 + \frac{\partial}{\partial t} \tag{4.3.17}$$

$$\left(\frac{d}{dt}\right)_3 = q\left(\frac{d}{d\sigma}\right)_2 + \frac{\partial}{\partial t} \tag{4.3.18}$$

Eqs. (4.3.14)–(4.3.16) are simple generalizations of the one dimensional unsteady equations. In the latter case, (4.3.15) is missing and the right side of (4.3.16) is zero. Note that (4.3.14)–(4.3.16) contain four directions which change from one point in the field to the next. The directions are illustrated in Fig. 4.3.

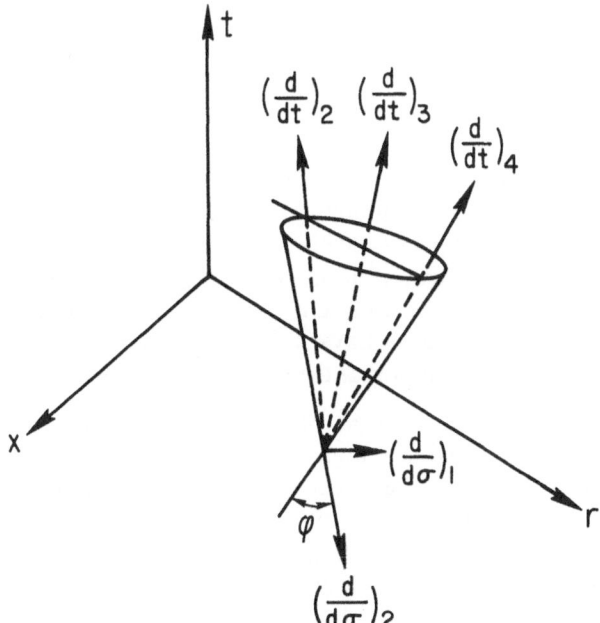

Fig. 4.3   Directions in Schaetz system for unsteady axisymmetric flow

The directions $(d/d\sigma)_2$, $(d/d\sigma)_1$ are parallel to the x, r plane. $(d/d\sigma)_2$ is in the direction of the resultant velocity; $(d/dt)_3$ is in the direction of the local particle path; $(d/dt)_1$, $(d/dt)_2$ are bicharacteristic directions determined by characteristic surfaces through an element in the $(d/d\sigma)_1$ direction (the latter is normal to $(d/d\sigma)_2$).

Schaetz System for Steady Flow in Three Dimensions: Using a similar procedure to that just described for unsteady axially symmetric flow, SCHAETZ has derived the following optimal system of equations for steady supersonic flow in three dimensions:

$$-(q^2 - a^2)^{1/2} p_{,1} - \varrho a q^2 \phi_{,1} = \varrho a^2 q \sin\phi\, \theta_{,4} \tag{4.3.19}$$

$$(q^2 - a^2)^{1/2} p_{,2} - \varrho a q^2 \phi_{,2} = \varrho a^2 q \sin\phi\, \theta_{,4} \tag{4.3.20}$$

$$\varrho q^2 \sin\phi\, \theta_{,3} = p_{,4} \tag{4.3.21}$$

$$\varrho q q_{,3} + p_{,3} = 0 \tag{4.3.22}$$

$$S_{,3} = 0 \tag{4.3.23}$$

Here, $,n$ denotes differentiation in the direction defined by some unit vector n_i. Direction 3 is the velocity direction.

Direction 4 is the binormal to the streamline.

Directions 1 and 2 are bicharacteristics determined by the binormal 4. Thus, directions 1, 2, and 3 are coplanar and 4 is normal to the common plane. (q,θ,ϕ) are spherical coordinates in hodograph space. SCHAETZ has solved (4.3.19)–(4.3.23) numerically in the case of three dimensional source flow. Locally one uses difference relations in three *coplanar* directions and in one direction normal to these. The method actually requires the use of only one more difference relation than the number needed in two dimensions. However, the direction of the common plane changes from point to point.

SCHAETZ has successfully applied the optimum characteristics method to the calculation of unsteady flow through a Laval nozzle. It is appropriate to conclude this section with a brief description of the difference scheme he used.

Application to Unsteady Flow through a Laval Nozzle: The shape of a two dimensional Laval nozzle is prescribed (see Fig. 4.4). It begins and ends with a section of constant area. At time $t=0$, an unsteady gas motion is initiated in the entry and up to a certain time t_1 (the first instant when the motion affects the variable area part of the nozzle) the flow is one-dimensional and can be determined precisely. This provides the initial data for calculating the unsteady two dimensional flow which develops at later instants.

The motion originates as one dimensional unsteady motion in region A, and becomes two dimensional on entering region B. The equations governing the motion, written in optimal characteristic form, are

$$\left(\frac{d}{dt}\right)_{1,2} p \pm a\varrho \left(\frac{d}{dt}\right)_{1,2} q = -a^2 q\varrho \left(\frac{d}{d\sigma}\right)_2 \theta \tag{4.3.24}$$

$$q\varrho \left(\frac{d}{dt}\right)_3 \theta = -\left(\frac{d}{d\sigma}\right)_2 p \tag{4.3.25}$$

$$\left(\frac{d}{dt}\right)_3 S = 0 \tag{4.3.26}$$

where (q,θ) are polar coordinates in the hohograph plane.

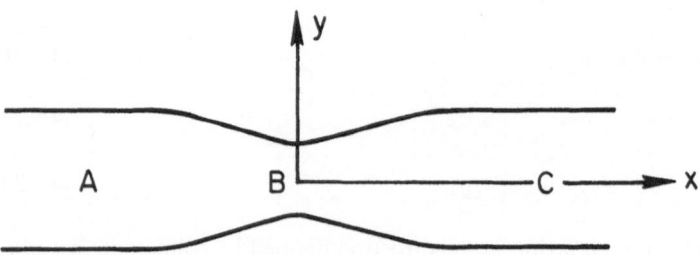

Fig. 4.4   Unsteady flow in a Laval nozzle

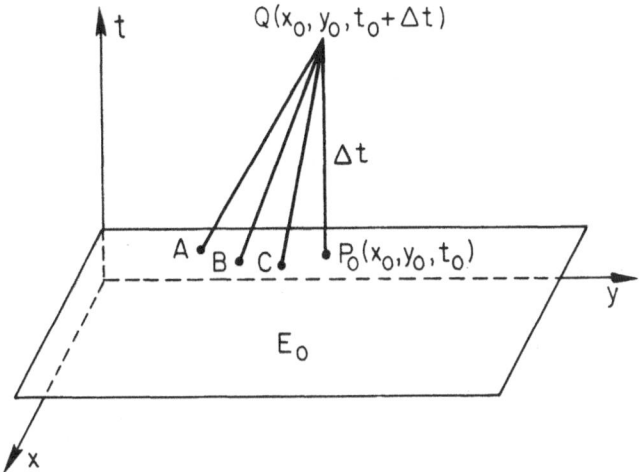

Fig. 4.5   Difference scheme in nozzle problem

The directions arising in (4.3.24)–(4.3.26) are shown in Fig. 4.5. $(d/dt)_1$ is along AQ, a bicharacteristic; $(d/dt)_2$ is along BQ also a bicharacteristic; $(d/dt)_3$ is along the stream direction CQ.

To solve (4.3.24)–(4.3.26), proceed as follows:

1) Assume all data known in plane E_0 (and in parallel planes corresponding to earlier times) at time t_0.

2) Select a point Q in a plane parallel to E_0 at time $t = t_0 + \Delta t$.

3) Project Q onto E_0 at P_0. Determine differentiation directions $(d/dt)_1$, $(d/dt)_2$, $(d/dt)_3$ at point P_0 and draw segments *in these directions* through Q, to intersect the plane E_0 at points A, B, C, respectively. Determine $(d/d\sigma)_2 p$, $(d/d\sigma)_2 \theta$ (inner derivatives in the E_0 plane) at points $A, B,$ and C.

4) Solve (4.3.24), (4.3.25), (4.3.26) as difference equations along the segments QA, QC, QB, and so determine values of p, q, θ, S, a at Q.

5) Repeat 3) and 4) over a network of points in the plane $t = t_0 + \Delta t$, and so determine the complete Laval nozzle flow field at a new time.

This method proved very successful in the Laval nozzle problem and could be applied to steady flow problems. Its only drawback is the fact that the normal to the plane containing the flow and bicharacteristic directions changes from point to point. The scheme to be described now avoids this disadvantage.

4.4 Near Characteristics Method (Sauer)

The basic idea of this method is to apply the same criterion for determining characteristics in the x, t plane in one dimensional unsteady flow to equations containing more than two independent variables.

Consider first the following pair of equations in two independent variables, x, t:

$$a_{\mu 1} u_t + a_{\mu 2} v_t + b_{11}^{(\mu)} u_x + b_{12}^{(\mu)} v_x = h_\mu \qquad (\mu = 1, 2) \tag{4.4.1}$$

The linear combination of (4.4.1) resulting from factoring by σ_1 $(\mu = 1)$ and σ_2 $(\mu = 2)$ and adding, is

$$(\sigma_1 a_{11} + \sigma_2 a_{21}) u_t + (\sigma_1 b_{11}^{(1)} + \sigma_2 b_{11}^{(2)}) u_x + (\sigma_1 a_{12} + \sigma_2 a_{22}) v_t$$
$$+ (\sigma_1 b_{12}^{(1)} + \sigma_2 b_{12}^{(2)}) v_x = \sigma_1 h_1 + \sigma_2 h_2 \tag{4.4.2}$$

or, in abbreviated form

$$A_1 u_t + B_1 u_x + A_2 v_t + B_2 v_x = H \tag{4.4.2*}$$

To get *characteristic* equations we require that u and v in (4.4.2), (4.4.2*) be differentiated in the *same* direction. Then,

$$\frac{dx}{dt} = \frac{B_1}{A_1} = \frac{B_2}{A_2} = \tau \tag{4.4.3}$$

$$B_1 = \tau A_1$$
$$B_2 = \tau A_2 . \tag{4.4.4}$$

Eqs. (4.4.4) are linear homogeneous equations for σ_1, σ_2. They have solutions only if the determinant formed from the coefficients is zero. This condition gives a quadratic in τ which has two roots, real if the equations are hyperbolic. When these values of τ are substituted in (4.4.4) we get two independent values of σ_1/σ_2 which determine the required characteristic combinations of (4.4.1).

Now consider the equations in n independent and m dependent variables

$$\sum_{k=1}^{m} \left\{ a_{\mu k} \frac{\partial u_k}{\partial t} + b_{1k}^{(\mu)} \frac{\partial u_k}{\partial x_1} + \cdots + b_{nk}^{(\mu)} \frac{\partial u_k}{\partial x_n} \right\} = h_\mu \tag{4.4.5}$$

We rewrite these equations with only $\partial/\partial t$ and $\partial/\partial x_1$ derivatives on the left hand side and seek linear combinations of these equations which will result in all variables $u_1, u_2, \ldots u_m$ being differentiated in the *same* direction in the x, t plane. If these factors are σ_μ, the linear combination formed is

$$\left\{ \sum_{\mu=1}^{m} \sigma_\mu a_{\mu 1} \frac{\partial u_1}{\partial t} + \sum_{\mu=1}^{m} \sigma_\mu b_{11}^{(\mu)} \frac{\partial u_1}{\partial x_1} \right\} + \cdots + \left\{ \sum_{\mu=1}^{m} \sigma_\mu a_{\mu m} \frac{\partial u_m}{\partial t} + \sum_{\mu=1}^{m} \sigma_\mu b_{1m}^{(\mu)} \frac{\partial u_m}{\partial x_1} \right\}$$
$$= H \left(\sigma_\mu, h_\mu \frac{\partial}{\partial x_2}, \ldots, \frac{\partial}{\partial x_n} u_1, \ldots, u_m \right) \tag{4.4.6}$$

where H is a linear combination of derivatives with respect to $x_2,...,x_n$ and also a function of the variables themselves. In abbreviated form (4.4.6) may be written

$$\left\{A_1\frac{\partial u_1}{\partial t} + B_1\frac{\partial u_1}{\partial x_1}\right\} + \cdots + \left\{A_m\frac{\partial u_m}{\partial t} + B_m\frac{\partial u_m}{\partial x_1}\right\} = H \tag{4.4.6*}$$

We now require that all the functions on the left of (4.4.6*) be differentiated in the same direction in, or parallel to, the $(t_1\,x_1)$ plane. Then this direction is

$$\frac{dx_1}{dt} = \frac{B_1}{A_1} = \cdots = \frac{B_m}{A_m} = \tau, \qquad dx_2 = \cdots = dx_n = 0 \tag{4.4.7}$$

Hence

$$B_1 = \tau A_1, \quad B_2 = \tau A_2,...,B_m = \tau A_m \tag{4.4.8}$$

Eqs. (4.4.8) are m homogeneous equations for the factors $\sigma_1, \sigma_2,...,\sigma_m$. They have solutions only if τ satisfies the m-th degree equation formed by the determinant of the coefficients.

The equations are hyperbolic if the equation has at least r real roots τ_ϱ $(\varrho=1,...,r)$ where $2\leq r\leq m$. In this case, we can form r linearly independent compatibility equations

$$A_1^{(\mu)}\left(\frac{du_1}{dt}\right)_\varrho + \cdots + A_m^{(\mu)}\left(\frac{du_m}{dt}\right)_\varrho = H_\varrho \qquad \varrho=1,2,...,\mathrm{r};\ \mu=1,...,m \tag{4.4.9}$$

where

$$\left(\frac{d}{dt}\right)_\varrho = \frac{\partial}{\partial t} + \tau_\varrho\frac{\partial}{\partial x_1}.$$

These equations are clearly equivalent to the original equations. In many respects, they are similar to the optimum characteristic equations derived by Bruhn and Haack, and Schaetz. However, all the characteristic directions are now found in the (x_1,t) plane and are therefore not bicharacteristic directions. Sauer calls the directions he defines *near characteristics*.

The details of Sauer's approach are most clearly understood in applications. Firstly, we shall consider two dimensional unsteady flow, which Sauer himself considers. Secondly, we work out the system for steady flow.

Application to Two-Dimensional Unsteady Flow: For isentropic flow (the extension to non-isentropic flow is simple) the equations of motion may be written

$$u_t + u u_x + v u_y + \frac{2}{\gamma - 1} a a_x = 0 \tag{4.4.10}$$

$$v_t + u v_x + v v_y + \frac{2}{\gamma - 1} a a_y = 0 \tag{4.4.11}$$

$$\frac{2}{\gamma - 1}(a_t + u a_x + v a_y) + a(u_x + v_y) = 0 \tag{4.4.12}$$

In these equations, place the y derivatives on the right, factor Eq. (4.4.10) by σ_1, (4.4.11) by σ_2, (4.4.12) by σ_3, and add.
Then we obtain

$$\left\{ \sigma_1 \frac{\partial}{\partial t} + (\sigma_1 u + \sigma_3 a) \frac{\partial}{\partial x} \right\} u + \left\{ \sigma_2 \frac{\partial}{\partial t} + \sigma_2 u \frac{\partial}{\partial x} \right\} v$$

$$+ \frac{2}{\gamma - 1} \left\{ \sigma_3 \frac{\partial}{\partial t} + (\sigma_3 u + \sigma_1 a) \frac{\partial}{\partial x} \right\} a \tag{4.4.13}$$

$$= -\sigma_1 v \frac{\partial u}{\partial y} - \sigma_2 \left(v \frac{\partial v}{\partial y} + \frac{2a}{\gamma - 1} \frac{\partial a}{\partial y} \right) - \sigma_3 \left(\frac{2v}{\gamma - 1} \frac{\partial a}{\partial y} + a \frac{\partial v}{\partial y} \right).$$

In order that the derivatives on the left of (4.4.13) are all in the same direction we require that

$$\frac{\sigma_1}{\sigma_1 u + \sigma_3 a} = \frac{\sigma_2}{\sigma_2 u} = \frac{\sigma_3}{\sigma_3 u + \sigma_1 a} = \frac{1}{\tau}.$$

This gives a cubic for τ with the roots

$$\tau_{1,2} = u \pm a, \qquad \tau_3 = u.$$

Corresponding values of $\sigma_1^{(\mu)}$, $\sigma_2^{(\mu)}$, $\sigma_3^{(\mu)}$ are then

$$\sigma_1^{(\mu)}, \sigma_2^{(\mu)}, \sigma_3^{(\mu)} = \begin{pmatrix} 1 & 0 & 1 \\ 1 & 0 & -1 \\ 0 & 1 & 0 \end{pmatrix}.$$

The near characteristic directions are therefore

$$\left(\frac{d}{dt}\right)_1 = \frac{\partial}{\partial t} + (u+a)\frac{\partial}{\partial x}$$

$$\left(\frac{d}{dt}\right)_2 = \frac{\partial}{\partial t} + (u-a)\frac{\partial}{\partial x}$$

$$\left(\frac{d}{dt}\right)_3 = \frac{\partial}{\partial t} + u\frac{\partial}{\partial x}.$$

The corresponding compatibility equations ((4.4.10)–(4.4.12) transformed in near characteristic form) are

$$\left(\frac{d}{dt}\right)_1\left(u + \frac{2}{\gamma-1}a\right) = -vu_y - av_y - \frac{2}{\gamma-1}va_y \tag{4.4.14}$$

$$\left(\frac{d}{dt}\right)_2\left(u - \frac{2}{\gamma-1}a\right) = -vu_y + av_y + \frac{2}{\gamma-1}va_y \tag{4.4.15}$$

$$\left(\frac{d}{dt}\right)_3 v = -vv_y - \frac{2}{\gamma-1}aa_y \tag{4.4.16}$$

The solution of (4.4.14)–(4.4.16) by a finite difference method is illustrated in Fig. 4.6. We assume that the flow field is known up to time t_0, represented by the plane E_0, and wish to determine the flow quantities at a typical point Q corresponding to time $(t_0+\Delta t)$. Let P_0 be the projection of Q on E_0.

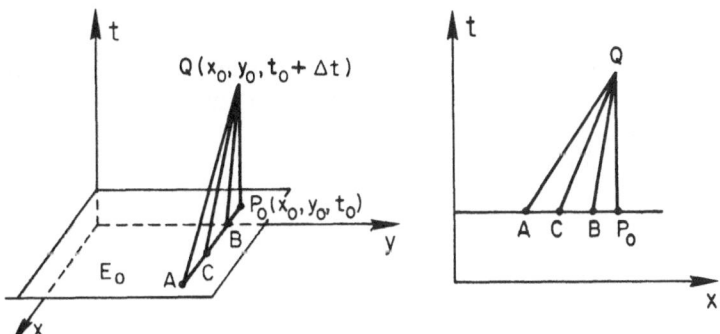

Fig. 4.6   Difference scheme in near characteristics method

QA, QB, QC are near characteristic directions $(d/dt)_1, (d/dt)_2, (d/dt)_3$, respectively.

We then proceed as follows:

1) Determine the directions $(d/dt)_1$, $(d/dt)_2$, $(d/dt)_3$ at P_0 and draw segments in these directions through Q to meet E_0 at A, B, C, respectively.

2) Evaluate the inner $(\partial/\partial y)$ derivatives occurring in (4.4.14) (on the right) at P_0.

3) Solve (4.4.14), (4.4.15), (4.4.16) as difference equations along AQ, BQ, CQ, to determine u, a, and v at Q.

4) Repeat 1), 2), 3) over a whole network of points in the plane $t = t_0 + \Delta t$.

This scheme is similar to Schaetz's scheme used in the optimal characteristics method. However, there is an important simplification. All the characteristic directions now lie in planes *parallel to the x, t plane*. Note that the network of points used in the difference scheme lie on a rectangular grid, facilitating interpolation.

Application to Three-Dimensional Steady Flow: For non-isentropic steady flow, the equations of motion may be written (using generalized coordinates)

$$u^j u_{i,j} + \frac{1}{\varrho} p_{,i} = 0 \tag{4.4.17}$$

$$g^{jk} u_{j,k} + \frac{u^k}{\varrho a^2} p_{,k} = 0 \tag{4.4.18}$$

$$u^j S_{,j} = 0 \tag{4.4.19}$$

$$p = p(\varrho, S) \tag{4.4.20}$$

We take Cartesian coordinates (x, y, z) as basic coordinates and assume that the resultant velocity component parallel to the x, y plane is supersonic.

We write the velocity in cylindrical polar components with

q the resultant component parallel to the x, y plane

w the component in the z direction

θ the angle between the direction of q and the x axis.

We introduce curvilinear coordinates in the x, y plane based on the projections of streamlines (element ds) and normals to these projections (see Fig. 4.7) (elements dn).

Referred to these coordinates, the equations of motion may be written

$$q \frac{\partial q}{\partial s} + \frac{1}{\varrho} \frac{\partial p}{\partial s} = -w \frac{\partial q}{\partial z} \tag{4.4.21}$$

$$q^2 \frac{\partial \theta}{\partial s} + \frac{1}{\varrho} \frac{\partial p}{\partial n} = -wq \frac{\partial \theta}{\partial z} \tag{4.4.22}$$

$$q \frac{\partial w}{\partial s} = -w \frac{\partial w}{\partial z} - \frac{1}{\varrho} \frac{\partial p}{\partial z} \tag{4.4.23}$$

$$\frac{\partial q}{\partial s} + q\frac{\partial \theta}{\partial n} + \frac{q}{\varrho a^2}\frac{\partial p}{\partial s} = -\frac{\partial w}{\partial z} - \frac{w}{\varrho a^2}\frac{\partial p}{\partial z} \tag{4.4.24}$$

$$q\frac{\partial S}{\partial s} = -w\frac{\partial S}{\partial z} \tag{4.4.25}$$

Eq. (4.4.25) can be considered separately from the other equations. Now apply Sauer's near characteristics method to (4.4.21)–(4.4.24).

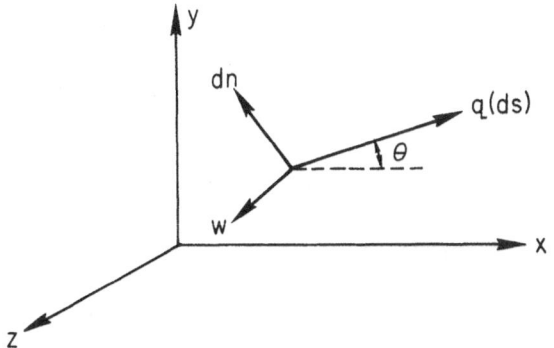

Fig. 4.7   Coordinate system in steady flow

Multiply these equations by factors $\sigma_1, \sigma_2, \sigma_3, \sigma_4$, respectively, and add. Then the condition that the resulting directional derivatives on the left should all be parallel leads to the condition

$$\frac{\sigma_1 + \sigma_4 q/a^2}{\sigma_2} = \frac{\sigma_1 q + \sigma_4}{0} = \frac{\sigma_2 q^2}{\sigma_4 q} = \frac{\sigma_3 q}{0} = \tau.$$

The near characteristic directions are given by

$$\tau = \pm\left(\frac{q^2}{a^2} - 1\right)^{1/2}, \qquad \sigma_1 = 1, \ \sigma_4 = -q, \ \sigma_2 = \tau, \ \sigma_3 = 0$$
$$\sigma_1 = \sigma_2 = \sigma_4 = 0, \ \sigma_3 = 1$$
$$\sigma_1 = 1, \ \sigma_2 = \sigma_3 = \sigma_4 = 0.$$

If we introduce $\sin\mu = a/q$, (4.4.21)–(4.4.26) may be written in near characteristic form

$$\frac{\cot\mu}{\varrho}\left(\frac{d}{dx}\right)_1 p - q^2\left(\frac{d}{dx}\right)_1 \theta$$

$$= \sec(\theta - \mu)\left[\cos\mu\, w q\frac{\partial \theta}{\partial z} + \sin\mu\left\{w\frac{\partial q}{\partial z} - q\frac{\partial w}{\partial z} - \frac{q w}{\varrho a^2}\frac{\partial p}{\partial z}\right\}\right] \tag{4.4.26}$$

$$\frac{\cot\mu}{\varrho}\left(\frac{d}{dx}\right)_2 p + q^2\left(\frac{d}{dx}\right)_2\theta$$
$$= \sec(\theta+\mu)\left[-\cos\mu\, w q\,\frac{\partial\theta}{\partial z} + \sin\mu\left\{w\frac{\partial q}{\partial z} - q\frac{\partial w}{\partial z} - \frac{qw}{\varrho a^2}\frac{\partial p}{\partial z}\right\}\right] \tag{4.4.27}$$

$$q\left(\frac{d}{dx}\right)_3 q + \frac{1}{\varrho}\left(\frac{d}{dx}\right)_3 p = -w\sec\theta\,\frac{\partial q}{\partial z} \tag{4.4.28}$$

$$q\left(\frac{d}{dx}\right)_3 w = -\sec\theta\left\{w\frac{\partial w}{\partial z} + \frac{1}{\varrho}\frac{\partial p}{\partial z}\right\} \tag{4.4.29}$$

$$q\left(\frac{d}{dx}\right)_3 S = w\sec\theta\,\frac{\partial S}{\partial z} \tag{4.4.30}$$

Here $(d/dx)_1$, $(d/dx)_2$, $(d/dx)_3$ are near characteristic directions. The solution of (4.4.26)–(4.4.30) is similar to that for the unsteady case. Assume data are known in some plane E_0, $x=x_0$ (parallel to the y, z plane) (see Fig. 4.8).

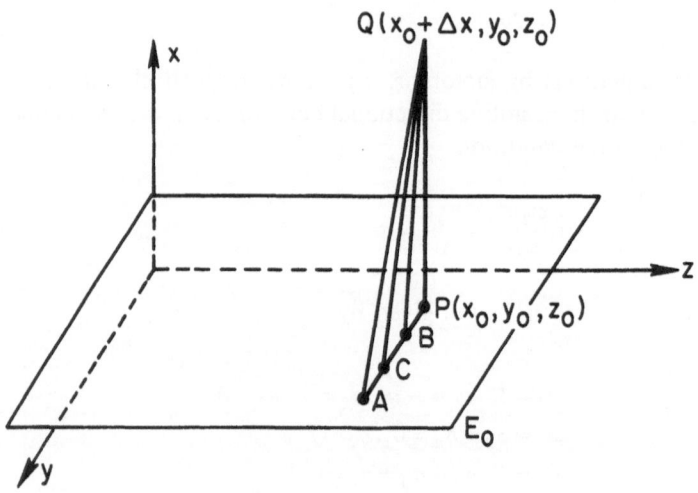

Fig. 4.8   Characteristics scheme in steady flow

To determine the flow quantities at a new point $Q(x=x_0+\Delta x)$ proceed as follows:

1) Project Q onto the plane E_0 at $P(x_0,y_0,z_0)$.

2) Determine the near characteristic directions $(d/dx)_1$, $(d/dx)_2$, $(d/dx)_3$ at P_0 and draw segments parallel to these direction, through Q to meet E_0 in A, B, C, respectively.

3) Evaluate the inner derivatives and coefficients in (4.4.26)–(4.4.30) at A, C, or B, as may be appropriate.

4) Solve (4.4.26), (4.4.27) as difference equations along AQ and BQ to determine p and θ at Q.

5) Solve (4.4.28)–(4.4.30) as difference equations along CQ to determine q, w, and S at Q.

6) Repeat 1)–5) over the whole plane $x = x_0 + \Delta x$.

This method is extremely simple to carry out on a high speed computer. Note that all difference equations are solved in directions *parallel to the x, y plane*. This simplification is very important for problems involving shocks, contact discontinuities, centered expansions, and so on.

The difference scheme proposed here would be modified for problems involving attached shocks (e. g., three dimensional flow past wings with supersonic edges, flow past pointed bodies, etc.). However, since all the work is done in parallel planes, we can simply extend the techniques for handling such problems in two dimensions.

The near characteristics method for steady flow has been worked out here for Cartesian coordinates. Similar derivations can be given in cylindrical or spherical polar coordinates.

References

Albrecht, R., Urich, W.: Numer. Math. **3**, 131 (1961).

Butler, D. S.: Proc. Roy. Soc. A **255**, 232 (1960).

Bruhn, G., Haack, W.: ZAMP **1 × b**, 173 (1958).

Coburn, N., Dolph, C. L.: Proc. 1st Symp. Appl. Math., Amer. Math. Soc. 1949.

Ferri, A., General Theory of High Speed Aerodynamics. Princeton 1954.

Fowell, L. R.: IAS-ARS Meeting, Los Angeles, June 1961. Preprint 61–208–1902.

Holt, M.: J. Fluid Mech. **1**, 409 (1956).

Holt, M.: J. Aero/Space Sci. **26**, 767 (1959).

Holt, M., Yim, B.: Proc. 4th U. S. Congress of Applied Mechanics, Berkeley 1962.

Ransom, V. H., Thompson, H. D., Hoffmann, J. D.: *Lecture Notes in Physics No. 8* (Ed. M. Holt). Berlin-Heidelberg-New York: Springer 1971.

Sauer, R.: Project Report, T. H. München 1962.

Schaetz, R.: Dissertation, T. H. München 1961.

Thornhill, C. K.: British Ministry of Supply, R & M 2615, 1952.

Titt, E. W.: Ann. Math. **2**, 40, 862 (1959).

Van Dyke, M. D.: NACA Tech Rept. 1194, 1954.

CHAPTER 5

The Method of Integral Relations

5.1 Introduction

The Method of Integral Relations was originally formulated by DORODNITSYN in 1956 and subsequently applied to a wide variety of current problems in fluid dynamics. It is designed for the solution of problems governed by partial differential equations of elliptic, mixed elliptic-hyperbolic, or parabolic type. It has been used mostly for problems in two independent variables but in recent years has been extended to deal with three dimensional problems.

The basic objective of the method is to integrate the governing partial differential equations with respect to one of the independent variables so that the original equations can be replaced by a system of equations containing independent variables reduced by one. Thus, in two dimensions, the reduced system consists of ordinary differential equations.

Dorodnitsyn's first presentation of the method (DORODNITSYN, 1956) dealt with inviscid flow problems. Its most notable application was to calculate supersonic or hypersonic flow past blunt-nosed cylinders or bodies of revolution; this important work is described in a series of papers by BELOTSERKOVSKII (1958, 1960, 1962a, 1962b). The inviscid formulation was also applied by CHUSHKIN (1956, 1957) to calculate subsonic critical flow past an ellipse or ellipsoid and sonic flow past an airfoil. Applications to inviscid transonic (supercritical) flow were carried out by HOLT (1962) and by HOLT and MASSON (1970). A complete account of the inviscid formulation is to be found in BELOTSERKOVSKII and CHUSHKIN (HOLT, 1965); these authors are currently preparing an updated version of the work. Generalised versions of the method are to be found in FINLAYSON (1972) and FLETCHER (1984).

In 1960 DORODNITSYN presented a new version of the Method of Integral Relations directed primarily to laminar boundary layer problems. Dorodnitsyn's papers (DORODNITSYN, 1960, 1962) dealt with incompressible flow but the treatment was extended to the compressible case by PAVLOVSKII (1962). Simultaneously S. T. LIU (1962a, 1962b) considered laminar boundary layers with wall suction or injection in both the incompressible and compressible cases. An analysis of the general method, especially when applied to attached flows, is given in ABBOTT and BETHEL (1968).

The viscous version of the Method of Integral Relations was extended by HOLT, NIELSEN and others to cover reversed flow regions in supersonic laminar streams, especially in boundary layers downstream of separation and in wakes (see, for example, HOLT and LU, 1975; NIELSEN et al., 1969).

More recently the viscous version was extended to three dimensions by MODARRESS and HOLT (1976) and to conical boundary layers by FLETCHER and HOLT (1975, 1976); in both cases applications have been made to flows extending up to separation.

Although this chapter deals mainly with viscous flow, it is instructive to emphasize the distinction between the inviscid and viscous formulations of the Method of Integral Relations, and we therefore compare the inviscid blunt body flow application with a simple incompressible boundary layer calculation.

In the general treatment of blunt-body problems, two coordinates, x, y are introduced, with x measured along the body surface and y along straight normals to the surface. The non-uniform, and unknown, part of the flow field is confined laterally at the lower end by the body surface and, at the upper end, by the detached shock wave. The shape of the shock wave, defined by its normal distance ε from the body as a function of x, is one of the unknowns in the problem. It is convenient to replace y by a dimensionless coordinate $\eta = y/\varepsilon$. The flow is then confined laterally in the x, η plane by the body $\eta = 0$, and the shock $\eta = 1$. The equations of motion, with x and η as independent variables, are written in divergence form. The flow field is then divided into a number of strips, parallel to the x-axis, at equal intervals along the η-axis. The transformed equations are now integrated with respect to η between the surface $\eta = 0$ and each successive strip boundary. To evaluate the integrals, it is assumed that the integrands can be represented by suitable interpolation formulas between the strip boundaries. The integrated equations then reduce to ordinary differential equations in certain combinations of the unknowns (those arising in the integrands) on the strip boundaries. The solutions for these equations are determined, in part, by boundary conditions on the axis of symmetry of the body and in part by smoothness conditions applied near the sonic line. The number of strips used depends on the order of the approximation. In the blunt-body problem up to three strips have been used, although two are adequate for most purposes. In this problem, a certain number of the ordinary differential equations have saddle-point singularities near the sonic line (one in the first approximation, two in the second, and so on). The solution is fixed by the requirement that integral curves must be regular at and near the saddle points.

Several changes are made in the method when it is applied to boundary-layer problems. Strictly speaking, there is no definite outer edge to the boundary layer in the physical plane. Accordingly, the space coordinate y is replaced by the x velocity component u, as the variable of integration, when integrating the governing equations across the boundary layer. The outer boundary can then be fixed at the position where $u = U$, the outside velocity. Further, the reciprocal of the skin friction enters into the integrals. This tends to infinity like $1/(u - U)$ at the edge of the boundary layer. It is therefore necessary to weight the equations of motion with factors vanishing at least as fast as $(u - U)$ at the boundary-layer edge.

In boundary-layer problems the initial or boundary conditions are simpler than in blunt body problems. No saddle points arise in the ordinary differential equations and, to start the method, it is only necessary to know the boundary-layer profile at one station. Usually, this is determined from a similarity solution near the stagnation point. However, the method can be started at any station where the boundary-layer solution is known completely.

In all the papers discussed, a great deal of attention is given to similar solutions corresponding to the outside velocity distribution $U = c x^m$. These are very easily handled by the method of integral relations; it is simply required to solve sets of algebraic equations for certain constants. The solutions obtained, particularly with the fourth approximation, agree very closely with the exact solutions of FALKNER and SKAN (1930), in the incompressible case, and those of COHEN and RESHOTKO (1956), in the compressible case.

We begin with a general description of the method and its application to a simple model problem. A discussion of the inviscid version follows with application to the blunt body problem, subsonic flow past elliptic cylinders, and transonic flow. We then describe the viscous version and its application to laminar boundary layer and wake problems in two dimensions.

Generalizations of this version for conical and three-dimensional laminar boundary layers are considered, together with certain improvements in the basic procedure. Finally, recent extensions of the method to deal with internal flow problems, both for laminar and turbulent boundary layers are described. Throughout, emphasis is placed on the value of the method in solving viscous interaction problems arising in separated flows.

5.2 General Formulation. Model Problem

To apply the Method of Integral Relations to the solution of a system of partial differential equations, it is convenient to write the system in divergence form, that is, arrange the system so that all coefficients of derivatives are equal to unity. There is no problem in carrying out this transformation for systems governing mechanics and physics expressing conditions of conservation of mass, momentum, energy, and so forth. The system is here assumed to be of elliptic or mixed elliptic-hyperbolic type. Parabolic equations are considered in Sect. 5.6.

In the case of two independent variables x, y such a system can be written

$$\frac{\partial}{\partial x} P_i(x, y, u_1, \ldots, u_k) + \frac{\partial}{\partial y} Q_i(x, y, u_1, \ldots, u_k) = F_i(x, y, u_1, \ldots, u_k),$$
$$i = 1, \ldots, k$$

(5.2.1)

where $u_1, \ldots, u_k$ are the k dependent variables and P_i, Q_i are known functions of $x, y, u_1, \ldots, u_k$.

We seek a solution of system (5.2.1) in a region bounded by $x=a$, $x=b$, $y=0$, $y=\Delta(x)$ (see Fig. 5.1).

We suppose that k conditions connect the unknowns on the boundaries $x=a$, $x=b$ and k similar conditions on the boundaries $y=0$, $y=\Delta(x)$. In some problems $\Delta(x)$ may also be an unknown and it is then necessary to add an extra boundary condition.

If system (5.2.1) is elliptic all these conditions are explicit relations between $u_1,\ldots,u_k$ on the boundaries themselves. If the system is mixed elliptic-hyperbolic, changing type within the quadrangle, some of the boundary conditions are replaced by conditions guaranteeing regularity of solutions at certain singular points.

We multiply each of (5.2.1) by some piecewise continuous function $f(y)$ and integrate the result with respect to y from $y=0$ to $y=\Delta(x)$. We then obtain k integral relations of the type (suffix i is dropped)

$$
\frac{d}{dx} \int_0^{\Delta(x)} f(y)P\,dy - \Delta'(x)f(\Delta)P_\Delta - f(0)Q_0 + f(\Delta)Q_\Delta - \int_0^\Delta f'Q\,dy
$$

$$
= \int_0^\Delta fF\,dy
$$

(5.2.2)

(suffix Δ refers to values on $y=\Delta$, suffix 0 to values on $y=0$).

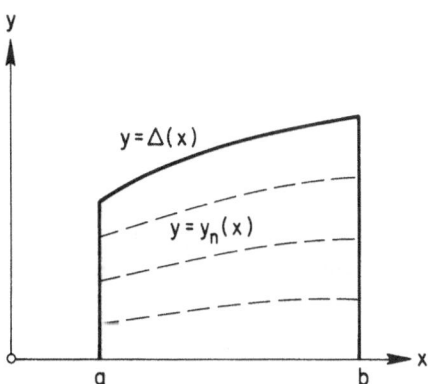

Fig. 5.1   The system of strips in the method of integral relations

To apply the method of integral relations in the N-th approximation we divide the quadrangle in Fig. 5.1 into N strips bounded by lines

$$
y = 0, \qquad y = y_n(x) = \frac{n}{N}\Delta(x), \qquad n = 1,2,3,\ldots,N-1
$$

$$
y = y_N = \Delta(x).
$$

To evaluate the unknown integrands we approximate the functions P, Q, F by suitable interpolation formulae containing their values P_n, Q_n, F_n on strip boundaries. For example,

$$P(x, y, u_1, \ldots, u_k) = \sum_{n=0}^{N} P_n(x) Z_n(y)$$

where $P_n(x) = P(x, y, u_{1n}, \ldots, u_{kn})$, $u_{1n} = u_1(x, y_n)$, etc., and $Z_n(y)$ are chosen interpolation functions depending on the problem in question (polynomials, trigonometric expressions, etc.). The integral in (5.2.2) will then be evaluated in the form

$$\int_0^{\Delta(x)} f P \, dy = \Delta(x) \sum_{n=0}^{N} A_n P_n(x)$$

where A_n are known numerical coefficients, with values depending on the choice of $f_n(y)$ and $Z_n(y)$.

Substituting N independent functions $f_n(y)$ in (5.2.2) we obtain a system of kN ordinary differential equations, supplemented by k algebraic relations in $y=0$ and $y=\Delta(x)$, to determine the $k(N+1)$ unknowns $u_{vn}(x)$, $(v=1, \ldots, k, n=0, \ldots, N)$. The boundary conditions on $x=a$ and $x=b$ (or combined boundary conditions and regularity conditions) provided the required kN conditions needed to fix the solutions of these equations. When $\Delta(x)$ is an unknown a geometric relation provides a differential equation connecting it with the other unknowns $u_1, \ldots, u_k$, and one more boundary condition on $x=a$ or $x=b$ is needed.

We consider two choices of the functions $f_n(y)$

1) Dirac delta function

$$f_n(y) = \delta(y - y_n) \quad (n = 1, 2, \ldots, N).$$

These lead to the well known method of lines (see Chapt. 6), in which y derivatives are replaced by finite difference expressions corresponding to the interpolation formulae selected.

2) Step functions

$$f_n(y) = \begin{cases} 0 & y < y_{n-1} \\ 1 & y_{n-1} \leq y \leq y_n \\ 0 & y > y_n. \end{cases}$$

This choice is equivalent to integrating (5.2.1) as they stand across successive strips and leads to the form of the method of integral relations usually applied to inviscid fluid flow problems.

A simple illustration of the method is provided by the following model problem, which can be solved both analytically and by the method of integral relations.

Model Problem

An elementary example of the method of integral relations to a mixed elliptic-hyperbolic system is provided by DORODNITSYN (1959).

We seek a solution of the system

$$\frac{\partial u}{\partial y} - \frac{\partial v}{\partial x} = 0$$

$$\frac{\partial}{\partial x}\left[(1-x)u\right] + \frac{\partial v}{\partial y} = 0$$

with boundary conditions

$$u = 0, \qquad x = 0$$

$$v = 0, \qquad y = 0, \qquad 0 \le x \le 1$$

$$u = x, \qquad y = 1, \qquad 0 \le x \le 1$$

given on three sides of the square AECB (see Fig. 5.2). The equations change from elliptic to hyperbolic type as x increases through value 1 and the boundary condition on $x=1$, $0 \le y \le 1$, is deliberately omitted. It is replaced by the condition that the solutions remain regular as x increases through 1. The equations show that this condition is $u = \partial v / \partial y$ when $x = 1$.

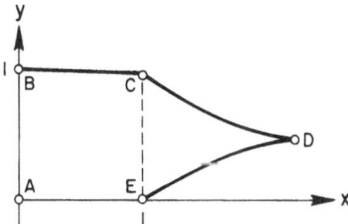

Fig. 5.2   Boundaries in model mixed type problem

Applying MIR in the first approximation we integrate across the strip $y=0$ to $y=1$ and represent the unknowns by the expressions

$$u = u_0 + (u_1 - u_0)y$$

$$v = v_1 y.$$

After substituting in the equations and integrating across $(0,1)$ in y we obtain the following ordinary differential equations for the unknowns u_0 and v_1 (note that $u_1 = x$)

$$\frac{dv_1}{dx} + 2u_0 = 2x$$

$$\frac{d}{dx}\left[(1-x)u_0\right] + 2v_1 = 2x - 1.$$

We seek a solution of these equations with boundary conditions $u_0 = 0$, $x = 0$ and regularity condition $2v_1 = 1 + u_0$, $x = 1$.

The reader should verify that the required solution of this system is

$$u_0 = (x-1) + \frac{1}{(1-x)^{1/2}} \frac{I_1[4(1-x)^{1/2}]}{I_1(4)}$$

$$v_1 = 2x - \frac{3}{2} + \frac{I_0[4(1-x)^{1/2}]}{I_1(4)}$$

where I_0 and I_1 are modified Bessel functions.

5.3 Flow Past Ellipses

One of the earliest applications of the method of integral relations in inviscid flow was carried out by CHUSHKIN (1957, 1958, 1959) in the case of subsonic flow past ellipses, ellipsoids and airfoils.

The two dimensional ellipse problem illustrates the technique used by CHUSHKIN.

The flow field is referred to elliptic coordinates ξ, η, related to the geometry of the ellipse and Cartesian coordinates (see Fig. 5.3) by

$$x + iy = c\cosh(\xi + i\eta).$$

In the case of irrotational flow the equations of motion can be written

$$\frac{\partial \chi}{\partial \xi} + \frac{\partial \omega}{\partial \eta} = 0$$

$$\frac{\partial \lambda}{\partial \xi} - \frac{\partial \mu}{\partial \eta} = 0$$

(5.3.1)

where

$$\chi = H\varrho u, \qquad \omega = H\varrho v; \qquad \lambda = Hv, \qquad \mu = Hu$$
$$\varrho = (1 - u^2 - v^2)^{1/(\gamma-1)}, \qquad H = (\sinh^2\xi + \sin^2\eta)^{1/2}.$$

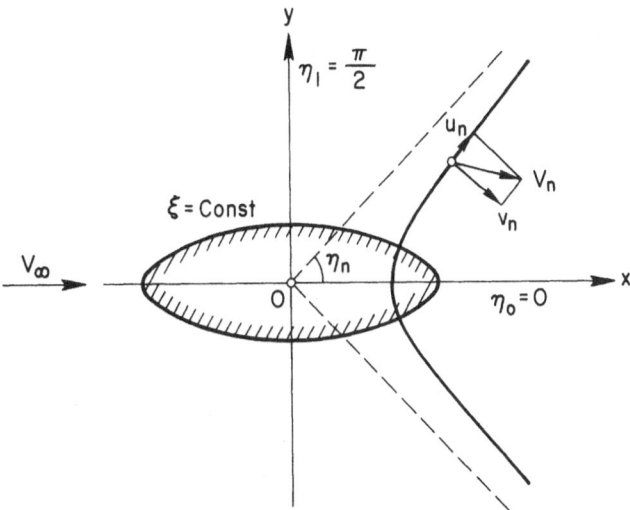

Fig. 5.3 Coordinate system for subsonic flow past ellipses

(H is the Lamé parameter), u, v are the velocity components in the ξ, η directions, respectively. Dimensionless variables are used.

Eqs. (5.3.1) are reduced to ordinary differential equations in the ξ direction. Because of symmetry it is sufficient to consider the first quadrant only. This is divided into N regions by $N-1$ equally spaced hyperbolas $\eta = \eta_n$ within $0 \leq \eta \leq \pi/2$. Eqs. (5.3.1) are integrated across successive strips as they stand (no weighting functions are needed) and the unknown integrands are represented by the following trigonometric expressions

$$\chi = \sum_{n=0}^{N-1} a_{2n+1}(\xi) \cos(2n+1)\eta$$

$$\lambda = \sum_{n=0}^{N-1} b_{2n+1}(\xi) \sin(2n+1)\eta$$

(5.3.2)

which, it will be observed, automatically satisfy the symmetry condition on the axes $x=0$, $y=0$. The coefficients a_{2n+1}, b_{2n+1} are linear combinations of the values χ_n and λ_n on the strip boundaries.

Substitution of (5.3.2) in the integral relations leads to $2N$ ordinary differential equations in a_{2n+1} and b_{2n+1} with boundary conditions provided by N conditions $u_n=0$ on the solid ellipses and N conditions at large distances stating that, on each strip boundary, $\eta=\eta_n$, $u \to V_\infty \cos\eta_n$, $v \to V_\infty \sin\eta_n$ as ξ approaches large values. At large distances the flow can be represented by the Prandtl-Glauert linearization of (5.3.1) and the solution can be given analytically. This means that certain linear relations must be satisfied by the coefficients a_{2n+1} and b_{2n+1} in (5.3.2) where ξ approaches a large value. To fulfill these conditions successive iterations on the values of u_n at the elliptic surface are carried out. The process converges very rapidly.

CHUSHKIN determines the critical Mach number (the free stream Mach number for which the maximum velocity on the body first becomes sonic) corresponding to a series of ellipses. He also carries out analogous calculations for ellipsoids of revolution. These results are shown in Fig. 5.4.

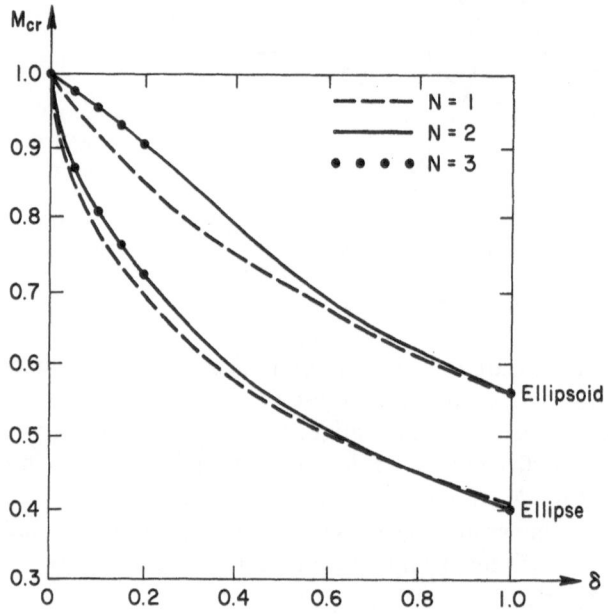

Fig. 5.4   Critical Mach numbers M_{cr} for ellipses and ellipsoids of revolution of various relative thicknesses δ and various N.

5.4 The Supersonic Blunt Body Problem

BELOTSERKOVSKII (1958, 1960) successfully applied the Method of Integral Relations to calculate the mixed subsonic-supersonic flow field adjacent to a circular cylinder or sphere in a supersonic stream. In fact, he was the first to solve this difficult problem correctly, requiring the determination of a detached shock wave and sonic line of unknown shapes.

We consider the two dimensional case and the geometry of the flow field for a general blunt body is shown in Fig. 5.5. The equations of motion are referred to polar coordinates (r, θ) with velocity components (u, v), and pressure, density, entropy denoted by p, ϱ, S', respectively. The unknowns are expressed in dimensionless form with velocities in terms of maximum velocities, pressures and densities in terms of respective stagnation values (far upstream). The stream function ψ is also introduced.

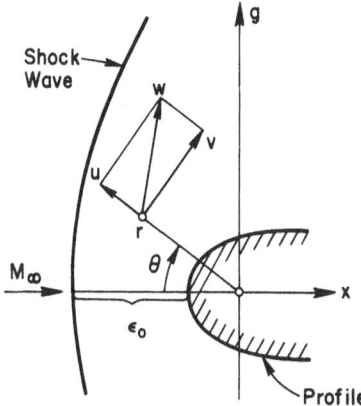

Profile   Fig. 5.5   Coordinate system in blunt body problem

The equations of motion consist of two components of the momentum equation, the continuity equation, conservation of entropy condition. These are combined to give the following equations,

$$\frac{\partial r H}{\partial r} + \frac{\partial S}{\partial \theta} = g \tag{5.4.1}$$

$$\frac{\partial r h}{\partial r} + \frac{\partial t}{\partial \theta} = 0 \tag{5.4.2}$$

$$\frac{\partial \psi}{\partial \theta} = \varrho \left(v \frac{\partial r}{\partial \theta} - r u \right) \tag{5.4.3}$$

$$\phi = \frac{p}{\varrho^{\gamma}} = \phi(\psi) \tag{5.4.4}$$

$$p = (1 - w^2) \varrho \quad \text{(BERNOULLI)} \tag{5.4.5}$$

Here

$$H = k p + \varrho u^2, \qquad S = \varrho u v, \qquad k = (\gamma - 1)/2\gamma$$
$$g = k p + \varrho v^2, \qquad h = \tau u, \qquad t = \tau v$$
$$\varrho = \tau \phi^{-1/(\gamma - 1)}, \qquad \tau = (1 - w^2)^{1/(\gamma - 1)}, \qquad w = (u^2 + v^2)^{1/2}.$$

Note that (5.4.1) and (5.4.2) are in divergence form.

Eqs. (5.4.1)–(5.4.5) are supplemented by a geometrical relation between the unknown distance, ε, measured along a radius from the body to the shock, and the shock angle σ. This is

$$\frac{d\varepsilon}{d\theta} = -(r_0+\varepsilon)\cot(\sigma+\theta) - \frac{dr_0}{d\theta} \tag{5.4.6}$$

(subscript 0 denotes values on the body).

The above equations are solved with boundary conditions on the body and on the shock

$$Body \quad u_0 = \frac{v_0}{r_0}\frac{dr_0}{d\theta}$$

$$\phi = \text{constant} \tag{5.4.7}$$

$$Shock \quad w_x = w_x(\sigma)$$

$$w_y = w_y(\sigma) \tag{5.4.8}$$

$$\phi = \phi(\sigma)$$

$$u = w_y\sin\theta - w_x\cos\theta$$

$$v = w_x\sin\theta + w_y\cos\theta$$

where $w_x(\sigma)$, $w_y(\sigma)$, $\phi(\sigma)$ are simple functions of σ only.

To apply the method of integral relations we divide the region between the body and shock into equal strips with boundaries

$$r = r_i(\theta) = r_0(\theta) + \xi_i\varepsilon(\theta)$$

where, in the N-th approximation,

$$\xi_i = \frac{N-i+1}{N} \qquad i=1,\ldots,N+1$$

$$i = 1 \quad \rightarrow \xi_i = 1 \quad \text{on the shock}$$

$$i = N+1 \rightarrow \xi_i = 0 \quad \text{on the body}$$

(suffix 1 is used on the shock).

To establish the basic integral relations we integrate (5.4.1), (5.4.2) with respect to r between the body and successice strip boundaries:

$$\frac{d}{d\theta}\int_{r_0}^{r_i} S(r,\theta)dr - \left(S_i\frac{dr_i}{d\theta} - S_0\frac{dr_0}{d\theta}\right) + r_iH_i - r_0H_0 = \int_{r_0}^{r_i} g(r,\theta)dr \tag{5.4.9}$$

$$\frac{d}{d\theta}\int_{r_0}^{r_i} t(r,\theta)dr - \left(t_i\frac{dr_i}{d\theta} - t_0\frac{dr_0}{d\theta}\right) + r_ih_i - r_0h_0 = 0 \tag{5.4.10}$$

In the N-th approximation we represent the unknown integrands S, t and g by N-th degree polynomials in ξ of the form

$$f(r,\theta) = \sum_{m=0}^{N} a_m(\theta)\,\xi^m.$$

The coefficients a_m are linear combinations of values of corresponding unknowns $(S, t$ or $g)$ on strip boundaries. The integral relations (5.4.9) and (5.4.10) then provide expressions for the derivatives of the unknowns, S_i', t_i' on strip boundaries.

The basic unknowns in the N-th approximation are $u_i, v_i, \psi_i, \sigma, \varepsilon \, (i=0,...,N)$. These, and their derivatives, are related to the auxiliary unknowns S_i, t_i, h_i by the identities listed below (5.4.1)–(5.4.5). Using these relations we obtain the following system of ordinary differential equations to determine $\varepsilon, \sigma, u_i, v_i, \psi_i$

$$\frac{d\sigma}{d\theta} = F$$

$$\frac{dv_0}{d\theta} = \frac{E_0}{a^{*2} - w_0^2}$$

$$\frac{d\psi_i}{d\theta} = \varrho_i \left[v_i \frac{dr_i}{d\theta} - r_i u_i \right] \qquad\qquad (5.4.11)$$

$$\frac{du_i}{d\theta} = G_i$$

$$\frac{dv_i}{d\theta} = \frac{E_i}{\{a^{*2} + 2u_i^2/(\gamma+1)\} - w_i^2}$$

Eqs. (5.4.11) are supplemented by (5.4.6) for ε. The coefficients F, E_0, G_i and E_i are known functions of S_i, t_i, h_i, H_i, g_i, the unknowns and derivatives S_i', t_i'. The latter are found from the integral relations (5.4.9), (5.4.10) and shock relations (which give S_1', t_1').

In the second approximation (5.4.7) and (5.4.11) provide seven ordinary differential equations for the unknowns $\varepsilon, \sigma, u_0, v_0, u_2, v_2, \psi_2$. There are five boundary conditions on the axis of symmetry $v_0 = v_2 = 0$, $\sigma = \pi/2$, $\psi_0 = \psi_2 = 0$ when $\theta = 0$. Two more conditions are needed. These are provided by the equations for v_0 and v_2 which have saddle point singularities at the sonic point on the body $(w_0 = a^*)$ and at a point near the sonic line on the half way line $(\xi = 1/2)$. For physically meaningful solutions we require that the solutions for v_0 and v_2 are regular at and near these saddle points and this imposes the two additional conditions required $(E_0 = 0, E_2 = 0$ at the corresponding points).

The positions of these points are not known in advance and have to be determined by iteration on the two unknown parameters $u_2(0)$ and $\varepsilon(0)$. For fixed $u_2(0)$ we vary $\varepsilon(0)$ until the first saddle point condition is satisfied. We then test the solution at the second saddle point, change the value of $u_2(0)$ if necessary, and repeat the iteration on $\varepsilon(0)$.

In the first approximation there is only one unknown parameter $\varepsilon(0)$ and a single saddle point on the body. In this case the calculation of the position of the saddle point and the construction of the solution passing through it is straightforward. Extending the process to the second approximation is difficult. Belotserkovskii's calculations are for the most part limited to the second approximation, which is sufficiently accurate for flow field information.

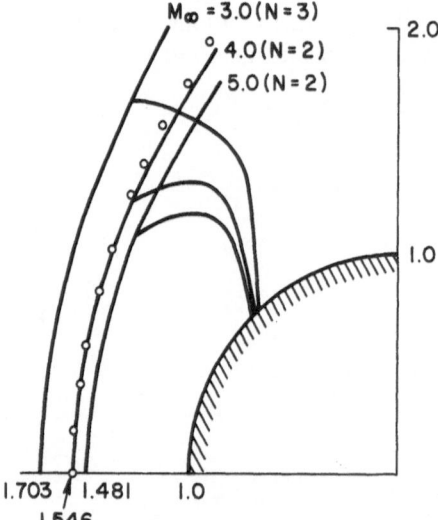

Fig. 5.6   Shapes of sonic line and shock for supersonic flow past a circular cylinder

Shapes of shock and sonic line for flow past a circle at three different Mach numbers are shown in Fig. 5.6. The circles represent experimental points.

BELOTSERKOVSKII has extended the calculations to reacting gas flows and non circular shapes. A full discussion is given in HOLT (1965) (article by BELOTSERKOVSKII and CHUSHKIN).

5.5 Transonic Flow

Several calculations of transonic flow through convergent-divergent nozzles have been carried out using the Method of Integral Relations. Since useful nozzle shapes are supposed to generate flows which are shock free, only irrotational flow conditions need to be considered, so that both a velocity potential Φ and stream function Ψ can be defined.

BELOTSERKOVSKII and CHUSHKIN (1962) formulate the problem in the Φ, Ψ plane. To apply the Method of Integral Relations they divide the region between the nozzle wall $\Psi = 1$ and the nozzle axis $\Psi = 0$ into N equally spaced strips. Integrating the governing equations across successive strips they obtain a series of ordinary differential equations between the velocity on the n-th strip boundary V_n, and the independent variable Φ. The latter may be replaced by s, the distance along the nozzle wall (since $d\Phi/ds = V_N$) so that the problem reduces to the solution of the $N + 1$ equations

$$\frac{dV_n}{ds} = \frac{\Delta_n}{\Delta} \quad (n = 0, 1, \ldots, N) \tag{5.5.1}$$

These equations have N saddle point singularities and the desired solution must be regular in the neighborhood of all of them, giving sufficient conditions to fix the solution. The first and second approximations were applied to both the direct problem (calculations of flow for a given nozzle shape) and design problem (determination of nozzle shape generating a prescribed axial velocity distribution). The applications, which include both plane and axi-symmetric flow are summarized in HOLT (1965) with fuller details to be found in BELOTSERKOVSKII and CHUSHKIN (1962) and ALIKHASHKIN et al. (1963).

HOLT (1964) considered the plane nozzle flow problem in the physical (x, y) plane and applied the first approximation to the hyperbolic channel flow problem. This originated with EMMONS (1948), who used a relaxation method to calculate the velocity field generated by a channel with hyperbolic walls. Subsequently CHERRY (1960) used his version of the hodograph method to calculate the nozzle wall corresponding to the axial velocity distribution found by EMMONS. The Cherry calculations then provide a suitable standard for testing other methods and the MIR calculation in this case agrees quite well with Cherry's results. In these calculations an attempt is made to treat the passage of solutions through critical points more systematically. This is done by starting solutions from a saddle point and determining derivatives there algebraically rather than trying to force solutions through such points from upstream or downstream. The approach was successfully carried out in the first approximation and partially completed recently in the second approximation.

In the physical plane the equations of motion for irrotational flow may be written

$$\frac{\partial(h)}{\partial x} + \frac{\partial(t)}{\partial y} = 0 \tag{5.5.2}$$

$$\frac{\partial v}{\partial x} - \frac{\partial u}{\partial y} = 0 \tag{5.5.3}$$

where (u, v) are the (x, y) velocity components

$$h = \tau u, \qquad t = \tau v, \qquad \tau = (1 - u^2 - v^2)^{1/(\gamma - 1)}.$$

We consider the design problem in which the channel height $Y(x)$ is an unknown, and change independent variables from x, y to x, η where $\eta = y/Y$. Applying MIR in the first approximation to the governing equations, after transforming to x, η, we arrive at the following three ordinary differential equations

$$v_1' = \frac{Y'}{Y} v_1 + \frac{2}{Y}(u_1 - u_0) \tag{5.5.4}$$

$$Y' = \frac{v_1}{u_1} \tag{5.5.5}$$

$$u_1' = E_1/D_1 \tag{5.5.6}$$

where $D_1 \equiv a^{*2} - a^{*2} v_1^2 - u_1^2$, E_1 is a given function of u_1, v_1, and values on the axis and suffix 1 refers to condition on the nozzle wall. Eqs. (5.5.4) and (5.5.5) have no singular points, while (5.5.6) has one saddle point at the position $x = x_1^*$, say, where $D_1 = 0$. Then, for regular behavior we must also satisfy $E_1 = 0$ at $x = x^*$.

To generate solutions of these equations we choose a value of x_1^* (close to the position of the sonic point on the axis) and prescribe $Y = Y(x_1^*)$ at this point. The saddle point conditions $D_1 = 0$, $E_1 = 0$ then provide the remaining two conditions needed to fix solutions of (5.5.4)–(5.5.6), and in fact determine the values u_1^*, v_1^* at $x = x_1^*$. To determine the slope of the regular integral u_1, x curve leaving the saddle point we write, for small $x - x_1^*$

$$u_1 = u_1^* + a(x - x_1^*)$$
$$v_1 = v_1^* + b(x - x_1^*)$$
$$Y = Y^* + Y'^*(x - x_1^*).$$

We immediately determine b and Y'^* from (5.5.4) and (5.5.5). We then substitute these expressions in (5.5.6), equate coefficients of $(x - x_1^*)$ and obtain a quadratic for a. The coefficients in the equations are long algebraic expressions but can all be calculated in terms of $u_1^*, v_1^*, b, Y^*, Y'^*$ and values on the axis (where the velocity is prescribed). Provided that x_1^* is sufficiently close to the sonic point on the axis the roots of this quadratic are real and of opposite sign. We choose the positive root (corresponding to accelerating flow in the x direction) and integrate (5.5.4)–(5.5.6) in both the upstream and downstream directions. If the $u_1(x)$ and $v_1(x)$ curves computed show local maxima or minima the estimated height $Y^*(x_1)$ leads to a nozzle flow with shocks. The calculation is repeated with different $Y^*(x_1)$ until the value (which is unique) giving shock free flow is found. Figure 5.7 shows a series of channel contours generated by Emmons' axial velocity distribution.

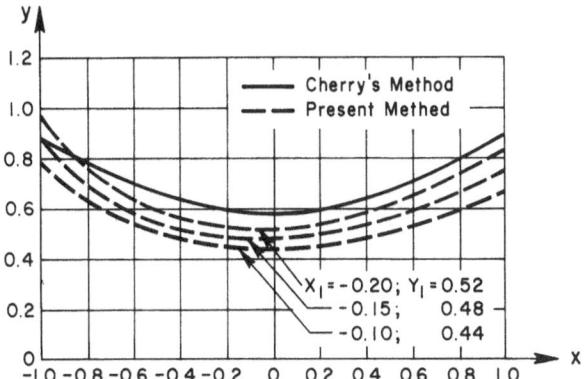

Fig. 5.7   Channel shapes corresponding to Emmons axial Mach number distribution

Less extensive calculations have been carried out for transonic flow past symmetric airfoils, notably by TAI (1975), MELNIK (1970), and HOLT and MASSON (1970). We describe briefly the latter formulation applied to calculate supercritical flow past a circle. Here the equations of motion are referred to polar coordinates r, θ based on the center of the circle

$$(r\tau u)_r + (\tau v)_\theta = 0 \tag{5.5.7}$$

$$(rv)_r - u_\theta = 0 \tag{5.5.8}$$

where (u, v) are the velocity components in the (r, θ) directions and $\tau = (1 - u^2 - v^2)^{1/(\gamma - 1)}$.

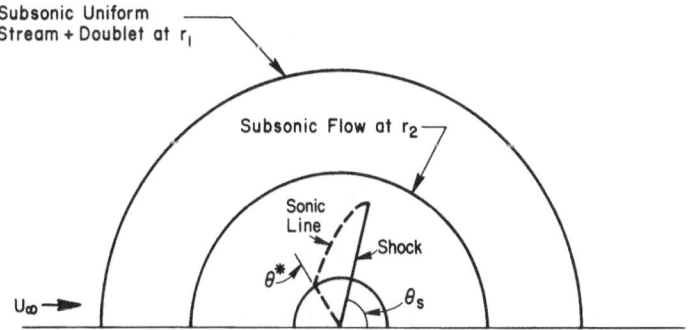

Fig. 5.8   Two strip MIR applied to transonic flow over a circular cylinder

We apply MIR over two circular strips concentric with the circle bounded by $r=r_0$ (circle radius), $r=r_2$ and $r=r_1$ (outer boundary) (see Fig. 5.8). On $r=r_1$ we represent the solution in dipole form

$$\phi = -U_\infty r_1 \cos\theta - \frac{D}{2\pi\lambda^{1/2}r_1}\left\{\frac{\cos\theta}{\cos^2\theta+\lambda\sin^2\theta}\right\} \tag{5.5.9}$$

where $\lambda=1-M_\infty^2$ and D is the dipole strength. In view of the behavior of u, v for large r we represent the integrands u, t, resulting from the application of MIR to (5.5.7), (5.5.8), by

$$t = t^{(0)}+t^{(1)}/r^2+t^{(2)}/r^3$$
$$u = u^{(0)}+u^{(1)}/r^2+u^{(2)}/r^3 \ .$$

The integral relations, together with (5.5.9), lead to a pair of non-linear equations

$$\frac{dv_0}{d\theta} = \frac{E_0(v_0,u_2,D,\theta)}{F_0(v_0,u_2,D,\theta)} \tag{5.5.10}$$

$$\frac{du_2}{d\theta} = G_2(v_0,u_2,D,\theta) \tag{5.5.11}$$

where E_0, F_0, G_2 are known functions.

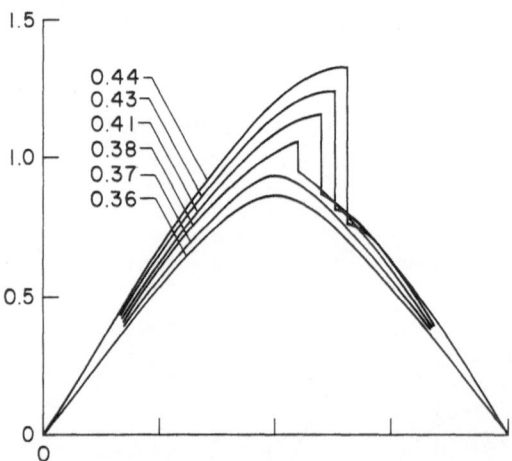

Fig. 5.9   Surface velocity for various near critical freestream Mach numbers

Eq. (5.5.10) has a saddle point where $F_0=0$ (the sonic point on the body for circle flow). We integrate (5.5.10), (5.5.11) starting at the stagnation point

$\theta = 0$ and iterate on the dipole constant D until we find a unique value leading to regular behavior at the saddle point $\theta = \theta^*$ ($E_0 = F_0 = 0$ at $\theta = \theta^*$). We then continue the solution over the aft portion of the cylinder introducing a shock point on the body at a position guaranteeing approach to a stagnation point at $\theta = \pi$ on its downstream side. Figure 5.9 shows the surface velocity for a series of supercritical Mach numbers.

TAI has carried out more ambitious calculations for airfoils (both symmetric and at angle of attack). This work is still continuing and its present status is surveyed briefly in HOLT (1975).

In Chapt. 6 both the transonic nozzle and airfoil problems are treated by Telenin's method and the Method of Lines. In these treatments saddle points are avoided so that, in principle, high order approximations can be carried out.

5.6 Incompressible Laminar Boundary Layer Equations. Basic Formulation

The application of the method of integral relations to the solution of the incompressible laminar boundary-layer equations was first formulated by DORODNITSYN in 1960 (DORODNITSYN, 1960). The approach used was subsequently extended to compressible flow and to flows with suction or injection at the wall. It is most easily understood in relation to the incompressible case.

The boundary layer equations, for plane incompressible flow, are

$$\bar{u}\frac{\partial \bar{u}}{\partial x} + \bar{v}\frac{\partial \bar{u}}{\partial y} = U U' + v\frac{\partial^2 \bar{u}}{\partial y^2}$$

$$\frac{\partial \bar{u}}{\partial x} + \frac{\partial \bar{v}}{\partial y} = 0 \tag{5.6.1}$$

with boundary conditions

$$\bar{u} = \bar{v} = 0 \quad \text{at} \quad y = 0, \qquad \bar{u} = U \quad \text{at} \quad y = \infty.$$

In (5.6.1) x, y are Cartesian coordinates (in plane flow), $\bar{u}, \bar{v}$ are the velocity components, U is the velocity outside the boundary layer, and v is the constant coefficient of viscosity.

The independent variables are changed to ξ, η where

$$\xi = \int_0^x U\,dx, \qquad \eta = \frac{1}{\sqrt{v}}\int_0^y U\,dy = \frac{U y}{\sqrt{v}}$$

and the velocity components are, written in dimensionless form,

$$u = \frac{\bar{u}}{U}, \qquad v = \frac{\bar{v}}{U\sqrt{v}}.$$

Further, v is replaced by w, where

$$w = v + \frac{\dot{U}}{U}\eta u \tag{5.6.2}$$

In terms of the new variables, (5.6.1) reduce to

$$u\frac{\partial u}{\partial \xi} + w\frac{\partial u}{\partial \eta} = \frac{\dot{U}}{U}(1-u^2) + \frac{\partial^2 u}{\partial \eta^2}$$

$$\frac{\partial u}{\partial \xi} + \frac{\partial w}{\partial \eta} = 0 \tag{5.6.3}$$

with boundary conditions

$$u = w = 0 \quad \text{at} \quad \eta = 0; \qquad u = 1 \quad \text{at} \quad \eta = \infty.$$

The method of integral relations, as applied to the boundary-layer equations, is an extension of von Kármán's integral method. In fact, if we multiply the second of (5.6.3) by $(1-u)$ and add to the first, integrating with respect to η from 0 to ∞, we recover the usual momentum integral.

As a generalization of this, DORODNITSYN replaces the factor $(1-u)$ by a weighting function $f(u)$, chosen to vanish at least as fast as $1-u$ as $u \to 1$ and $\eta \to \infty$. The second of (5.6.3) is factored by $f(u)$, the first by $f'(u)$; the two are added and the resulting equation is integrated with respect to η from 0 to ∞. We then obtain the equation

$$\frac{d}{d\xi}\int_0^\infty uf(u)\,d\eta = \frac{\dot{U}}{U}\int_0^\infty (1-u^2)f'(u)\,d\eta - f'(0)\left(\frac{\partial u}{\partial \eta}\right)_{\eta=0} - \int_0^\infty \left(\frac{\partial u}{\partial \eta}\right)^2 f''(u)\,d\eta \tag{5.6.4}$$

We next introduce the quantity

$$\theta = \frac{1}{(\partial u/\partial \eta)} \tag{5.6.5}$$

and change the variable of integration in (5.6.4) from η to u. The resulting equation is

$$\frac{d}{d\xi}\int_0^1 \theta uf(u)\,du = \frac{\dot{U}}{U}\int_0^1 \theta(1-u^2)f'(u)\,du - \frac{f'(0)}{\theta_0} - \int_0^1 \frac{f''(u)\,du}{\theta} \tag{5.6.6}$$

where suffix 0 refers to quantities evaluated at the wall. When $f(u) = 1 - u$ (5.6.6) reduces to von Kármán's momentum integral equation.

Eq. (5.6.6) is the basic equation used in the method of integral relations. Its further treatment depends on the order of approximation used.

To evaluate the integrals in (5.6.6) it is necessary to represent θ and $1/\theta$ as interpolation functions of u. These representations must satisfy the known behavior of θ at the edge of the boundary layer, namely, that $1/\theta$ (the dimensionless shearing stress) approaches zero like $1 - u$ as $u \to 1$,

$$\theta = O\left(\frac{1}{1-u}\right) \quad \text{as} \quad u \to 1 \tag{5.6.7}$$

In the n-th approximation, we then write θ and $1/\theta$, respectively,

$$\theta = \frac{1}{(1-u)}(c_0 + c_1 u + \cdots + c_{n-1} u^{n-1}) \tag{5.6.8}$$

$$\frac{1}{\theta} = (1-u)(b_0 + b_1 u + \cdots + b_{n-1} u^{n-1}) \tag{5.6.9}$$

We now introduce (5.6.8) and (5.6.9) into (5.6.6), assign to $f(u)$ the successive forms

$$(1-u), (1-u)^2, \ldots, (1-u)^n$$

and evaluate the resulting integrals. The integrands are all polynomials and we obtain n equations with coefficients which are known combinations of $c_0, c_1, \ldots, c_{n-1}, b_0, b_1, \ldots, b_{n-1}$. These in turn can be expressed in terms of $\theta_0, \theta_1, \ldots, \theta_{n-1}$ or their reciprocals.

In summary, the method of integral relations consists in replacing the original boundary-layer equations by n ordinary differential equations, derived from (5.6.6), to determine values of θ on the boundaries of n equally spaced strips in u. The procedure is equivalent to satisfying the exact boundary-layer equations along these n boundaries.

To make the procedure clear, we now give the details of the first and second approximations.

First Approximation

Eqs. (5.6.8) and (5.6.9) reduce to

$$\theta = \frac{\theta_0}{1-u}, \qquad \frac{1}{\theta} = \frac{1}{\theta_0}(1-u).$$

Eq. (5.6.6) is integrated once with $f(u)=(1-u)$ to give the single integral relation

$$\dot{\theta}_0 + 3\frac{\dot{U}}{U}\theta_0 = \frac{2}{\theta_0} \tag{5.6.10}$$

(Here the dot denotes differentiation with respect to ξ.)

Second Approximation

The interpolation formulas (5.6.8) and (5.6.9) now become

$$\theta = \frac{1}{(1-u)}[\theta_0(1-2u)+\theta_1 u]$$

$$\frac{1}{\theta} = (1-u)\left[\frac{1}{\theta_0}(1-2u)+\frac{1}{\theta_1}4u\right] \tag{5.6.11}$$

where θ_1 denotes the value of θ at $u=1/2$. We now substitute (5.6.11) in (5.6.6) and evaluate the integrals in (5.6.6), firstly, with $f(u)=(1-u)$; secondly, with $f(u)=(1-u)^2$. We then obtain two integral equations

$$\dot{\theta}_0 + \frac{\dot{U}}{U}(9\theta_0+7\theta_1) = \frac{34}{\theta_0} - \frac{32}{\theta_1} \tag{5.6.12}$$

$$\dot{\theta}_1 + \frac{\dot{U}}{U}(4\theta_0+6\theta_1) = \frac{20}{\theta_0} - \frac{16}{\theta_1} \tag{5.6.13}$$

Eqs. (5.6.12) are two ordinary differential equations to determine θ_0 and θ_1, respectively.

The third approximation leads to three ordinary differential equations in θ_0, θ_1, and θ_2, where θ_1 corresponds to $u=1/3$ and θ_2 to $u=2/3$. The fourth approximation introduces four such quantities, θ_0, θ_1, θ_2, and θ_3, defined in an analogous way. The detailed equations are given in DORODNITSYN (1960, 1962), and are too lengthy to quote here.

To establish boundary conditions for these equations we need to know the boundary-layer profile (or at least some key properties connected with it) at a particular ξ or x station.

DORODNITSYN considers the case when the outside velocity is proportional to some power of x, namely,

$$U = cx^m = c^{1/(m+1)}(m+1)^{m/(m+1)}\xi^{m/(m+1)} \tag{5.6.14}$$

and correspondingly,

$$\frac{\dot{U}}{U} = \frac{\beta}{2\xi} \tag{5.6.15}$$

where

$$\beta = \frac{2m}{m+1}.$$

In this case, it is well known that the flow field depends on a single similarity variable

$$\zeta = \frac{\eta}{\sqrt{2\xi}}.$$

Then

$$u = u(\zeta)$$

and

$$\frac{\partial u}{\partial \eta} = \frac{1}{\sqrt{2\xi}} u'(\zeta).$$

Since u is a function of ζ only, when u is constant ζ is constant, so that we may write, on the i-th u boundary

$$\theta_i = A_i \sqrt{\xi} \tag{5.6.16}$$

where $A_0, A_1, \ldots$, are constants.

When the method of integral relations is applied to the flow defined by (5.6.14), the problem reduces to the solution of algebraic equations.

From (5.6.10) we have, in the first approximation

$$A_0(1 + 3\beta) = \frac{4}{A_0} \tag{5.6.17}$$

and in the second approximation,

$$(1 + 9\beta) A_0 + 7\beta A_1 = \frac{68}{A_0} - \frac{64}{A_1}$$

$$\tag{5.6.18}$$

$$4\beta A_0 + (1 + 6\beta) A_1 = \frac{40}{A_0} - \frac{32}{A_1}$$

with corresponding algebraic equations in succeeding approximations.

In any approximation the most interesting quantity is $1/A_0$, which determines the skin friction coefficient at the wall. In fact,

$$c_f = \frac{\tau_0}{\frac{1}{2}\varrho U^2} = \sqrt{2\nu}\,\frac{\phi''(0)}{\sqrt{\xi}} = 2\sqrt{\frac{\nu}{\xi}}\,\frac{1}{A_0} \qquad\qquad (5.6.19)$$

In the following table the values of $1/A_0$, determined by the method of integral relations, using the first four approximations, are compared with corresponding values found from the exact similarity solutions. The first approximation is generally too crude, but the second approximation gives good agreement with exact values in negative pressure gradient regions ($m>0$). The method is severely tested in regions of rising pressure ($m<0$), particularly as the separating flow corresponding to $\beta = -0.1988$ is approached. However, the fourth approximation closely agrees with the exact solution even in this range.

No other approximate boundary-layer solution gives such good agreement with the exact Falkner-Skan solution as this one. If the fourth approximation is used, the method of integral relations is uniformly accurate both in accelerated and retarded regions. On the other hand, earlier integral methods, such as Pohlhausen's, were definitely less accurate in the retarded region.

Table of $1/A_0$

β	First Approx.	Second Approx.	Third Approx.	Fourth Approx.	Exact
−0.19	0.32787	*	0.14253	0.08514	0.06060
−0.15	0.37081	*	0.18072	0.14999	0.15299
−0.10	0.41833	*	0.23246	0.22255	0.22576
0.00	0.50000	0.31649	0.32968	0.33191	0.33206
+0.50	0.79057	0.65628	0.65416	0.65586	0.65597
1.00	1.00000	0.87247	0.87056	0.87164	0.87157
1.50	1.17260	1.04538	1.04386	1.04470	1.04456
2.00	1.32288	1.19371	1.19252	1.19321	1.19304

* Separation at $\beta \approx -0.05$.

The above analysis provides the information needed to start the method of integral relations near a two-dimensional stagnation point. We take the origin $x=0$, $\xi=0$ at this point and use the solution for $m=1$, determining the number of parameters $\theta_0, \theta_1,...$, equal to the order of approximation. We then have sufficient initial conditions to start the integration of the ordinary differential equations in $\theta_0, \theta_1, \theta_2,...$.

DORODNITSYN applied his method to the velocity distribution $U = c\sqrt{\xi}(1-\xi)$ corresponding to incompressible flow around a typical wing profile with rounded leading edge. The third and fourth approximations agree very closely, showing that the method converges in this severe case.

5.7 The Method in the Compressible Case

This was worked out by PAVLOVSKII (1962) for a perfect gas with constant specific heats c_p, c_v, gas constant and constant Prandtl number Pr, not necessarily equal to 1. The coefficient of viscosity μ is assumed to be an arbitrary function of temperature.

The equations of motion are, in the usual notation,

$$\varrho \bar{u} \frac{\partial \bar{u}}{\partial x} + \varrho \bar{v} \frac{\partial \bar{u}}{\partial y} = -\frac{dp_e}{dx} + \frac{\partial}{\partial y}\left(\mu \frac{\partial \bar{u}}{\partial y}\right)$$

$$\frac{\partial \varrho \bar{u}}{\partial x} + \frac{\partial \varrho \bar{v}}{\partial y} = 0, \qquad p_e = \varrho R T,$$

$$\varrho \bar{u} \frac{\partial}{\partial x}\left(I c_p T + \frac{1}{2}\bar{u}^2\right) + \varrho \bar{v} \frac{\partial}{\partial y}\left(I c_p T + \frac{1}{2}\bar{u}^2\right)$$

$$= \frac{1}{\mathrm{Pr}} \frac{\partial}{\partial y}\left[\mu \frac{\partial}{\partial y}\left(I c_p T + \frac{1}{2}\bar{u}^2\right)\right] - \left(\frac{1}{\mathrm{Pr}} - 1\right)\frac{\partial}{\partial y}\left(\mu \frac{\partial}{\partial y}\frac{\bar{u}^2}{2}\right)$$

(5.7.1)

The boundary conditions are

$$\bar{u} = \bar{v} = 0, \qquad y = 0$$

$$\bar{u} \to U, \qquad y \to \infty$$

$$I c_p T + \tfrac{1}{2}\bar{u}^2 \to \tfrac{1}{2} V_{\max}^2, \qquad y \to \infty$$

(5.7.2)

$$\frac{\bar{u}}{U} \to 1, \qquad y \neq 0, \qquad x \to 0$$

$$(I c_p T + \tfrac{1}{2}\bar{u}^2)/\tfrac{1}{2} V_{\max}^2 \to 1, \qquad y \neq 0, \qquad x \to 0$$

Here $U(x)$ denotes the velocity outside the boundary layer related to p_e by the relation

$$p_e(x) = p_0 \left(1 - \frac{U^2}{V_{\max}^2}\right)^{\gamma/(\gamma-1)}$$

(5.7.3)

Also,

$$V_{\max}^2 = 2 I c_p T_0$$

where T_0 is the stagnation temperature.

Finally, there is some heat-balance condition at the wall, which is considered in three alternative forms

(i) $$\left(\lambda \frac{\partial T}{\partial y}\right)_{y=0} = \varepsilon \sigma (T^4)_{y=0}$$

(5.7.4)

where λ is the thermal conductivity, σ the Stefan-Boltzmann constant, and ε the emissivity.

Eq. (5.7.4) states that the heat transfer from the gas due to conduction equals the amount of heat radiated by the body per unit time.

(ii) $\lambda \left(\dfrac{\partial T}{\partial y} \right)_{y=0} = \phi(x)$ (5.7.5)

In this case the heat transfer is a specified function of x. This includes the special case of zero wall heat transfer $\phi(x)=0$.

(iii) $(T)_{y=0} = \phi_1(x)$ (5.7.6)

In this case the wall temperature is specified.

Eqs. (5.7.1) are reduced to incompressible form (STEWARTSON, 1949) and further modified to correspond to (5.6.3) used in the basic formulation of the method. The combined transformation of coordinates and dependent variables is

$$s = \frac{1}{V_{max} L} \int\limits_0^x \frac{p_e(x)}{p_0} U(x)\,dx$$

$$t = \frac{U(x)}{\sqrt{v_0 \, V_{max} L_0}} \int\limits_0^y \frac{\varrho}{\varrho_0}\,dy, \qquad 1-h = \frac{I\,c_p\,T + \tfrac{1}{2} u^2}{I\,c_p\,T_0} \qquad (5.7.7)$$

$$\bar{u} = u\,U$$

$$w = \frac{u\,\partial t/\partial x}{U\,p_e/p_0} + \sqrt{V_{max} L}\,\frac{\bar{v}}{U\sqrt{v_0}}\,\frac{T_0}{T}$$

The resulting equations of motion are

$$u \frac{\partial u}{\partial s} + w \frac{\partial u}{\partial t} = \frac{\dot{\alpha}_e}{\alpha_e(1-\alpha_e^2)}(1-h-u^2) + \frac{\partial}{\partial t}\left(b \frac{\partial u}{\partial t} \right) \left.\vphantom{\begin{array}{c}1\\1\\1\end{array}}\right\}$$

$$\frac{\partial u}{\partial s} + \frac{\partial w}{\partial t} = 0 \qquad\qquad (5.7.8)$$

$$u \frac{\partial h}{\partial s} + w \frac{\partial h}{\partial t} = \frac{1}{Pr} \frac{\partial}{\partial t}\left(b \frac{\partial h}{\partial t} \right) + \left(\frac{1}{Pr} - 1 \right) \alpha_e^2 \frac{\partial}{\partial t}\left(b \frac{\partial u^2}{\partial t} \right)$$

where

$$b(\tau) = \frac{\mu}{\mu_0} \frac{1}{\tau}, \qquad \alpha_e = \frac{U(x)}{V_{max}}, \qquad \tau = \frac{T}{T_0}.$$

Boundary conditions corresponding to (5.7.2), (5.7.4), (5.7.5), and (5.7.6) can be expressed in terms of the transformed variables. The most important conditions for the velocity and enthalpy are

$$u = w = 0, \qquad t = 0$$
$$u \to 1, \qquad h \to 0, \qquad t \to \infty \tag{5.7.9}$$

One of three thermal conditions is applied at the wall.

To apply the method of integral relations we multiply the first of (5.7.8) by $f'(u)$, the second by $f(u)$ (where $f(u)$ is a continuous function which vanishes when $u = 1$), add the results, and integrate with respect to t from 0 to ∞. We change the variable of integration from η to u and introduce the quantities

$$\theta = 1/(\partial u/\partial t), \qquad \kappa = \theta h, \qquad \beta = \frac{\dot{\alpha}_e}{\alpha_e(1 - \alpha_e^2)}.$$

Then we obtain

$$\frac{d}{ds} \int_0^1 \theta(u) f(u) du = \beta \int_0^1 \theta f'(u)(1 - u^2) du$$

$$- \beta \int_0^1 \kappa f'(u) du - \frac{f'(0) b_0}{\theta_0} - \int_0^1 \frac{b}{\theta} f''(u) du \tag{5.7.10}$$

where suffix zero refers to values on the wall.

Now multiply the first of (5.7.8) by $h f'(u)$, the second by $h f(u)$, the third by $f(u)$ and add, then integrate with respect to η from 0 to ∞, or with respect to u from 0 to 1. We then obtain

$$\frac{d}{ds} \int_0^1 \kappa(u) f(u) du = \beta \int_0^1 \kappa f'(u)(1 - u^2) du - \beta \int_0^1 \frac{\kappa^2}{\theta} f'(u) du$$

$$- \frac{\kappa_0 b_0}{\theta_0^2} f'(0) - \int_0^1 \frac{b \kappa}{\theta_0^2} f''(u) du - \left(\frac{1}{Pr} + 1\right) \int_0^1 q f'(u) du$$

$$- \frac{q_0 f(0)}{Pr} - 2\alpha_e^2 \left(\frac{1}{Pr} - 1\right) \int_0^1 \frac{b}{\theta}(u) f'(u) du \tag{5.7.11}$$

where

$$q = b \frac{\partial h}{\partial t} = \frac{b}{\theta} \frac{\partial}{\partial u} \frac{\kappa}{\theta}.$$

Eqs. (5.7.10) and (5.7.11) are the basic equations to determine the integral relations. We choose $f(u)$ so that all the integrals in these equations are finite. In fact, we take $f(u) = (1-u), (1-u)^2, \ldots, (1-u)^n$ in the n-th approximation.

In the first-order approximation, we take $f(u) = (1-u)$ in (5.7.10). We then obtain one integral relation. In the second approximation we take $f(u) = (1-u)$ in (5.7.11) and $(1-u), (1-u)^2$ in (5.7.10) and so obtain three integral relations. In general, we establish $2n-1$ integral relations in the n-th approximation. However, we need $2n$ equations to determine $2n$ unknowns $\theta_0, \theta_1, \ldots, \theta_{n-1}$, $\kappa_0, \kappa_1, \ldots, \kappa_{n-1}$. The remaining equation is supplied by the wall heat balance equation.

The first approximation results in the single ordinary differential equation

$$\frac{1}{2} \theta_0 = -\frac{2}{3} \beta \theta_0 + \beta \kappa_0 + \frac{b_0}{\theta_0} \tag{5.7.12}$$

where we have written

$$q_0 = -\frac{b_0 \kappa_0}{\theta_0^2}.$$

The second approximation contains the equations:

$$-\frac{1}{6} \theta_0 + \frac{1}{3} \theta_1 = -\beta \left(-\frac{1}{6} \theta_0 + \frac{5}{6} \theta_1 \right) + \beta \kappa_1 + \frac{b_0}{\theta_0}$$

$$\frac{1}{12} \theta_1 = -\beta \left(\frac{1}{3} \theta_0 + \frac{1}{2} \theta_1 \right) + \beta \left(\frac{1}{3} \kappa_0 + \frac{2}{3} \kappa_1 \right) + \frac{5}{3} \frac{b_0}{\theta_0} - \frac{4}{3} \frac{b_1}{\theta_1}$$

$$\frac{1}{6} \kappa_1 = -\beta \left(\frac{1}{6} \kappa_0 + \frac{1}{2} \kappa_1 \right) + \left(\frac{1}{6} \frac{\kappa_0^2}{\theta_0} + \frac{2}{3} \frac{\kappa_1^2}{\theta_1} \right) + \frac{\kappa_0 b_0}{\theta_0^2} \tag{5.7.13}$$

$$\qquad\qquad + \left(\frac{1}{Pr} + 1 \right) \left(\frac{1}{6} q_0 + \frac{2}{3} q_1 \right) - \frac{q_0}{Pr} + \frac{2}{3} \alpha_e^2 \left(\frac{1}{Pr} - 1 \right) \frac{b_1}{\theta_1}$$

where

$$q_0 = -\frac{b_0}{\theta_0} \left(3 \frac{\kappa_0}{\theta_0} - 4 \frac{\kappa_1}{\theta_1} \right), \qquad q_1 = -\frac{b_1}{\theta_1} \frac{\kappa_0}{\theta_0}.$$

Succeeding approximations become increasingly complicated and are discussed in Pavlovskii's (1962) paper.

The way in which various terms in the integrands of (5.7.10) and (5.7.11) are represented by interpolation formulas is not made very clear in Pavlovskii's paper, and it may be helpful to explain the procedure more fully.

In the n-th approximation we write

$$\theta = \frac{1}{(1-u)}\left[c_0 + c_1(1-u) + \cdots + c_{n-1}(1-u)^{n-1}\right]$$

and evaluate $c_0, c_1, \ldots, c_{n-1}$ from the conditions $\theta = \theta_0$, $u=0$, $\theta = \theta_1$, $u=1/n$, $\theta = \theta_2$, $u=2/n$, etc. We write $\kappa = (d_0 + d_1(1-u) + \cdots + d_{n-1}(1-u)^{n-1})$ and determine the coefficients in the same manner. Combinations of κ and θ are treated in a similar manner, always allowing for the singularity in θ at $u=0$.

To evaluate q, for example, we use the definition

$$q = \frac{b}{\theta}\frac{\partial}{\partial u}\left(\frac{\kappa}{\theta}\right)$$

where

$$\frac{\kappa}{\theta} = (1-u)\left[e_0 + e_1(1-u) + \cdots + e_{n-1}(1-u)^{n-1}\right].$$

In the second approximation we find that

$$\frac{\kappa}{\theta} = (1-u)\left[\frac{4\kappa_1}{\theta_1} - \frac{\kappa_0}{\theta_0} + (1-u)\left(\frac{2\kappa_0}{\theta_0} - \frac{4\kappa_1}{\theta_1}\right)\right]$$

so that

$$\frac{\partial}{\partial u}\frac{\kappa}{\theta} = \left(\frac{4\kappa_1}{\theta_1} - \frac{\kappa_0}{\theta_0}\right) + 2(1-u)\left(\frac{2\kappa_0}{\theta_0} - \frac{4\kappa_1}{\theta_1}\right).$$

Put $u=0$ in this expression to give

$$q_0 = -\frac{b_0}{\theta_0}\left(\frac{3\kappa_0}{\theta_0} - \frac{4\kappa_1}{\theta_1}\right).$$

With $u=1/2$ we obtain

$$q_1 = -\frac{b_1}{\theta_1}\left(\frac{\kappa_0}{\theta_0}\right).$$

These agree with the expressions stated below (5.7.13).

The boundary conditions are incorporated in (5.7.12) and (5.7.13). The solution of the equations in each approximation requires initial conditions. In general these are determined from a similarity solution near the stagnation point. The discussion of this is not as simple as in the incompressible case.

Assume that near $s=0$ the dimensionless outside velocity distribution can be written

$$\alpha_e(s) = C s^m [1 + 0(s)] \tag{5.7.14}$$

Then

$$\beta(s) = \frac{m}{s} [1 + 0(s)] \tag{5.7.15}$$

Eq. (5.7.14) is equivalent to assuming that

$$U = C' x^{m/(m-1)}.$$

Hence, it follows that

$$0 \le m \le 1.$$

If the leading terms only in (5.7.14) and (5.7.15) are retained, it can be shown that the reduced boundary-layer equations (5.7.8) have a similarity solution with similarity variable $\lambda = t/s^{1/2}$. The equations then reduce to ordinary differential equations in λ equivalent to those obtained by COHEN and RESHOTKO (1956).

The method of integral relations will lead, in this case, to equations which are approximately equivalent to the similarity equations. The approximate equations are algebraic equations which can be solved to determine the shearing stress and heat transfer on strip boundaries $u = $ constant.

Near a stagnation point we may write, at the i-th u strip boundary

$$\theta_i = A_{ni}^0 \sqrt{s} \, [1 + 0(s)]$$
$$\kappa_i = B_{ni}^0 \sqrt{s} \, [1 + 0(s)].$$

In the n-th approximation there are sufficient algebraic equations to determine the constants A_{ni}^0, B_{ni}^0. These coefficients fix the leading terms in θ_i and κ_i when expanded in powers of $s^{1/2}$.

The whole procedure is best understood when applied to an example. PAVLOVSKII considered the flow past a slab with a cylindrical leading edge and used the inviscid velocity distribution for this shape determined by BELOTSERKOVSKII (1958) and CHUSHKIN (1960) (see Fig. 5.10). He took $M_\infty = 4$,

$Pr = 3/4$, and used Sutherland's formula for the variation of coefficient of viscosity

$$b(\tau) = \tau^{1/2} \frac{C/T_0 + 1}{C/T_0 + \tau} \quad (C = 117\,^\circ C) \tag{5.7.16}$$

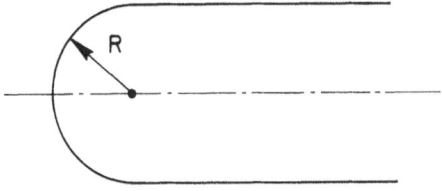

Fig. 5.10 Two-dimensional configuration considered in the example of PAVLOVSKII (1962)

The stagnation temperature is $T_0 = 1000\,^\circ C$. The wall is assumed to be insulated and nonradiating so the wall heat balance condition is

$$\left(\frac{\partial T}{\partial y}\right)_{y=0} = 0 \tag{5.7.17}$$

The outside velocity distribution is shown in Fig. 5.11 with α_e increasing linearly with $\bar{x}$ up to $\bar{x} = \pi/2$ ($\bar{x} = $ arc length/radius) and remaining constant thereafter.

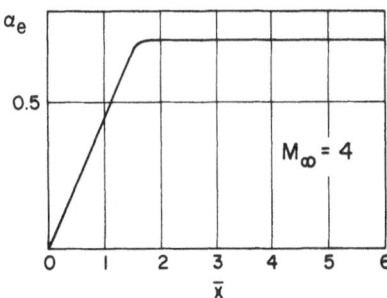

Fig. 5.11 Variation of edge-velocity assumed in PAVLOVSKII (1962) for the cylindrically capped plate of Fig. 5.10

The solution near the stagnation point corresponds to $m = 1/2$. In the first approximation we have

$$\theta_0 = A_{10}^0 \sqrt{s} + \cdots$$

while

$$\beta = m/s + \cdots .$$

Substituting in (5.7.12) we obtain

$$\tfrac{1}{4} A_{10}^0 = \tfrac{1}{3} A_{10}^0 + 1/A_{10}^0$$

or

$$A_{10}^0 = 1.309307 \,.$$

From (5.7.17)

$$q_0 = \kappa_0 = 0 \,.$$

In the second approximation we obtain algebraic equations for A_{10}^0 and A_{21}^0 and B_{20}^0, B_{21}^0 (we write $\kappa_i = B_{ni}^0 \sqrt{s}$ near the stagnation point). It is found that

$$A_{20}^0 = 1.1461666 \,, \qquad A_{21}^0 = 1.823133 \,, \qquad B_{20}^0 = 0 \,, \qquad B_{21}^0 = 0 \,.$$

The B coefficients are zero in all cases (this follows from (5.7.17)). Corresponding coefficients can be found in the third and fourth approximations.

The value of A_{40}^0 is compared with the corresponding value determined from the exact similarity solutions A_0^0. We find

$$A_{40}^0 = 1.1477527$$
$$A_0^0 = 1.147354 \,.$$

There is no question, therefore, that the method of integral relations provides a simple accurate means of calculating the similarity solution near the stagnation point.

However, it is found that if only the leading terms in expansions near the stagnation point are retained, the initial conditions for $\theta_0, \theta_1, \dots$, are not sufficiently accurate to start the integration of the ordinary differential equations further along the body. The next higher order terms (at least in κ_i) should be calculated.

Thus we write

$$\alpha_e(s) = C_0 s^{1/2} + C_1 s + \cdots$$
$$\kappa_i = B_{ni}' s^{3/2} + \cdots$$

substitute in the integral relations, compare powers of $s^{3/2}$, and obtain algebraic equations for B_{ni}'. The values in the second approximation are

$$B_{20}' = 0.1838 \, C^2$$
$$B_{21}' = 0.2193 \, C^2$$

where

$$C = \lim_{s \to 0} \frac{\alpha_e}{\sqrt{s}} = 0.4697659 .$$

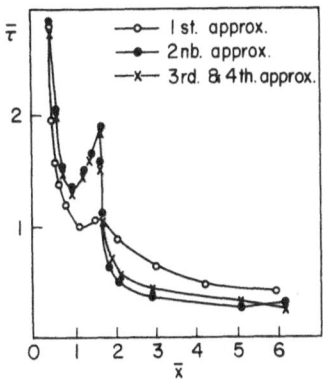

Fig. 5.12 Variation in normalized wall shear stress, $\bar{\tau}$, for an insulated, cylindrically capped plate. $M_\infty = 4$, $Pr = 3/4$ (PAVLOVSKII, 1962)

With the revised initial values, the ordinary differential equations were integrated along the body up to $s = 2.422$ or $\bar{x} = 7.02$, using the first four approximations. From the results of the integrations the variations in skin friction coefficient and recovery enthalpy around the body were calculated.

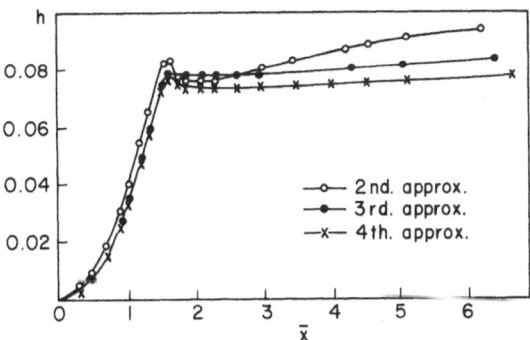

Fig. 5.13 Variation in the normalized difference between stagnation and recovery enthalpies, $h = (H_{stag} - H_{rec})/H_{stag}$, for an insulated, cylindrically capped plate. $M_\infty = 4, Pr = 3/4$ (PAVLOVSKII, 1962)

Results are shown in Figs. 5.12 and 5.13. The fourth approximation can be expected to give accurate results under all conditions and the third approximation is adequate to determine the skin friction coefficient. In this particular application some inaccuracies were apparently introduced at the cylinder-plate junction where the velocity gradient is discontinuous.

5.8 Laminar Boundary Layers with Suction or Injection

Dorodnitsyn's (1960) treatment of the incompressible laminar boundary layer
was extended by S. T. Liu (1962a) to include wall suction or injection. The
main change comes in the boundary condition where, instead of

$$v = 0, \qquad y = 0$$

we have

$$v = v_0(x), \qquad y = 0$$

$v_0(x)$ being a prescribed function. In the transformed $\bar{u}, w$ variables we must
also introduce

$$w_0(\xi) = \frac{v_0(\xi)}{U\sqrt{v}}$$

with

$$w = w_0(\xi), \qquad \eta = 0.$$

The general integral relation (5.6.6) now has an extra term $-w_0(\xi)f(0)$
on the left side. Correspondingly, the ordinary differential equations in the
successive approximations contain extra terms.

Thus (5.6.10) now reads:

$$\dot{\theta}_0 = 2w_0(\xi) - 3\frac{\dot{U}}{U}\theta_0 + \frac{2}{\theta_2} \tag{5.8.1}$$

In the second approximation (5.6.12) has an extra term $18w_0(\xi)$ on the right,
while (5.6.13) has an additional term $12w_0(\xi)$.

The method was applied to flow with uniform suction or injection. In
this case

$$w_0(\xi) = \frac{m}{\sqrt{2\xi}}, \quad \text{where} \quad m = \frac{v_0}{\sqrt{vc}}.$$

The outside velocity distribution is assumed to be

$$U = cx.$$

Then near the stagnation point (where $U = cx$) the solution is fixed by constants $A_0, A_1, \ldots$. The equations to determine these are

$$A_0 + \frac{m}{\sqrt{2}} = \frac{1}{A_0} \qquad\qquad \text{1st approximation}$$

$$\left.\begin{aligned}
10\,A_0 + 7\,A_1 + \frac{36\,m}{\sqrt{2}} &= \frac{68}{A_0} - \frac{64}{A_1} \\[2mm]
4\,A_0 + 7\,A_1 + \frac{24\,m}{\sqrt{2}} &= \frac{40}{A_0} - \frac{32}{A_1}
\end{aligned}\right\} \quad \text{2nd approximation.}$$

Using the fourth approximation with $m = -0.5$, we find

$$1/A_0 = 1.09001$$

while the value from the exact similarity solution is

$$1/A_0 = 1.09018 \,.$$

These results are used to start the integration of the ordinary differential equations for the airfoil flow defined by

$$U = c\sqrt{\xi}\,(1 - \xi)$$

with suction corresponding to a value $m = -0.5$. Full results are given in LIU (1962a).

LIU (1962b) also extended Pavlovskii's treatment of the compressible laminar boundary-layer to include suction or injection (at a constant rate).

The extension results in the addition of new terms to (5.7.10) and (5.7.11), the general integral relations, namely a term $w_0(s)\,f(0)$ on the left of (5.7.10) and a term $w(s)\,f(0)\,\kappa_0/\theta_0$ on the left of (5.7.11). These introduce additional terms in the ordinary differential equations in the various approximations and considerably complicate them.

In contrast to the incompressible case, the wall suction or injection condition now affects the temperature; as a result, the analysis is not a simple extension of Pavlovskii's treatment of boundary layer without suction.

The transformation of independent variables used by PAVLOVSKII, is also applied in this case and the transformed equations of motion still have the form (5.7.8). However, the wall boundary condition for w is now

$$w_0(s) = \sqrt{\frac{l}{V_{\max}\,v_\infty}\,\frac{m_0}{\varrho_\infty\,\alpha_0(1 - \alpha_0^2)^{\gamma/(\gamma - 1)}}}$$

where m_0 is the suction or injection parameter ($m_0 > 0$ for injection, $m_0 < 0$ for suction).

The heat balance condition is stated in case (i) only (defined in Sect. 5.7), namely, when the wall is radiating. This now has the form

$$b\left(\frac{\partial h}{\partial t}\right)_{t=0} = -\sqrt{\frac{l}{V_{max} \, v_\infty}} \, \frac{Pr}{\varrho_\infty \alpha_0 (1 - \alpha_0^2)^{\gamma/(\gamma-1)}}$$

$$\times \left[m_0 (1 - h_w - \tau_k) + \frac{\varepsilon \sigma}{C_p} (1 - h_w)^4 \, T_\infty^3 \right]$$

where τ_k is the dimensionless temperature of the injected fluid. The extra term, with factor m_0, results from the presence of injected fluid.

The method of integral relations is applied to the governing equations, exactly as in Pavlovskii's treatment, with additional terms arising at each stage.

The ordinary differential equations governing the fourth approximation are written out in full by LIU and consist of four equations for shearing stress coefficients $\theta_0, \theta_1, \theta_2, \theta_3, \kappa_0, \kappa_1, \kappa_2, \kappa_3$. The missing equation is supplied by the heat balance condition (this is not a differential equation but an algebraic relation between $\kappa_0, \kappa_1, \kappa_2, \kappa_3$).

Attention is again given to the flow defined by $U = cx^m$. The characteristics of this are worked out by the method of integral relations using the first four approximations. Specific results are given for flow near a stagnation point, $m = 1$, with incident Mach number $M_\infty = 5$, Prandtl number $Pr = 0.736$, $\gamma = 1.4$, and assigned values of $C_p, R, \sigma, \varepsilon$. The temperature of the injectant is taken as $T_k = 273.16°$ K. The outside gas temperature corresponds to air at 20,000 meters; this leads to incident stream data $T_\infty = 1299.96°$ K, $\varrho_\infty = 0.18047 \times 10^7$ bar, $\mu_\infty = 0.48366 \times 10^{-3}$ poise.

The similarity solution for flow with suction or injection near the stagnation point is also worked out (the writer is not aware that this had been done previously) and the constant A_0, related to the skin friction coefficient, is again compared with corresponding values determined by the method of integral relations. Results are very satisfactory.

These starting conditions are used to integrate the ordinary differential equations in the fourth approximation in the case of supersonic flow past a circular cylinder with detached shock wave. The outside velocity distribution is based on Belotserkovskii's calculations in this case (1962). The integration is carried out for three values of m_0, $m_0 = -0.1$ (suction), $m_0 = 0$ (no suction or injection), $m_0 = 0.1$ (injection). The values of $\theta_0, \theta_1, \theta_2, \theta_3$ and corresponding enthalpy coefficients are tabulated over a complete range of x. From these results, graphs of velocity and temperature profiles, and skin friction variation with x are derived.

5.9 Extension to Separated Flows

The above formulation of the method of integral relations for laminar boundary layer flows needs to be modified in regions of positive pressure gradient, for example, near a separation point or in wake flows. Two principal changes need to be made. Firstly, the representation of the velocity gradient $\partial u/\partial \eta$ as a function of u must be extended to permit flow reversal near the wall of a retarded boundary layer or near the axis of a wake. Secondly, in contrast to accelerated flow regions, the pressure outside the boundary layer is no longer a given function determined from a prior inviscid calculation, but depends on the boundary layer growth and is found by considering the interaction between the outer inviscid flow and the boundary layer flow.

In subsonic flow this interaction process is difficult to investigate because a change in the effective outer boundary of the viscous flow affects the whole inviscid flow and a complete iteration between the viscous and inviscid regions must be carried out. The general procedure in this case is described in CATHERALL and MANGLER (1966). A recent example of this type of calculation is to be found in FLETCHER and HOLT (1975), in treatment of conical boundary layers (on the windward side of a yawed cone, the inviscid cross flow is always subsonic). In supersonic flow, on the other hand, the influence of the effective outer edge of a boundary layer or wake on the inviscid flow is purely local; in fact, when the slope of the outer boundary is small the pressure can be found simply from the Ackeret rule of proportionality to local slope.

The application of the method of integral relations to the problem of laminar separation in supersonic flow over a flat plate with a compression corner is considered in a series of papers by HOLT (1966, 1967, 1968) and NIELSEN et al. (1969). Most of these are listed and briefly reviewed in NIELSEN et al. (1969). The approaches of HOLT and NIELSEN agree substantially upstream of separation but differ significantly in reversed flow regions. NIELSEN treats the latter in two parts; in the outer (forward moving region) he uses a Dorodnitsyn type method of integral relations with $\partial u/\partial \eta$ represented as a function of u, in the inner (reversed velocity region) he uses quadratic and cubic profiles for the velocity as a function of η. In the Holt formulation $\partial u/\partial \eta$ is represented as a function of u across the whole viscous layer although the form is changed near the wall.

The first paper by HOLT (1966) investigates the effect of introducing a small pressure increment into a uniform supersonic flow past a flat plate under adiabatic conditions. It is shown that, immediately downstream of the pressure increase location the boundary layer thickens, causing a change in the displacement thickness and a change in slope of the effective boundary between the viscous and inviscid regions. As a consequence the pressure increases further, leading to continued boundary layer thickening a little further downstream. This interaction between pressure increase and boundary layer growth is very strong in supersonic flow and causes separation to occur within very few boundary layer thicknesses of the initial pressure jump position. The process is called free interaction. The second paper (HOLT, 1966) adds consideration of thermal effects to this analysis, especially wall cooling, and shows that the extent of separation is reduced by increased cooling. In a third paper (HOLT, 1967) the calculations are continued along the inclined face of the corner up to reattachment. The final paper in the series (HOLT and LU, 1975) deals with the entire problem of laminar separation in a supersonic corner and establishes a connection between location of initial pressure rise and regular reattachment, leading to a unique solution for given wedge angle and upstream flow conditions. NIELSEN et al. (1969) have also found this unique criterion for reattachment using their formulation of the reversed flow analysis.

Mathematical Details: The equations of motion governing compressible laminar boundary layer flow in two dimensions were discussed in Sect. 5.7, following Pavlovskii's analysis. The same transformations (STEWARTSON and DORODNITSYN) are applied in the separated flow analysis but a more convenient notation is adopted. The dimensionless enthalpy function is now denoted by

$$s = 1 - \frac{C_p T + \dfrac{\bar{u}^2}{2}}{C_p T_\infty + \dfrac{\bar{u}_\infty^2}{2}} \tag{5.9.1}$$

and the actual transverse velocity gradient

$$z = \partial u / \partial \eta \tag{5.9.2}$$

is used in place of its reciprocal θ. When wall suction (represented by a prescribed function $w_0(\xi)$) is included in the analysis, the basic integral relations, derived from suitable combinations of the continuity, momentum and energy equations are

$$\frac{d}{d\xi} \int_0^1 \frac{u f(u)}{z} du = f(0) w_0(\xi) + \frac{\dot{U}_1}{U_1} \int_0^1 \frac{(1-u)^2 f'(u)}{z} du - \frac{\dot{U}_1}{U_1} \int_0^1 \frac{s f'(u)}{z} du$$

$$- f'(0) z_0 - \int_0^1 f''(u) z \, du \tag{5.9.3}$$

$$\frac{d}{d\xi} \int_0^1 \frac{s u f(u)}{z} du = s_w f(0) w_0(\xi) + \frac{\dot{U}_1}{U_1} \int_0^1 \frac{s(1-u^2) f'(u)}{z} du$$

$$-\frac{\dot{U}_1}{U_1} \int_0^1 \frac{s^2 f'(u)}{z} du - s_w f'(0) z_0^2 - \left(\frac{1}{\sigma}+1\right) \int_0^1 \frac{\partial s}{\partial \eta} f'(u) du$$

$$-\int_0^1 s f''(u) z \, du - \frac{1}{\sigma} f(0) \left(\frac{\partial s}{\partial \eta}\right)_0$$

$$-2\left(\frac{1}{\sigma}-1\right)\left(\frac{m_1}{1+m_1}\right) \int_0^1 u f'(u) z \, du$$

(5.9.4)

where $f(u) = (1-u), \ldots, (1-u)^n$ as in 5.6.

The representations of z and $1/z$ used in attached accelerated flow (see (5.6.8) and (5.6.9)) must be modified in regions of retarded and separated flows. Figure 5.14 shows the evolution of the boundary layer profiles (η, u), $\left(\frac{\partial u}{\partial \eta}, u\right)$ and the thermal profiles (s, η), (s, u) in regions of positive pressure gradient moving downstream from attached flow to reversed, separated flow. Throughout the region the velocity profile has a point of inflexion, while downstream of separation the velocity has a minimum at the point of maximum reversed flow velocity, denoted by $u = -\alpha$. To take account of this behavior we add a factor $(u+\alpha)^{1/2}$ to the representation z in terms of u, throughout the positive pressure gradient zone. In addition, downstream of separation we note that z is a double valued function of u when $u < 0$, and therefore we need different (z, u) representations in the two regions $z < 0$ and $z > 0$, matching these at $z = -\alpha$ to satisfy continuity of z and $\partial z/\partial u$ at $z = 0$.

As shown by COHEN and RESHOTKO (1956) in the case of their similarity profiles, s is a monotonically decreasing function of η in both accelerated and retarded flow regions. However, s is a double valued function of u (since η is double valued in u) when $u < 0$, so again two representations (s, u) are needed in reversed flow regions.

The general representations of (z, u) and (s, u) are taken as follows

$$z = (u+\alpha)^{1/2}(1-u)(b_0 + b_1 u + \cdots + b_{n-1} u^{n-1})$$
$$s = [(1+\alpha)^{1/2} - (u+\alpha)^{1/2}][d_0 + d_1 u + \cdots + d_{n-1} u^{n-1}]$$

(5.9.5)

in the n-th approximation, with modifications in reversed flow regions when $z < 0$.

The presence of the additional factor $(u+\alpha)^{1/2}$ in retarded flow regions immediately adds one unknown parameter to the number of parameters in the corresponding representation in accelerated flow and, more seriously, makes the algebraic expressions for the factors of the unknown coefficients $b_0, \ldots, b_{n-1}$, etc. quite complicated. Furthermore, in such regions, the pressure at the edge of the boundary layer is no longer prescribed but is determined from a free interaction condition (represented by an ordinary differential equation). In other words, the pressure is an additional unknown.

The analysis of separated flow has therefore been confined to calculations using the second order approximation (two parameter representations of (z, u), and (s, u) with α common to both). This leads to a system of four first order ordinary differential equations for the three unknown profile parameters and the outside pressure (or Mach number). This approximation is sufficiently

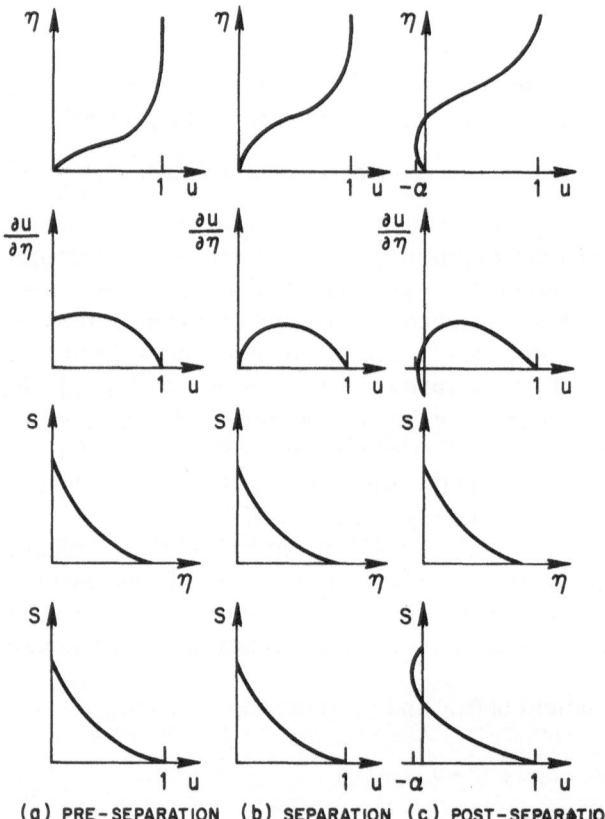

(a) PRE-SEPARATION   (b) SEPARATION   (c) POST-SEPARATION

Fig. 5.14   Velocity, velocity gradient and enthalpy profiles near a separation or reattachment point

accurate for most applications, based on a comparison of calculations with observations, and any improvement should be in the treatment of the viscous-inviscid interaction condition rather than in the use of higher order integral relations.

In the second approximation the representations of z, $1/z$ and s needed to evaluate the integrals in (5.9.3) and (5.9.4) are as follows:

Attached region

$$z = \frac{1}{c_0} \sqrt{(u+\alpha)(1-u)}$$

$$\frac{1}{z} = \frac{c_0}{\sqrt{(u+\alpha)(1-u)}}$$

(5.9.6)

$$s = (\sqrt{[1+\alpha]} - \sqrt{[u+\alpha]})\left[\frac{s_w}{(\sqrt{[1+\alpha]} - \sqrt{\alpha})} + \frac{d_1}{c_0}u\right]$$

Separated region

$z < 0$ $z > 0$

$$z = -\frac{(1+\alpha)\sqrt{(u+\alpha)}}{c_0} \qquad z = \frac{1}{c_0}\sqrt{(u+\alpha)(1-u)}$$

$$\frac{1}{z} = -\frac{c_0}{(1+\alpha)\sqrt{(u+\alpha)}} \qquad \frac{1}{z} = \frac{c_0}{\sqrt{(u+\alpha)(1-u)}}$$

(5.9.7)

$$s = s_w \left(\frac{\sqrt{(1+\alpha)} + \sqrt{(u+\alpha)}}{\sqrt{(1+\alpha)} + \sqrt{\alpha}}\right)(z<0)$$

$$s = (\sqrt{[1+\alpha]} - \sqrt{[u+\alpha]})\left[\frac{s_w}{(\sqrt{[1+\alpha]} + \sqrt{\alpha})} + \frac{d_1}{c_0}(u+\alpha)\right](z>0)$$

These are substituted in (5.9.3) with $f(u)=(1-u),(1-u)^2$, and then in (5.9.4) with $f(u)=(1-u)$. This yields three ordinary differential equations in the unknowns c_0, d_0, α and m_1. This is related to the unknown Mach number at the edge of the boundary layer and is connected with boundary layer geometry through an interaction condition.

The Interaction Equation: The matching of flow conditions inside the boundary layer with the inviscid flow outside is the subject of several studies, some of them very sophisticated (see BROWN and STEWARTSON, 1969; STEWARTSON and WILLIAMS, 1969). The matching condition adopted in the method of integral relations calculations is the comparatively simple one derived from CROCCO and LEES and CHAPMAN, KUEHN and LARSON (1958). This states that the pressure at the outer edge of the boundary layer is that generated by inviscid supersonic flow past the solid wall enlarged by an amount equal to the local displacement thickness. This leads to the equation

$$\frac{d\delta^*}{dx} + \frac{dH}{dx} = \tan\theta$$

(5.9.8)

in the physical plane where δ^* is the displacement thickness, H is the wall height, and θ the flow inclination at the displacement thickness edge. The angle θ is connected with local Mach number or m_1 by the Prandtl-Meyer relation, while δ^* is a known function of m_1, α, b_0, d_0.

With this interaction equation the analysis of separated flow reduces to the solution of the following system of four ordinary differential equations,

$$
\begin{bmatrix}
A_{11} & A_{12} & A_{13} & A_{14} \\
A_{21} & A_{22} & A_{23} & A_{24} \\
A_{31} & A_{32} & A_{33} & A_{34} \\
A_{41} & A_{42} & A_{43} & A_{44}
\end{bmatrix}
\begin{bmatrix}
\dot{c}_0 \\
\dot{\alpha} \\
\dot{U}_1 \\
d_1
\end{bmatrix}
=
\begin{bmatrix}
b_1 \\
b_2 \\
b_3 \\
b_4
\end{bmatrix}
\tag{5.9.9}
$$

where A_{ij} and b_i are algebraic expressions depending on $c_0, \alpha, m_1, d_1, s_w, R_0$, and g_n, with

$$
g_n(\alpha) = \frac{2}{2(n+1)}\left[(1+\alpha)^{1/2} - n g_{n-1}(\alpha)\right]
$$

$$
g_0(\alpha) = 2\left[(1+\alpha)^{1/2} - \alpha^{1/2}\right].
$$

R_0 is the Reynolds number based on free stream conditions and distance from the plate leading edge. The expressions for A_{ij}, b_i in the pre-separation region are given in HOLT and LU (1975). Those in the post separation region are given in LU (1970).

Application: The analysis just described has been applied to calculate separated flow in a concave corner. Far upstream on the incident wall the flow can be assumed uniform with a laminar boundary layer of Blasius type. At a certain distance upstream of the corner the presence of the corner induces an infinitesimal increase in pressure outside the boundary layer. This distance is an unknown in the problem and the reciprocal of its ratio to the distance of the corner from the leading edge of the plate is denoted by ω. For given free stream conditions, corner angle, and assumed ω (5.9.9) are integrated from the beginning of interaction, across the corner up to reattachment. Downstream of reattachment, the outside pressure distribution should approach a constant value, but this will only occur if a unique correct value is selected for ω, corresponding to the simultaneous fulfillment of the conditions $\dot{\alpha}=0$, $\dot{m}_1=0$, downstream of reattachment. For a general (incorrect) value of ω the pressure distribution downstream of reattachment is one of two types, illustrated in Fig. 5.15. The correct value of ω is determined by iterating until the difference between ω (type 1) and ω (type 2) is less than some assigned small quantity.

Figure 5.16 shows a comparison between calculated and measured pressure distributions for supersonic flow over an $11°$ corner with incident Mach number 2.64 and Reynolds number 1.4×10^5. These conditions correspond to

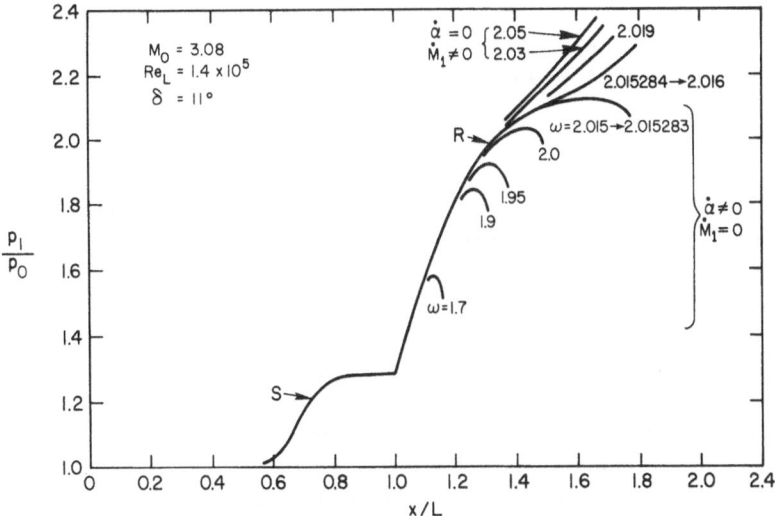

Fig. 5.15   Criterion for locating initial point of interaction region

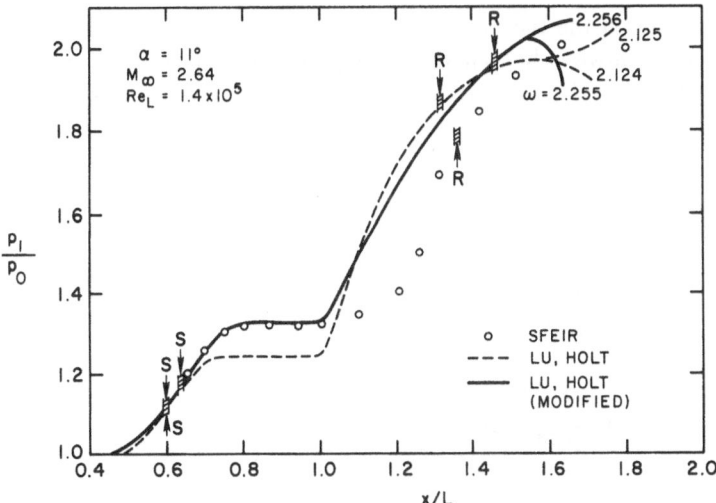

Fig. 5.16   Pressure distribution. Comparison between theory and experiment

those of experiments performed in the $15.2\,cm \times 15.2\,cm$ supersonic wind tunnel at the University of California, Berkeley. The open circles correspond to observed values of pressure over the corner. The dashed curve shows the calculated pressure distribution using the simple interaction condition represented by (5.9.8). The full curve shows the calculated pressure when this condition is modified in the reversed flow region. The modification consists in taking the effective wall as the dividing streamline in the recirculating zone and shows considerable improvement in the first part of the separated region. Further calculations on this problem include the effect of wall cooling and wall suction on the separated flow characteristics and are described in HOLT and LU (1975).

5.10 Application to Supersonic Wakes and Base Flows

HOLT and MENG (1970, 1973) adapted the formulation of the Method of Integral Relations for separated laminar boundary layers to the study of supersonic base flow and supersonic laminar wakes.

The analysis is confined to adiabatic conditions so that only the continuity and momentum equations need to be considered. The wake equations are identical to the laminar boundary layer equations in the adiabatic case. The boundary conditions at the outer edge of the wake are the same as for a boundary layer but the wall condition is replaced by the condition

$$\eta = 0, \qquad u = -\alpha, \qquad z = 0 \quad \text{(symmetrical wake)}.$$

A sketch of the flow in the base and near wake region is shown in Fig. 5.17. The velocity profile always has an inflexion point but α changes sign from positive to negative as we move downstream from the recirculating zone near the base to the wake region.

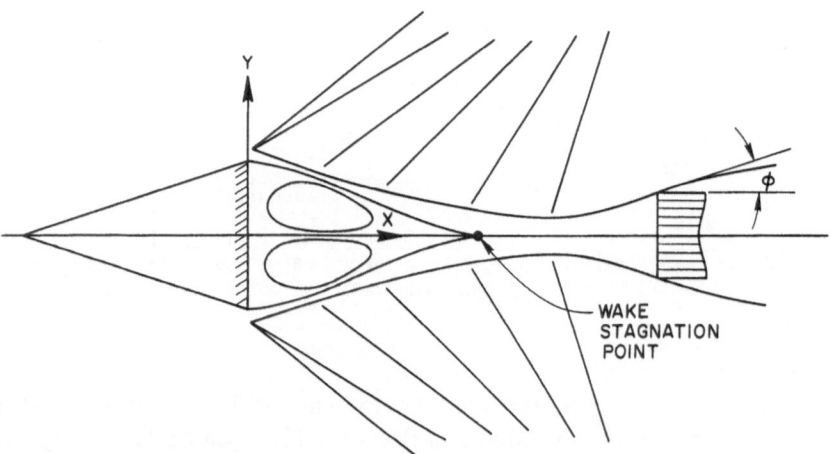

Fig. 5.17   Physical diagram of the near wake region

The velocity gradient z is represented in the same form as for separated flow, given by (5.9.5). Since $z \geq 0$, always, a single z, u representation can be used in both the recirculating zone and wake region. The basic integral relation in this case is

$$
\frac{d}{d\xi} \int_{-\alpha}^{1} \frac{uf(u)\,du}{z} + [wf(u)]_{-\alpha}^{1} = \frac{\dot{U}_1}{U_1} \int_{-\alpha}^{1} \frac{(1-u^2)f'(u)\,du}{z}
$$

$$
+ [zf'(u)]_{-\alpha}^{1} - \int_{-\alpha}^{1} zf''(u)\,du
$$

(5.10.1)

Use of weighting functions $f(u)=(1-u), (1-u)^2,\ldots$, guarantees that the term $[wf(u)]_{-\alpha}^{1}$ always vanishes.

The Method of Integral Relations is again applied in the second approximation, so that the $z(u)$ function contains two unknown parameters c_0, α. The interaction condition (5.9.8) is used, reducing here to

$$
\frac{d\delta^*}{dx} = \tan\theta
$$

(5.10.2)

We arrive at the following three ordinary differential equations for α, m, and c_0.

$$
\frac{d\alpha}{d\xi} = \frac{D_1}{D}
$$

(5.10.3)

$$
C_2 \frac{dm_1}{d\xi} + A_2 \frac{d\alpha}{d\xi} = D_2
$$

(5.10.4)

$$
C_3 \frac{dm_1}{d\xi} + B_3 \frac{dc_0}{d\xi} + A_3 \frac{d\alpha}{d\xi} = D_3
$$

(5.10.5)

where

$$
C_2 = -\frac{(1+\alpha)(4\alpha^2-7\alpha+4)}{m_1}
$$

$$
A_2 = 1+8\alpha-8\alpha^2
$$

$$
D_2 = \frac{2}{c_0^2}(1-2\alpha)(1+\alpha)^2
$$

$$
C_3 = -\frac{2c_0(1+\alpha)(\alpha-2)}{m_1}
$$

$$
B_3 = 2(1+\alpha)(1-2\alpha)
$$

$$A_3 = -3c_0(1+2\alpha)$$

$$D_3 = 0$$

$$D_1 = 2(1+\alpha)^2 \left\{ [3+2\bar{m}_1(2-\alpha)] \left[5-4\alpha - \frac{\lambda\theta\bar{m}_1}{1+\bar{m}_1}(1-2\alpha) \right] \right.$$

$$\left. -4(1-2\alpha)(2-\alpha)\theta\bar{m}_1 \right\} - 3c_0\sqrt{1+\alpha}\, R_0^{1/2} \frac{1+\bar{m}_1}{1+\bar{m}_0} \tan\phi(4\alpha^2 - 7\alpha + 4)$$

$$D = c_0^2 \left[A(1) + A(2)\alpha + A(3)\alpha^2 + A(4)\alpha^3 \right]$$

$$A(1) = -33 - 28\bar{m}_1 - 8\theta\bar{m}_1 - (3+4\bar{m}_1)\frac{\lambda\theta\bar{m}_1}{1+\bar{m}_1}$$

$$A(2) = 78 + 144\bar{m}_1 - 60\theta\bar{m}_1 - (24+30\bar{m}_1)\frac{\lambda\theta\bar{m}_1}{1+\bar{m}_1}$$

$$A(3) = -24 - 96\bar{m}_1 + 96\theta\bar{m}_1 + 24(1+2\bar{m}_1)\frac{\lambda\theta\bar{m}_1}{1+\bar{m}_1}$$

$$A(4) = 32\bar{m}_1 - 32\theta\bar{m}_1 - 16\frac{\lambda\theta\bar{m}_1^2}{1+\bar{m}_1}$$

θ is related to ϕ by the equation

$$\theta = \frac{m_1 + \cos^2\phi}{m_1 + \cos^2\phi + \sin\phi\sqrt{5m_1 - \cos^2\phi}}.$$

Solutions of (5.10.3), (5.10.4), (5.10.5) are sought which will generate regular flow at a critical point downstream of wake reattachment and connect with constant pressure conditions near the base.

With typical values, corresponding to conditions in the wakes of hypersonic wedges and cones, assigned to the free stream Mach number M_0 and Reynolds number R_0, the singularities of (5.10.3), (5.10.4) and (5.10.5) are analyzed. The positions of the singular points are determined from the roots of the cubic equation $D(\alpha)=0$, for chosen values of the local Mach number M_1. The singular point of greatest physical significance is located a short distance downstream of the wake reattachment point ($\alpha=0$). This singularity is only of the required saddle point type if M_1 exceeds M_0 and lies within a narrow range of values. Of the other two values of α at the remaining singular points, one is physically unacceptable (outside the range $|\alpha|\leq 1$), while the other is a positive value close to unity and is unattainable in practice.

With the values of the free stream parameters taken as

$$M_0 = 6, \qquad R_0 = 10^6$$

the variation of D with α for various M_1 is shown in Fig. 5.18.

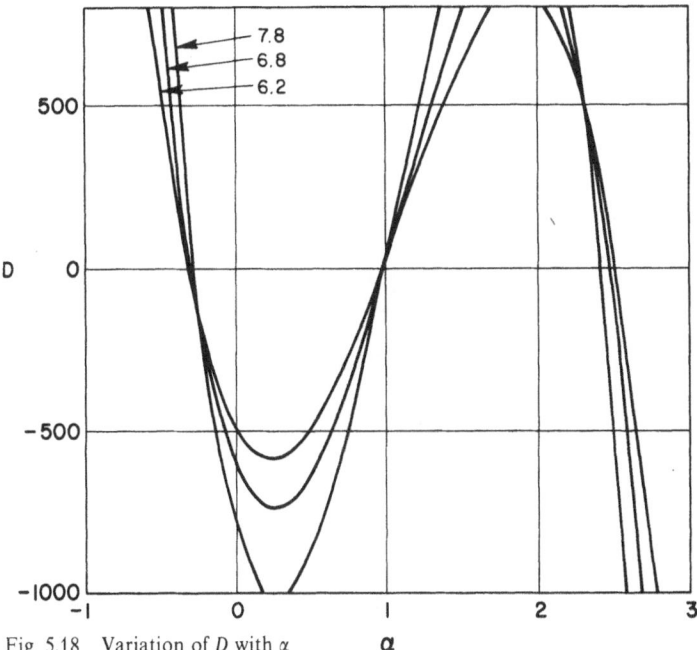

Fig. 5.18   Variation of D with α α

It is found that a saddle point (as required) downstream of reattachment only exists (for the above M_0, R_0) if M_1 lies in the range

$$6.0 \le M_1 \le 8.2 \,.$$

For given M_1 in this range we integrate the (α, ζ) equation (together with the (M_1, ζ), (c_0, ζ) equations) upstream from the saddle point towards the base. Several such curves are shown in Fig. 5.19. It is observed that if M_1 is too large (~ 6.8) the integral curve does not reach the wake reattachment point ($\alpha = 0$). Of the curves which can be continued through this point there is one which leads to uniform conditions in the base, giving the solutions of the constant pressure base mixing problem for the given values of M_0 and R_0. This curve corresponds to $M_1 = 6.01$ at the saddle point and is shown in Fig. 5.20.

HOLT and MENG (1973) use this basic approach to generate details of base and near wake solutions for a supersonic wedge of semi-angle 10°, Mach number 6 and two Reynolds numbers (based on base height) of 5.5×10^4 and 4.1×10^4.

5.11 Application to Three-Dimensional Laminar Boundary Layers

The theory of laminar boundary layers changes significantly when extended from two to three dimensions. Since the boundary layer equations of motion must be referred to coordinates tied to the surface generating viscous effects,

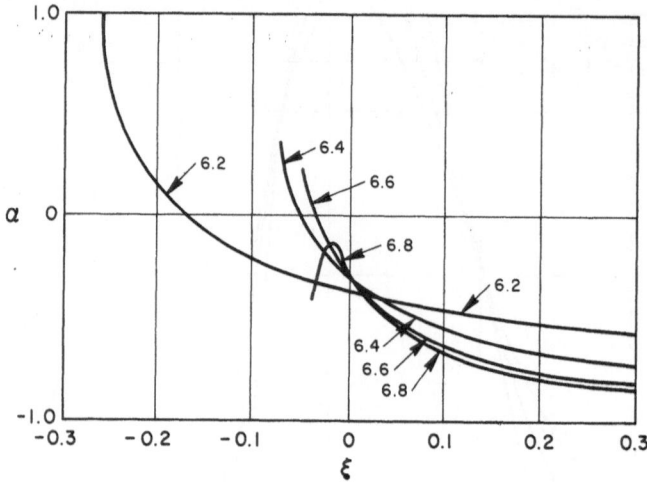

Fig. 5.19   Integral curves for α vs. ξ

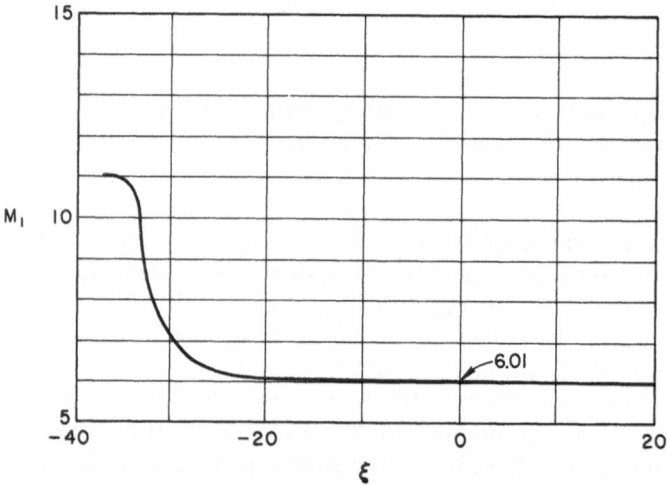

Fig. 5.20   Integral curve connecting recompression region with base region

they become much more complicated in the three dimensional case. Furthermore, the additional degree of freedom permits quite different and frequently
less severe physical behavior in the boundary layer. This is especially true in
regions where the pressure increases in the streamwise direction. In two dimensions a positive pressure gradient causes complete flow reversal near the
wall and a change in sign in the skin friction coefficient. On the other hand,
in three dimensions, fluid particles encountering unfavorable pressure gradients
can move in the transverse as well as in the streamwise direction so that, at
separation, only one component of fluid velocity needs to be reversed rather

than the whole velocity vector. As a result, three dimensional boundary layer separation is frequently a combination of retarded flow in one direction and accelerated flow in a transverse direction (an experimental investigation of this behavior is given in BACHALO (1975) and BACHALO and HOLT (1975)).

Several finite difference methods have been developed for laminar flow in three dimensions. Some are designed for the full Navier Stokes equations and usually are limited to a moderate Reynolds number range. Other methods apply to the boundary layer equations and are valid at high values of Reynolds number. These have been developed by DER and RAETZ (1962), DWYER (1968, 1974), KRAUSE (1969, 1973), WANG (1969, 1970), among others.

All the finite difference methods demand a great deal of computer time and it is therefore worth while to investigate the use of the Method of Integral Relations in this connection with the object of eliminating finite difference calculations normal to the solid surface.

BASHKIN (1968) investigated the application of MIR to three dimensional laminar boundary layer flow over conical bodies and later (BASHKIN, 1970) calculated the flow near a line of symmetry.

A more complete generalization of the Method of Integral Relations to laminar boundary layers in three dimensions was carried out by HOLT and MODARRESS (1977) and MODARRESS and HOLT (1976). This formulation was applied to several model problems previously treated by finite difference methods, with satisfactory results. It was then applied to calculate the supersonic laminar flow over a swept back wedge, investigated experimentally by BACHALO (1975).

Derivation of the Basic Integral Relations: The laminar boundary layer equations for compressible flow of a perfect gas in three dimensions (referred to Cartesian coordinates) are

$$\frac{\partial(\varrho u)}{\partial x} + \frac{\partial(\varrho v)}{\partial y} + \frac{\partial(\varrho w)}{\partial z} = 0, \tag{5.11.1}$$

$$u\frac{\partial u}{\partial x} + v\frac{\partial u}{\partial y} + w\frac{\partial u}{\partial z} = -\frac{1}{\varrho}\frac{\partial p}{\partial x} + v\frac{\partial^2 u}{\partial z^2}, \tag{5.11.2}$$

$$u\frac{\partial v}{\partial x} + v\frac{\partial v}{\partial y} + w\frac{\partial v}{\partial z} = -\frac{1}{\varrho}\frac{\partial p}{\partial y} + v\frac{\partial^2 v}{\partial z^2}, \tag{5.11.3}$$

$$\frac{\partial p}{\partial z} \approx 0, \tag{5.11.4}$$

where z is the coordinate normal to the wall. The wall is assumed to be thermally insulated so that the energy equation reduces to an algebraic relation. x, y are body coordinates. The velocity components are (u, v, w) in the (x, y, z) directions, respectively. The pressure $p(x, y)$ is independent of z, ϱ is the density, v is the kinematic viscosity and is assumed to be proportional to the square of the temperature.

At the wall $(u, v, w) = 0$ while at the outer edge of the boundary layer $u \to u_e$, $v \to v_e$, where (u_e, v_e) are the velocity components of the outside inviscid flow.

To reduce (5.11.1)–(5.11.4) to quasi incompressible form we apply the Howarth-Dorodnitsyn transformation in the form adopted by MOORE (1951). This is defined by

$$\frac{\partial \psi}{\partial z} = \frac{\varrho}{\varrho_\infty} u, \qquad \frac{\partial \phi}{\partial z} = \frac{\varrho}{\varrho_\infty} v$$

$$\frac{\partial \psi}{\partial x} + \frac{\partial \phi}{\partial y} = -\frac{\varrho}{\varrho_\infty} w \tag{5.11.5}$$

$$\hat{u} = \frac{\partial \psi}{\partial z}, \qquad \hat{v} = \frac{\partial \phi}{\partial z}, \qquad \hat{w} = -\left(\frac{\partial \psi}{\partial x} + \frac{\partial \phi}{\partial y}\right) \tag{5.11.6}$$

$$\bar{v} = C v_\infty \frac{p(x, y)}{p_\infty}$$

$$\bar{z} = \int_0^z \frac{\varrho}{\varrho_\infty} dz' \tag{5.11.7}$$

$$\bar{w} = \frac{\varrho}{\varrho_\infty} w + u \frac{\partial \bar{z}}{\partial x} + v \frac{\partial \bar{z}}{\partial y}$$

Eqs. (5.11.5), (5.11.6) are from MOORE (1951), while (5.11.7) is based on HOWARTH (1948). In the $\bar{v}$, v_∞ relation C is the Chapman-Rubesin constant.

The effect of these transformations is to reduce (5.11.1)–(5.11.4) to incompressible equations for u, v, $\bar{w}$ in x, y, $\bar{z}$. We now replace these variables by the following dimensionless variables

$$U = u/u_e, \qquad V = v/v_e, \qquad W = \sqrt{\frac{u_\infty L}{v_\infty}} \frac{\bar{w}}{u_e}$$

$$\xi = x/L, \qquad \eta = y/L, \qquad \zeta = \sqrt{\frac{v_\infty L}{v_\infty}} \frac{\bar{z}}{L}.$$

The equations of motion then reduce to

$$\frac{\partial U}{\partial \xi} + \frac{\partial V}{\partial \eta} + \frac{\partial W}{\partial \zeta} = -\frac{1}{U_e}\left(U \frac{\partial U_e}{\partial \xi} + V \frac{\partial U_e}{\partial \eta}\right), \tag{5.11.8}$$

$$U \frac{\partial U}{\partial \xi} + V \frac{\partial U}{\partial \eta} + W \frac{\partial U}{\partial \zeta} = \frac{1}{U_e} \frac{\partial U_e}{\partial \eta} (R - U^2)$$

$$+ \frac{1}{U_e} \frac{\partial U_e}{\partial \eta} (R V_e - U V) + \bar{b} \frac{\partial^2 U}{\partial \zeta^2},$$

(5.11.9)

$$U \frac{\partial V}{\partial \xi} + V \frac{\partial V}{\partial \eta} + W \frac{\partial V}{\partial \zeta} = \frac{1}{U_e} \frac{\partial U_e}{\partial \xi} (R V_e - U V) + \frac{1}{U_e} \frac{\partial U_e}{\partial \eta} (R V_e^2 - V^2)$$

$$+ R \left(\frac{\partial V_e}{\partial \xi} + V_e \frac{\partial V_e}{\partial \eta} \right) + \bar{b} \frac{\partial^2 V}{\partial \zeta^2},$$

(5.11.10)

where

$$R = \frac{\varrho}{\varrho_e} = 1 + H_e \left[(1 - U^2) + (V_e^2 - V^2) \right]$$

$$\bar{b} = C \frac{p_e/p_\infty}{U_e}$$

(5.11.11)

$$H_e = U_e^2 \left(\frac{U_\infty^2}{2 c_p T_e} \right)$$

The boundary conditions are

$$U = V = W = 0 \quad \text{at} \quad \zeta = 0.$$

$$U = 1, \quad V = V_e$$

(5.11.12)

$$\frac{\partial U}{\partial \zeta} = \frac{\partial V}{\partial \zeta} = \frac{\partial^2 U}{\partial \zeta^2} = \frac{\partial^2 V}{\partial \zeta^2} = \cdots = 0 \quad \text{as} \quad \zeta \to \infty$$

Eqs. (5.11.8)–(5.11.10) are of hyperbolic type and are solved with Cauchy data given on two oblique planes normal to the surface.

To apply the Method of Integral Relations we introduce complete sets of weighting functions $f(U)$, $g(U)$ satisfying the conditions

$$\underset{\zeta \to \infty}{\text{Lt}} \ f(U) = 0, \qquad \underset{\zeta \to \infty}{\text{Lt}} \ g(U) = \begin{cases} 1 \ \text{(lowest order)}, \\ 0 \ \text{(all higher orders)}. \end{cases}$$

We derive one basic integral relation by adding (5.11.8) factored by $f(U)$ to (5.11.9) factored by $f'(U)$, integrating the result with respect to ζ across the boundary layer (0 to ∞) and then replacing ζ by U as the variable of integration. The second basic integral relation is obtained by changing the independent variables in (5.11.10) from ξ, η, ζ to ξ, η, U, factoring by a weighing function $g(U)$ and integrating the result from 0 to 1 with respect to U.

The resulting equations are

$$\frac{\partial}{\partial \xi} \int_0^1 U f \theta \, dU + \frac{\partial}{\partial \eta} \int_0^1 V f \theta \, dU = \frac{1}{U_e} \frac{\partial U_e}{\partial \xi} \int_0^1 [(R - U^2) f' - U f] \theta \, dU$$

$$(5.11.13)$$

$$+ \frac{1}{U_e} \frac{\partial U_e}{\partial \eta} \int_0^1 [(V_e R - U V) f' - V f] \theta \, dU - \bar{b} \frac{f'(0)}{\theta_0} - \bar{b} \int_0^1 \frac{f''}{\theta} \, dU,$$

$$\int_0^1 U \frac{\partial V}{\partial \xi} g(U) \, dU + \int_0^1 V \frac{\partial V}{\partial \eta} g(U) \, dU = \left(\frac{\partial V_e}{\partial \xi} + V_e \frac{\partial V_e}{\partial \eta} \right) \int_0^1 g(U) R \, dU$$

$$+ \frac{1}{U_e} \frac{\partial U_e}{\partial \xi} \int_0^1 \left[R V_e - U V - \frac{\partial V}{\partial U} (R - U^2) \right] g(U) \, dU \qquad (5.11.14)$$

$$+ \frac{1}{U_e} \frac{\partial U_e}{\partial \eta} \int_0^1 \left[R V_e^2 - V^2 - \frac{\partial V}{\partial U} (R V_e - U V) \right] g(U) \, dU + \bar{b} \int_0^1 \frac{\partial^2 V}{\partial U^2} \frac{g(U)}{\theta^2} \, dU.$$

In the applications made so far we take

$$f(U) = (1 - U), (1 - U)^2 \dots (1 - U)^n$$
$$g(U) = 1, (1 - U), (1 - U)^2 \dots (1 - U)^{n-1}.$$

For attached flow we represent the unknowns in the integrands in (5.11.13) and (5.11.14) as follows

$$\theta = \frac{1}{1 - U} \sum_{n=1}^N a_{n-1} U^{n-1}$$

$$(5.11.15)$$

$$\frac{1}{\theta} = (1 - U) \sum_{n=1}^N \bar{a}_{n-1} U^{n-1}$$

$$V = U \left[V_e + \sum_1^M b_n (1 - U)^n \right] \qquad (5.11.16)$$

In regions of rising pressure where separation may occur the factor $(1 - U)$ in (5.11.15) is preceded by a factor $(U + \alpha)^{1/2}$, permitting an inflexion point in the U, z profile. Also the first term in (5.11.16) is replaced by

$$V_e \{(U + \alpha)^{1/2} - \alpha^{1/2}\} / \{(1 + \alpha)^{1/2} - \alpha^{1/2}\}.$$

In attached, incompressible flow, the lowest order approximation requires the solution of the following partial differential equations in θ_0 and b_1 [here $\theta = \theta_0/(1-U)$, $V = U[V_e + b_1(1-U)]$] with two independent variables ξ, η

$$
3\frac{\partial\theta_0}{\partial\xi} + (3+b_1)\frac{V_e}{U_e}\frac{\partial\theta_0}{\partial\eta} + \theta_0\frac{V_e}{U_e}\frac{\partial b_1}{\partial\eta}
$$
$$
= -\theta_0\left[\frac{3}{U_e}\frac{\partial U_e}{\partial\xi} + (b_1+3)\frac{1}{U_e}\frac{\partial V_e}{\partial\eta} + (9-2b_1)\frac{V_e}{U_e^2}\frac{\partial U_e}{\partial\eta}\right] + \frac{6\bar{b}}{\theta_0}
$$

(5.11.17)

$$
4\frac{\partial b_1}{\partial\xi} + (b_1+4)\frac{V_e}{U_e}\frac{\partial b_1}{\partial\eta}
$$
$$
= \frac{1}{U_e}\left\{-2(b_1^2-3b_1+9)\frac{V_e}{U_e}\frac{\partial U_e}{\partial\eta} + (18-8b_1-b_1^2)\frac{\partial V_e}{\partial\eta} - (18-2b_1)\frac{\partial U_e}{\partial\xi}\right\}
$$
$$
+ 2(9-2b_1)\frac{1}{V_e}\frac{\partial V_e}{\partial\xi} - \frac{1}{2}\frac{b_1\bar{b}}{\theta_0^2}
$$

(5.11.18)

In the second order approximations four partial differential equations for the unknown coefficients θ_1, θ_0, b_1, b_2 exist. These are given in HOLT and MODARRESS (1977).

As in two dimensional flow, when regions of separation are approached the pressure distribution outside the boundary layer must be determined from an analysis of the interaction between the inner viscous and outer inviscid flow. The interaction analysis considerably modifies the pressure distribution corresponding to entirely inviscid flow. The analysis is naturally more complicated than in the two dimensional case and is not yet complete.

Several applications of the three dimensional Method of Integral Relations have been worked out and we shall discuss two of them. The first is to parabolic flow over a flat plate with outside velocity

$$
u_e = U_\infty, \qquad v_e = U_\infty(1-x)
$$

or, in dimensionless form

$$
U_e = 1, \qquad V_e = 1-\xi.
$$

This flow has an exact similarity solution, expressed in terms of Prandtl's variable $\bar{\zeta} = 1/2(U_e/\nu x)^{1/2} z$, of the form

$$
U = g(\bar{\zeta})
$$
$$
V = g(\bar{\zeta}) - \xi h(\bar{\zeta})
$$

where $g(\bar{\zeta})$, $h(\bar{\zeta})$ are known functions, tabulated in ROSENHEAD (1966). The functions are solutions of ordinary differential equations, calculated by SOWERBY (1955) and LOOS (1954).

When the method of integral relations is applied to the same problem in the second approximation it is reduced to the solution of the following four ordinary differential equations (similarity eliminates one independent variable and MIR eliminates a second)

$$\overset{\circ}{\theta}_0 = \frac{34}{\theta_0} - \frac{32}{\theta_1} \tag{5.11.19}$$

$$\overset{\circ}{\theta}_1 = \frac{20}{\theta_0} - \frac{16}{\theta_1} \tag{5.11.20}$$

$$-20\overset{\circ}{b}_1 + 5\overset{\circ}{b}_2 = 5(18 + 4b_1 - b_2)\frac{\overset{\circ}{V}_e}{V_e} + \frac{4(5b_1 - 7b_2)}{\theta_0^2}$$
$$+ \frac{32(5b_1 - b_2)}{\theta_1^2} - \frac{48b_2}{\theta_0\theta_1} \tag{5.11.21}$$

$$-5\overset{\circ}{b}_1 + 2\overset{\circ}{b}_2 = (5b_1 - 2b_2 + 40)\frac{\overset{\circ}{V}_e}{V_e} + \frac{2(8b_1 - 13b_2)}{\theta_0^2}$$
$$+ \frac{32(2b_1 - b_2)}{\theta_1^2} + \frac{16(b_1 - 2b_2)}{\theta_0\theta_1} \tag{5.11.22}$$

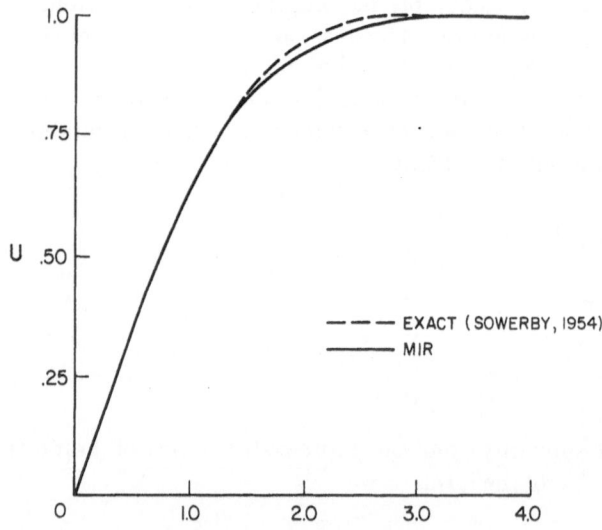

Fig. 5.21   Comparison between the exact (Sowerby) and the approximate (MIR 2nd approximation) solution for the streamwise velocity profile

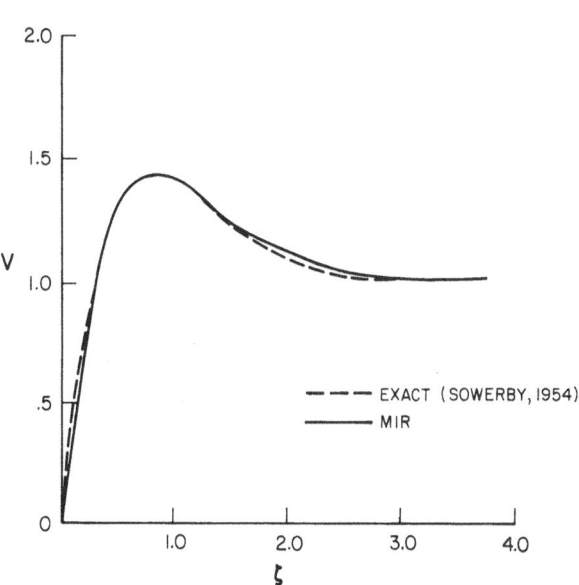

Fig. 5.22   Comparison between the exact and the approximate solution for crossflow velocity profile

In addition, since $\theta_i = \xi^{1/2}/\beta_i$, again from similarity, these equations reduce to a set of algebraic relations giving $\beta_0 = 0.3169$, $\beta_1 = 0.2874$. The MIR solution for the U and V profiles is compared with the corresponding exact solution in Figs. 5.21 and 5.22, showing very good agreement. Notice that the reversed cross flow is accurately represented.

The second application is to supersonic flow past a swept back wedge mounted on a flat plate (Fig. 5.23). This configuration was selected for investigation of three dimensional supersonic separation effects because it represents the simplest generalization of two dimensional flow in a plane concave corner (see HOLT and LU (1975) and SFEIR (1970)). The main advantage of this model from the theoretical point of view is that all surfaces are plane so that the Cartesian formulation of the MIR derivation can be used without any change. Experiments on the configuration are described in BACHALO (1975) and BACHALO and HOLT (1975).

The second approximation of MIR for separating flow is applied to this problem leading to three partial differential equations for α, c_0 (U profile) and d_1 (V profile). These equations are hyperbolic and are solved with data given on two planes. The first of these is normal to the incident flow far upstream (where two dimensional Blasius flow may be assumed). The second is the central plane of symmetry and the data over this plane are determined from a separate two dimensional calculation. The method of lines (described in Chapt. 6) is used to integrate the equations numerically.

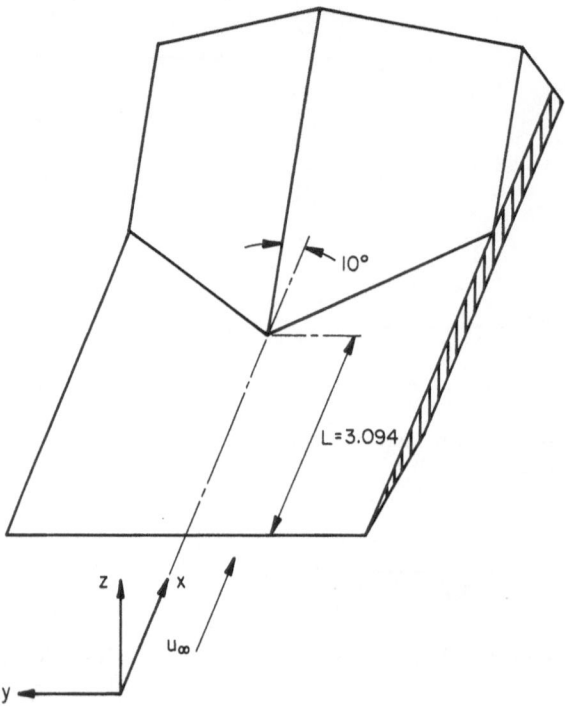

Fig. 5.23   Model configuration and coordinate system for the swept back wedge

In the absence of an interaction equation (still to be established) the experimental pressure distribution found in Bachalo's experiments was applied at the outer edge of the boundary layer. The data used are for a wedge angle of 10° and cross slope angle of 7.5°. Figure 5.24 shows the position of the separation line calculated with this pressure and a comparison with the position found experimentally using oil injection at the model surface. Figure 5.25 shows the U, ζ profile at $\eta = 0.242$, some distance from the wedge center plane. Figure 5.26 shows the V, ζ profile in the same plane for a series of values of ξ near the separation line. The U, ζ profiles are all of attached type and the departure from BLASIUS behavior develops slowly. The V, ζ profiles show cross flow reversal at all stations with this feature dominating at $\xi = 0.808$. The calculated results are in qualitative accord with the behavior observed in Bachalo's experiments.

Three other laminar boundary layer problems in three dimensions have been treated by this MIR formulation, principally with the object of testing MIR against other methods for known solutions. The first problem concerns flow over a yawed cylinder of infinite length, treated by SEARS (1948) and JONES (1947). The second problem deals with flow round a circular cylinder mounted normal to a flat plate, which was solved using an implicit finite difference scheme by DWYER (1968). Finally, MIR has been applied to calculate

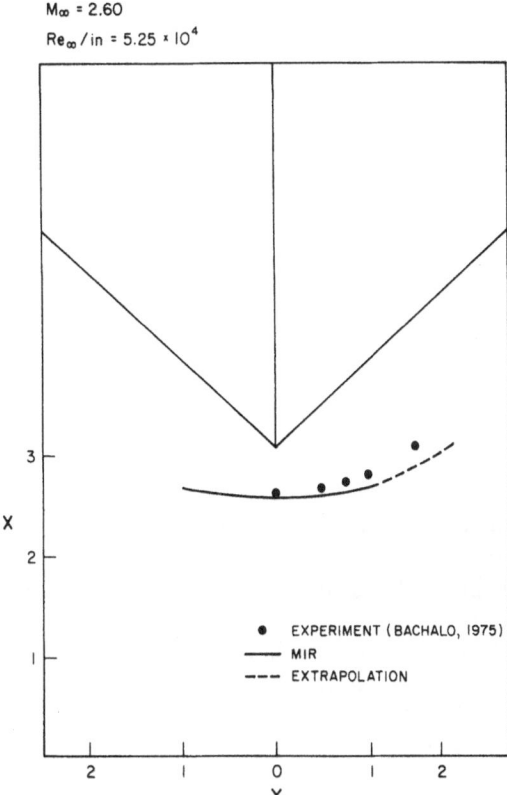

Fig. 5.24 $10°/7.5°$ wedge with separation lines

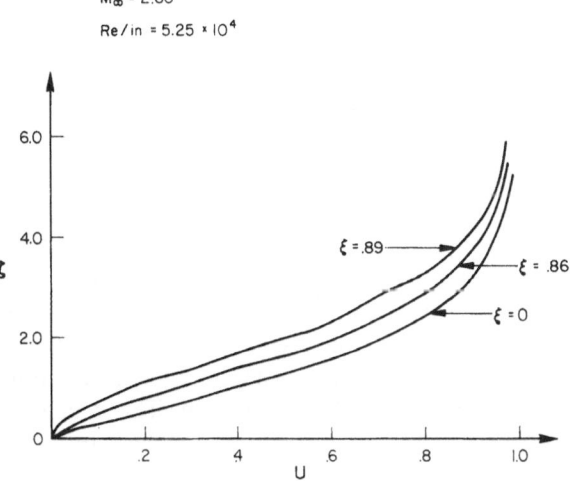

Fig. 5.25 U profile for $\eta = 0.242$

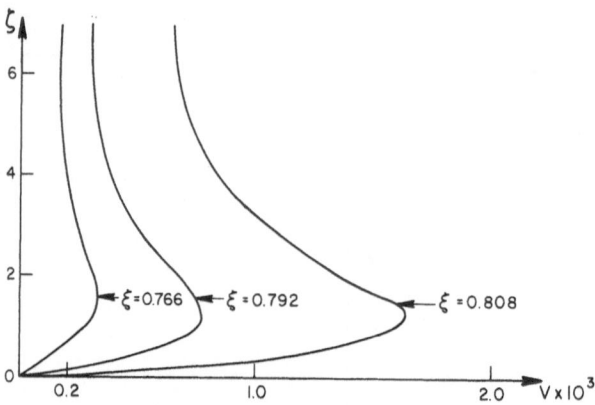

Fig. 5.26   Crossflow velocity variation at $\eta = 0.2424$

the growth of the laminar boundary layer on a spheroid at angle of attack in a uniform incompressible stream. This is a difficult problem treated by WANG (1969, 1970) by a finite difference method. To calculate the same flow field by MIR the formulation of the basic integral relations (see (5.11.13), (5.11.14)) was generalized for spheroidal coordinates.

Further details of the parabolic plate problem, the yawed cylinder problem the platecylinder problem, and the supersonic swept back wedge problem are given in HOLT and MODARRESS (1977). The spheroidal flow applications are described in MODARRESS and HOLT (1976).

5.12 A Modified Form of the Method of Integral Relations

As the previous section shows, the original formulation of the method of integral relations for laminar boundary layers becomes increasingly complicated as we proceed to higher orders of approximation. For attached flows analysis has not been carried out beyond the fourth approximation, while calculations concerning separated flows are restricted to the second approximation (albeit requiring the determination of four unknowns). One cause of the difficulty is the derivation of algebraic expressions arising in integrals in the basic integral relations. For attached flows the integrands concerned are rational functions (combinations of powers of $(1-u)$ with polynomials in u), while in separated flow square root factors are added to these rational functions. There is no systematic way of expressing these integrands in the general case. A more serious contribution to the complexity of the analysis for higher approximations comes from the nature of the first order differential equations for the n unknown coefficients in the n-th approximation. The coefficient matrix in this system of equations consists entirely of non-zero terms, in general. As n increases it becomes progressively more ill conditioned and more difficult to reduce to triangular form.

A simple change in the MIR formulation overcomes these difficulties. Instead of using a complete set of functions $f_i(u)$ as weighting functions we employ an orthonormal set $g_i(u)$ formed from appropriate linear combinations of $f_i(u)$. Furthermore, we now expand the unknowns arising in the integrands of the basic integral relations in series of $g_i(u)$ rather than in polynomials in u. As a result, the coefficient matrix in the resulting system of ordinary differential equations for the unknown coefficients reduces to diagonal form and problems of matrix inversion are eliminated.

We define the orthonormal weighting functions g_j by

$$g_j = \sum_{k=1}^{j} b_{kj} f_k \tag{5.12.1}$$

where f_k are the Dorodnitsyn weighting functions considered in Sect. 5.6 (e. g., $(1-u)^k$) and the coefficients b_{kj} are constants determined by the Gram-Schmidt orthonormalization process (ISAACSON and KELLER, 1966). We define the inner product of g_i, g_j with respect to a base weighting function w, namely

$$(g_i, g_j) = \int_a^b g_i(x) g_j(x) w(x) dx \tag{5.12.2}$$

Then the set of functions $g_i(s)$ is orthonormal if

$$(g_i, g_j) = 1, \quad i = j$$
$$= 0, \quad i \neq j.$$

We now introduce these functions into the solution of the problem of incompressible laminar boundary layer flow considered in Sect. 5.6. In the basic integral relation (5.6.6), the reciprocal of the velocity gradient $\theta = 1/(\partial u/\partial \eta)$ is now approximated by

$$\theta = \frac{1}{(1-u)} \left[b_0 + \sum_{j=1}^{N-1} b_j g_j \right] \tag{5.12.3}$$

where b_0 is retained to ensure the correct behavior of θ at the outer edge of the boundary layer. Replacing f by g (5.6.6) then yields

$$\frac{d}{d\xi} \int_0^1 \left[b_0 + \sum_{j=1}^{N-1} b_j g_j \right] g_k \frac{u}{1-u} du = C'(k), \quad k = 1, \dots, N \tag{5.12.4}$$

where $C'(k)$ is independent of ξ derivatives. Comparing (5.12.2) and (5.12.4) we see that the correct choice of the basic weighting function w in this case is

$$w(u) = u/(1-u).$$

This function is not the same in all problems; for example, in viscous conical flow (Sect. 5.13) we take $w(u) = 1/(1-u)$. With g_j determined by the Gram-Schmidt process (5.12.4) reduces to

$$\frac{db_0}{d\xi} \int_0^1 g_k w\, du + \frac{db_k}{d\xi} = C'(k), \quad k = 1, \ldots, N-1 \tag{5.12.5}$$

and, for $k = N$

$$\frac{db_0}{d\xi} \int_0^1 g_N w\, du = C'(N) \tag{5.12.6}$$

From (5.12.5) and (5.12.6)

$$\frac{db_k}{d\xi} = C'(k) - C'(N) \frac{\left[\int_0^1 g_k w\, du \right]}{\left[\int_0^1 g_N w\, du \right]} \tag{5.12.7}$$

From (5.6.6) we find that

$$C'(k) = \frac{\dot{V}}{V} \int_0^1 \theta g'_k(u)(1-u^2)\, du - [g'_k(u)\tau]_{\text{wall}} - \int_0^1 g''_k(u)\tau\, du \tag{5.12.8}$$

where V now denotes the dimensionless outside velocity and $\tau = 1/\theta$. The integrals on the right of (5.12.8) are evaluated by a complete Simpson's rule with m evaluations of the integrands at equal intervals in the range $u=0$ to $u=1$. For $m \geq 30$ it is found that $C'(k)$ is accurate to six decimal places.

To compare the orthonormal MIR with the original formulation both versions are applied to Falkner-Skan solutions for a) $\beta = 0.5$ (favorable pressure gradient), b) $\beta = 0$ (Blasius flow), c) $\beta = -0.14$ (flow about to separate). The solutions are started with Falkner-Skan data provided at $\xi = 1$ and the ordinary differential equations in each version are integrated numerically for increasing ξ instead of reducing them to algebraic equations with the aid of similarity properties. By this device we can observe how rapidly each version approaches the similarity Falkner-Skan solution as ξ increases and we can also verify whether or not the Falkner-Skan value of skin friction is approached. Figure 5.27 in the case of accelerated flow shows how the difference between the approximate value of τ and the exact Falkner-Skan value develops in each version of MIR for ascending order of approximations. Figure 5.28 shows

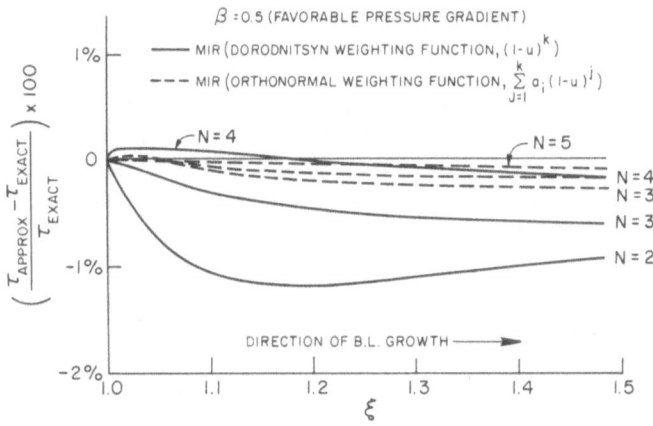

Fig. 5.27 Comparison of orthonormal MIR and conventional MIR for Falkner-Skan solutions

corresponding results for retarded flow. In the first case the orthonormal MIR is always more accurate than the standard version and approaches the Falkner-Skan condition more rapidly. In the separating flow case the orthonormal MIR departs more from the exact solution than the standard version for initial

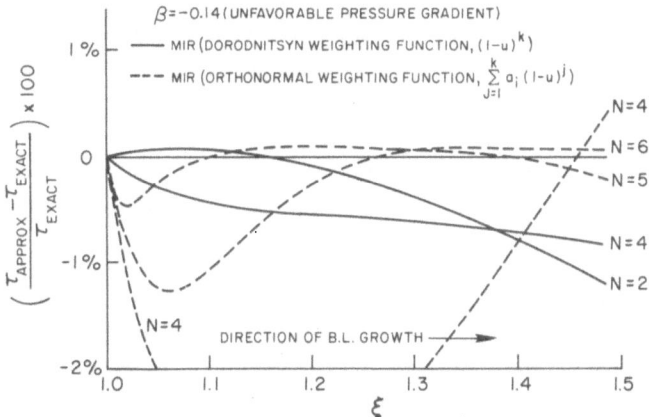

Fig. 5.28 $\beta = -0.14$ (unfavorable pressure gradient)

values of ξ but, at sufficiently high order $(N = 6)$ reproduces the exact solution very closely as ξ is increased.

The new version of MIR is also applied to the problem of viscous flow over a cone, discussed in the next section. There is no serious limitation on the order of approximation used in this version nor on the number of intervals used for the evaluation of the coefficients $C'(k)$. Calculations have been performed up to $m = 128$ and $N = 14$ with little expense in computer time.

A fuller discussion is given in FLETCHER and HOLT (1975).

5.13 Application to Viscous Supersonic Conical Flows

FLETCHER and HOLT (1976) have recently employed the orthonormal version
of the method of integral relations to calculate the laminar boundary layer
growth on a circular cone in a uniform supersonic stream at angle of attack.
Both moderate and large angles of attack are considered. When the angle of
attack is less than the semi-cone angle the component of flow normal to the
cone axis (the cross flow) is subsonic everywhere and the boundary layer is
everywhere attached. For angles of attack larger than the semi-cone angle the
cross flow pattern is similar to that for two dimensional supersonic flow past
a blunt body, being subsonic with detached shock wave on the windward side
of the cone, accelerating to supersonic speeds over the central region and
compressing to subsonic speeds again as the leeward side of the cone is approached. The cross flow profile of the boundary layer in this range develops
an inflection point as the outside pressure increases and becomes of reversed
flow type towards the leeward side. In FLETCHER and HOLT (1976) such profiles
are calculated as far as the line of cross flow reversal. The calculations are
presently being extended to the leeward side.

The variable of integration used in the basic integral relations is the radial
velocity component. Since this quantity is everywhere positive in the field of
disturbance and only the cross flow velocity component is reversed there is
no need to use square root factors in the approximating expressions for unknown integrands. This simplifies the formulation of the method of integral
relations in this application and also permits the use of the orthonormal version of the method.

The boundary layer flow adjacent to the cone is closely linked with the
surrounding inviscid flow and the two flow fields must be calculated simultaneously. The complete inviscid flow around supersonic cones at high angles of
attack has been calculated by FLETCHER (1975), while the corresponding flow
at angles of attack smaller than the semi-cone angle has been determined by
HOLT and NDEFO (1970). Fletcher's high angle calculations used Telenin's
method in the subsonic windward region and the method of characteristics in
the supersonic portion. HOLT and NDEFO use Telenin's method over the whole
cone. When account is taken of boundary layer growth the inviscid calculations are applied to a cone enlarged by an amount equal to the cross flow
displacement thickness rather than to the solid cone itself.

The three dimensional boundary layer equations are simplified by the
introduction of the conical similarity variable $\eta = (u_\infty/2\mu_\infty)^{1/2} y_2/x_2^{1/2}$ where
x_2 is the distance from the cone apex and y_2 is normal to the cone surface.
The Howarth, Mangler and Crocco transformations are also applied. After
factoring the resulting equations by appropriate factors containing orthonormal weighting functions g_k and their derivatives the following basic integral
relations are obtained

$$\frac{\partial}{\partial \zeta} \int_0^1 g_k \frac{w}{\tau} du = \sin \theta_b \left[l_5 \int_0^1 g_k \frac{w}{\tau} du - 1.5 \int_0^1 g_k \frac{u}{\tau} du - l_2 \{g_k' \tau\}_{\text{wall}} \right.$$

$$\left. - l_2 \int_0^1 g_k'' \tau du - \int_0^1 g_k' \frac{w}{\tau} \left\{ u \frac{w_e}{u_e} - w \right\} du \right]. \tag{5.13.1}$$

$$\frac{\partial}{\partial \zeta} \int_0^1 g_k \frac{w^2}{\tau} du = \sin \theta_b \left[l_7 \int_0^1 g_k \frac{w^2}{\tau} du - 2.5 \int_0^1 g_k \frac{uw}{\tau} du - l_2 \int_0^1 g_k'' \tau w du \right.$$

$$- 2 l_2 \int_0^1 g_k' \tau \frac{\partial w}{\partial u} du - \int_0^1 g_k' \frac{w^2}{\tau} \left\{ u \frac{w_e}{u_e} - w \right\} du \tag{5.13.2}$$

$$\left. + l_6 \frac{B}{u_e^2} \int_0^1 g_k \frac{(1-s)}{\tau} du - l_2 \left\{ g_k \frac{\partial w}{\partial u} \tau \right\}_{\text{wall}} + l_8 \int_0^1 g_k \frac{u^2}{\tau} du \right].$$

$$\frac{\partial}{\partial \zeta} \int_0^1 g_k \frac{sw}{\tau} du = \sin \theta_b \left[l_5 \int_0^1 g_k \frac{sw}{\tau} du - 1.5 \int_0^1 g_k \frac{su}{\tau} du - l_2 \{g_k' s \tau\}_{\text{wall}} \right.$$

$$- l_2 \int_0^1 g_k'' \tau s du - (l_2 + l_3) \int_0^1 g_k' \tau \frac{\partial s}{\partial u} du \tag{5.13.3}$$

$$- \int_0^1 g_k' \frac{sw}{\tau} \left\{ u \frac{w_e}{u_e} - w \right\} du - l_3 \left\{ g_k \frac{\partial s}{\partial u} \tau \right\}_{\text{wall}}$$

$$\left. - l_4 \int_0^1 g_k' \tau \left\{ u + \frac{\partial w}{\partial u} w \right\} du \right].$$

Here u is the velocity component parallel to a cone generator, w is the circumferential velocity component, $\tau = \partial u/\partial \eta$ and s is the dimensionless enthalpy used previously (Sect. 5.7). The factors $l_2, \ldots, l_7$, depend on inviscid flow conditions at the outer edge of the boundary layer, while θ_B is the semi-cone angle and $B = (\gamma + 1)/2(\gamma - 1)$ where γ is the ratio of specific heats in the gas (assumed perfect). The coordinate ζ is in the circumferential direction.

The unknown terms appearing in the integrands in (5.13.1), (5.13.2) and (5.13.3) are represented as follows

$$y_1 = \frac{w}{\tau} = \left\{ b_{01} + \sum_{j=1}^{N-1} b_{j1} g_j(u) \right\} \bigg/ (1-u) \tag{5.13.4}$$

$$y_2 = \frac{w^2}{\tau} = \left\{ b_{02} + \sum_{j=1}^{N-1} b_{j2} g_j(u) \right\} \bigg/ (1-u) \tag{5.13.5}$$

$$y_3 = \frac{sw}{\tau} = \left\{ b_{03} + \sum_{j=1}^{N-1} b_{j3} g_j(u) \right\} \bigg/ (1-u) \tag{5.13.6}$$

where $g_k(u)$ are orthonormal functions with base weighting function $1/(1-u)$ (see (5.12.2)). These coefficients b_{k1}, b_{k2}, b_{k3} are related by seven boundary conditions at the wall and outer edge of the boundary layer. Substitution of (5.13.4)–(5.13.6) in (5.13.1)–(5.13.3) leads to $3N-7$ ordinary differential equations for b_{k1}, b_{k2}, b_{k3} required to close the system ($N-2$ equations from (5.13.1), $N-3$ from (5.13.2) and $N-2$ from (5.13.3)). The system can be diagonalized to yield explicit equations for the unknown coefficients. The integration is started near the windward plane of symmetry, where the boundary layer profile can easily be calculated.

The outer edge values of u, w (and hence $l_2, \ldots, l_7$) depend on the inviscid solution for a cone angle increased by the local displacement thickness. Near the windward plane of symmetry the outer flow is elliptic and the displacement thickness is determined by iteration between the boundary layer computation and inviscid computation. The effect of displacement thickness on the outer pressure distribution is first neglected; then the boundary layer equations are integrated to get a first estimate of this thickness. This estimate is then added to the cone thickness and the outer pressure distribution is recalculated, with subsequent recalculation of displacement thickness. Convergence is achieved after three iterations.

Further around the cone, in the shoulder region, the outer flow equations are hyperbolic so that the displacement thickness calculated from the boundary layer computation can immediately be introduced into conical inviscid equations with no iteration required.

Calculations have been completed for cones with semi-angles up to $30°$ set at angles of incidence ranging from $12°$ to $50°$. The pressure distribution, heat transfer and skin friction in each case are determined as functions of the circumferential angle ϕ between $\phi = 0°$ (the windward plane of symmetry) and $\phi = 150°$, a position ahead of the cross flow separation station. Beyond this station the pressure is almost constant, also the heat transfer and skin friction are essentially constant there except for local peaks near the cores of the cross flow vortices which form near the leeward plane of symmetry.

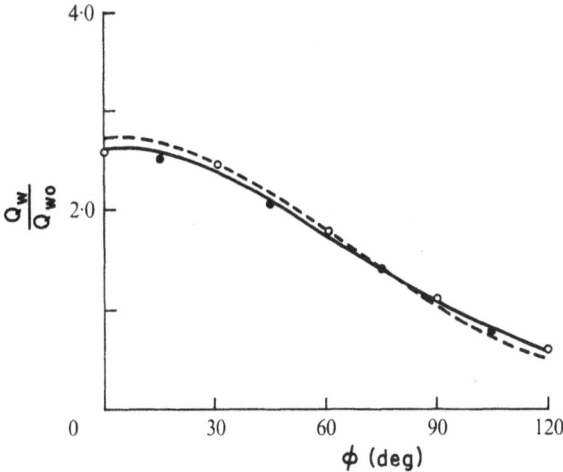

Fig. 5.29   Variation of surface heat transfer on a yawed circular cone

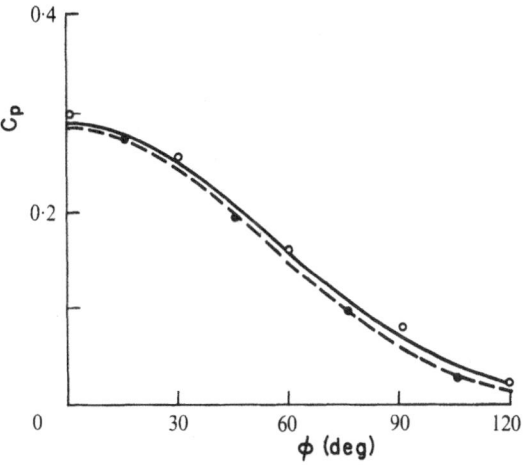

Fig. 5.30   Surface pressure coefficient on a yawed circular cone

Figure 5.29 shows the heat transfer variation with ϕ for $M_\infty = 7.95$, $\theta_b = 10°$, $Re_x = 4.2 \times 10^5$, $T_w/T_0 = 0.41$, $\alpha = 12°$. Figure 5.30 shows the pressure distribution for the same free stream condition. In each figure open circles represent results from a finite difference solution of the modified Navier-Stokes equations by LUBARD and HELLIWELL (1974), filled circles denote experimental points obtained by TRACY (1963), the dashed curve shows MIR results with no correction for displacement thickness effect and the full curve shows MIR results with this correction included. The corrected curves agree closely with experiment. The discrepancy between the present results and those of LUBARD and HELLIWELL is small. The advantages of the use of MIR in this problem are firstly, the small cost in computer time (40–60 seconds for each configuration on the CDC 7600 computer) and, secondly, the treatment of viscous effects by boundary layer theory coupled with an outer inviscid calculation. This has no Reynolds number restrictions and can in principle be extended to turbulent flows.

Recently, HOLT and CHAN (1979) improved the efficiency of the Fletcher-Holt program and extended the shoulder region calculations further round the cone.

5.14 Extension to Unsteady Laminar Boundary Layers

It has recently been shown by HOLT and CHAN (1975) that the method of integral relations can be extended to apply to unsteady laminar boundary layer flow in two dimensions. The effect of the method now is to reduce the unsteady boundary layer equations to a system of first order partial differential equations of hyperbolic type. These can be integrated by a time dependent finite difference process.

We shall first outline the extension and then illustrate it by application to the special case of impulsive motion past a circular cylinder, previously considered by SCHUH (1953) using a Pohlhausen type approach. Both the first and second MIR approximations are worked out. The first approximation has an analytical representation and gives a wall shearing stress with the correct qualitative behavior. The second approximation is computed by the method of characteristics and is in excellent agreement with Schuh's results.

The governing equations start from the time dependent form of the two dimensional boundary layer equations,

$$\frac{\partial u'}{\partial t} + u' \frac{\partial u'}{\partial x} + v' \frac{\partial u'}{\partial y} = \frac{\partial U}{\partial t} + U \frac{\partial U}{\partial x} + v \frac{\partial^2 u'}{\partial y^2} \tag{5.14.1}$$

$$\frac{\partial u'}{\partial x} + \frac{\partial v'}{\partial y} = 0 \tag{5.14.2}$$

using the same notation as in Sect. 5.6. The boundary conditions are as stated in Sect. 5.6, but U, the outer edge velocity, is now a function of x and t. We apply the same Dorodnitsyn transformation to arrive at the reduced equations

$$\frac{1}{U^3} \frac{\partial}{\partial \tau}(u\,U) + \frac{h}{U^3} \frac{\partial}{\partial \xi}(u\,U) + u \frac{\partial u}{\partial \xi} + w \frac{\partial u}{\partial \eta} + \left(\frac{1}{U^3} \dot{U} + \frac{h}{U^3} U'\right) \eta \frac{\partial u}{\partial \eta}$$

$$= \frac{1}{U^3} \dot{U} + \frac{h}{U^3} U' + \frac{1}{U} U'(1 - u^2) + \frac{\partial^2 u}{\partial \eta^2} \tag{5.14.3}$$

$$\frac{\partial u}{\partial \xi} + \frac{\partial w}{\partial \eta} = 0$$

where

$$h = \int_0^x \frac{\partial U}{\partial t} dx' \quad \text{and} \quad \dot{U} = \frac{\partial U}{\partial \tau} \tag{5.14.4}$$

with boundary conditions

$$u = w = 0, \qquad \eta = 0, \qquad u = 1, \qquad \eta = \infty.$$

The basic integral relation is derived from (5.14.3) and (5.14.4) by the same procedure as that in steady flow, after the introduction of weighting functions $f_i(u) = (1-u)^i$. Of course, time dependence introduces several additional terms, and in fact the full basic relation is now

$$\left[\frac{1}{U^2}\frac{\partial}{\partial \tau} + \frac{h}{U^2}\frac{\partial}{\partial \xi} - \left(\frac{1}{U^3}\dot{U} + \frac{h}{U^3}U'\right)\right]\int_0^1 f_i(u)\frac{du}{z} + \frac{\partial}{\partial \xi}\int_0^1 f_i(u)u\frac{du}{z}$$

$$= \left(\frac{1}{U^3}\dot{U} + \frac{h}{U^3}U'\right)\int_0^1 (1-u)f_i'(u)\frac{du}{z} + \frac{1}{U}U'\int_0^1 f_i'(u)(1-u^2)\frac{du}{z} \tag{5.14.5}$$

$$- f_i'(1)z_0 - \int_0^1 f_i''(u)z\,du$$

We consider non-interacting boundary layers and adopt the same re-presentations for the unknowns in the integrands of (5.14.5) as in Sect. 5.6 ((5.6.8), (5.6.9) with $\theta = 1/z$).

Applying the first and second approximations, we then obtain the following partial differential equations to determine the unknown coefficients in the (z, u), $(1/z, u)$ relations (here a replaces c used in Sect. 5.6).

First Approximation

$$z = a_0(1-u), \qquad \theta \equiv 1/z = b_0/(1-u), \qquad a_0 = 1/b_0$$

$$\left[\frac{1}{U^2}\frac{\partial}{\partial \tau} + \left(\frac{h}{U^2} + \frac{1}{2}\right)\frac{\partial}{\partial \xi}\right]b_0 + \frac{3}{2}\frac{U'}{U}b_0 = \frac{1}{b_0} \tag{5.14.6}$$

Second Approximation

$$z = (1-u)(a_0 + a_1 u), \qquad \theta = \frac{b_0 + b_1 u}{1-u}$$

$$\left[\frac{1}{U^2}\frac{\partial}{\partial\tau} + \left(\frac{h}{U^2} + \frac{1}{2}\right)\frac{\partial}{\partial\xi}\right]b_0 + \left[\frac{1}{2}\frac{1}{U^2}\frac{\partial}{\partial\tau} + \left(\frac{1}{2}\frac{h}{U^2} + \frac{1}{3}\right)\frac{\partial}{\partial\xi}\right]b_1 \qquad (5.14.7)$$

$$+ \frac{U'}{U}\left(\frac{3}{2}b_0 + \frac{5}{6}b_1\right) = a_0$$

associated with $f_1(u) = (1-u)$.

$$\left[\frac{1}{2}\frac{1}{U^2}\frac{\partial}{\partial\tau} + \left(\frac{1}{2}\frac{h}{U^2} + \frac{1}{6}\right)\frac{\partial}{\partial\xi}\right]b_0 + \left[\frac{1}{6}\frac{1}{U^2}\frac{\partial}{\partial\tau} + \left(\frac{1}{6}\frac{h}{U^2} + \frac{1}{12}\right)\frac{\partial}{\partial\xi}\right]b_1$$

$$- \frac{1}{2}\left(\frac{\dot{U}}{U^3} + \frac{h}{U^3}U'\right)\left(b_0 + \frac{b_1}{3}\right) + \frac{U'}{U}\left(\frac{4}{3}b_0 + \frac{b_1}{2}\right) = a_0 - \frac{a_1}{3} \qquad (5.14.8)$$

associated with $f_2(u) = (1-u)^2$.

In the general case (5.14.6) (first approximation) or (5.14.7) and (5.14.8) (second approximation) are integrated numerically using characteristic curves with initial conditions

$$b_0 = B_0(X), \qquad \xi = X, \qquad t = t_0 \qquad \text{(first approximation)}$$

$$b_0 = B_0(X), \qquad b_1 = B_1(X), \qquad \xi = X, \qquad t = t_0 \quad \text{(second approximation)}.$$

The outside velocity distribution $U = U(x, t)$ is prescribed. In each approximation we integrate an equivalent set of first order partial differential equations of hyperbolic type.

As an example, the method is applied to the flow generated by a circular cylinder started impulsively with velocity distribution

$$U = 0, \qquad t < 0$$

$$U = 2U_\infty \sin 2(x/d), \qquad t \geq 0$$

where U_∞ is the free stream velocity, x is the arc length round the cylinder measured from the stagnation point and d is the cylinder diameter.

In terms of dimensionless variables

$$X = x/d, \qquad T = tU_\infty/d, \qquad \Theta_0 = \theta_0/(U_\infty d)^{1/2}$$

the first approximation requires the solution of the equation

$$\left[\frac{1}{U^2}\frac{\partial}{\partial\tau}+\left(\frac{h}{U^2}+\frac{1}{2}\right)\frac{\partial}{\partial\xi}\right]\theta_0+\frac{3}{2}\frac{U'}{U}\theta_0=\frac{1}{\theta_0}\tag{5.14.9}$$

For $U=2U_\infty\sin 2(x/d)$,

$$h=0,\qquad\frac{\partial}{\partial\tau}=\frac{\partial}{\partial t},\qquad\frac{\partial}{\partial\xi}=\frac{1}{U}\frac{\partial}{\partial x}$$

with initial condition $\Theta_0(0,X)=0$.

The solution to this problem is

$$\Theta_0=\frac{2}{(\sin\phi)^3}\left[\frac{35}{64}(\cos\phi_0-\cos\phi)+\frac{7}{64}(\cos 3\phi-\cos 3\phi_0)\right.$$
$$\left.+\frac{7}{320}(\cos 5\phi_0-\cos 5\phi)+\frac{1}{448}(\cos 7\phi-\cos 7\phi_0)\right]^{1/2}\tag{5.14.10}$$

where $\phi=2X$.

The characteristic curves arising in the second approximation have slopes $(1/2)U(1+1/\sqrt{3})$ (ζ^+ curves) and $(1/2)U(1-1/\sqrt{3})$ (ζ^- curves) along which the compatibility relations are

ζ^+ curves

$$\frac{dX}{dT}=\left(1+\frac{1}{\sqrt{3}}\right)\sin(2X)\tag{5.14.11}$$

$$\left(1+\frac{1}{\sqrt{3}}\right)\Theta_{0\zeta}-\left(1+\frac{2}{\sqrt{3}}\right)\Theta_{1\zeta}=2X_\zeta\left\{\sin(2X)\left[\left(14-\frac{6}{\sqrt{3}}\right)\frac{1}{\Theta_0}-\frac{16}{\Theta_1}\right]\right.$$
$$\left.-\cot(2X)\left[\left(5+\frac{1}{\sqrt{3}}\right)\Theta_0+\left(1-\frac{5}{\sqrt{3}}\right)\Theta_1\right]\right\}$$

ζ^- curves

$$\frac{dX}{dT}=\left(1-\frac{1}{\sqrt{3}}\right)\sin(2X)\tag{5.14.12}$$

$$\left(1-\frac{1}{\sqrt{3}}\right)\Theta_{0\zeta}-\left(1-\frac{2}{\sqrt{3}}\right)\Theta_{1\zeta}=2X_\zeta\left\{\sin(2X)\left[\left(14+\frac{6}{\sqrt{3}}\right)\frac{1}{\Theta_0}-\frac{16}{\Theta_1}\right]\right.$$
$$\left.-\cot(2X)\left[\left(5-\frac{1}{\sqrt{3}}\right)\Theta_0+\left(1+\frac{5}{\sqrt{3}}\right)\Theta_1\right]\right\}$$

respectively, where Θ is the dimensionless form of θ.

Eqs. (5.14.11) and (5.14.12) are solved by the method of characteristics with initial conditions obtained from an analytical solution of the linearized version of (5.14.1) valid for early times.

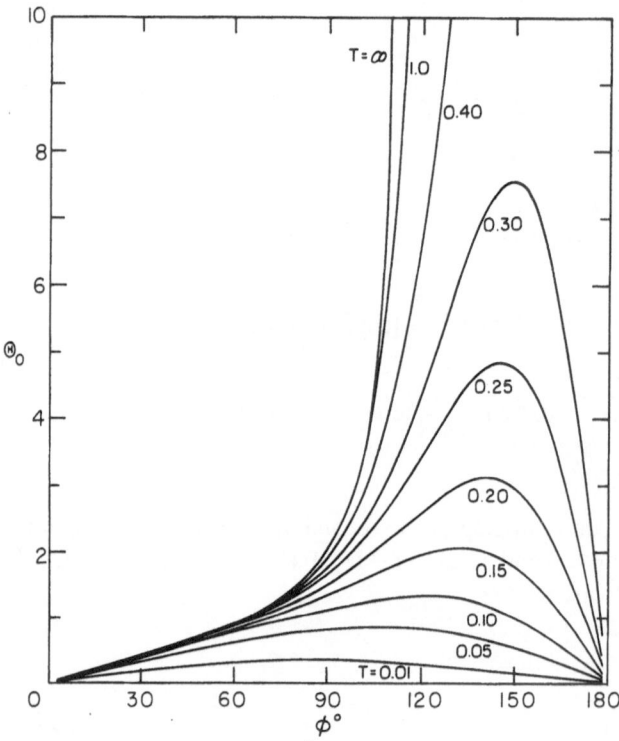

Fig. 5.31 Variation of Θ_0 with ϕ

Figure 5.31 shows the variation of Θ_0 with $\phi°$ obtained by the second approximation. In this calculation the steady separation position is found at $\phi = 110°$, which agrees very well with Schuh's result (SCHUH, 1953) and the result given in SCHLICHTING (1968, p. 161). Also, the behavior of the steady wall shear stress agrees very well with that shown in Fig. 9.7 of SCHLICHTING (1968).

5.15 Application to Internal Flow Problems

In recent years, the Method of Integral Relations, in conjunction with Telenin's Method and the Method of Lines (see Chap. 6), has been applied to a series of problems connected with coal gasification. Systems producing coal gas can suffer damage due to particle erosion in the curved pipe leading from the combustion chamber and from corrosion (principally due to oxides and sulfides) in the heat exchange system further downstream. The erosion problem led to an investigation of gas particle flow at high values of Reynolds number and Dean number near the entry section of a curved pipe of circular cross section with a 90° bend. This was carried out by YEUNG (1980) and YEUNG and HOLT (1983) and concentrated on the three dimensional laminar boundary layer flow near the pipe walls, matching this with an inviscid flow in the core of the pipe. A separate formulation of the Method of Integral Relations was used to treat the viscous part of this flow.

To handle the corrosion problem, it is proposed to introduce a normal or parallel jet of steam at the entry section of each pipe in the heat exchange system and so reduce the concentrations of sulfides or oxides at the pipe wall to levels at which their corrosive effects are unimportant. Strong injection is used, that is, the flow rate in the incident jet is sufficiently high to displace the pipe boundary layer away from the wall. The effect of such injection, in the laminar case, was investigated by YEUNG and HOLT (1982), using both laminar and normal injection. Flow in the laminar mixing and boundary layers is calculated using the Method of Integral Relations, adapted to equations of motion including heat transfer and non-equilibrium chemical reactions between the injected steam and the corrosive species in the main pipe flow. The flow is incompressible but the algebraic details of the MIR formulation are similar to those described in Sect. 5.7.

In the turbulent case, the effect of parallel injection on pipe flow has recently been investigated by YANG and HOLT (1983). Here, the viscous flow is calculated by a new version of the Method of Integral Relations, developed for turbulent boundary layers by YEUNG and YANG (1981). The essentials of this formulation are summarized.

Laminar Boundary Layer Flow in a Curved Circular Pipe: The coordinate system used to analyze laminar boundary layer flow in a curved circular pipe is shown in Fig. 5.32 where R is the radius of curvature of the pipe axis,

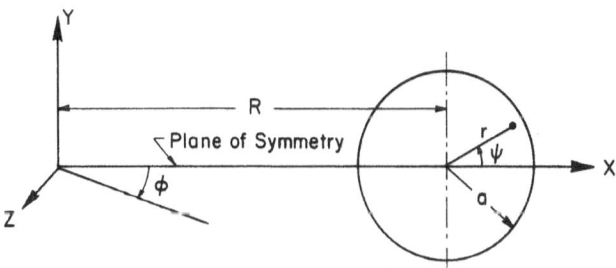

Fig. 5.32   Toroidal coordinate system

a is the radius of the cross section, u, v, w are the velocity components in the direction of increasing r, ψ, ϕ, respectively. As usual, p, ϱ and ν denote pressure, density and kinematic viscosity coefficient.

The boundary layer equations are derived from the Navier Stokes equations, written in terms of toroidal coordinates. Here we use the following dimensionless variables

$$U = \frac{u\,D^{1/2}}{w_e}, \quad V = \frac{v}{w_e}, \quad W = \frac{w}{w_e}, \quad P = \frac{p}{\varrho\,W_i^2}, \quad W_e = \frac{w_e}{W_i},$$

$$\eta = (a-r)D^{1/2}/a, \quad \psi = \psi, \quad s = \frac{\phi}{\alpha}, \tag{5.15.1}$$

where D is the Dean number given by

$$D = \mathrm{Re}\sqrt{\alpha}, \quad \mathrm{Re} = \frac{W_i\,a}{\nu} \tag{5.15.2}$$

and α is the curvature ratio

$$\alpha = a/R. \tag{5.15.3}$$

The boundary layer equations in toroidal coordinates are then

$$\frac{\partial P}{\partial \eta} = 0, \tag{5.15.4}$$

$$-U\frac{\partial V}{\partial \eta} + V\frac{\partial V}{\partial \psi} + \frac{W}{J}\frac{\partial V}{\partial s} = V_e\frac{\partial V_e}{\partial \psi} + \frac{1}{J}\frac{\partial V_e}{\partial s} + \frac{\alpha}{J}(1-W^2)\sin\psi$$

$$+ \frac{1}{W_e}\left\{\frac{\partial W_e}{\partial \psi}(V_e^2 - V^2) + \frac{1}{J}\frac{\partial W_e}{\partial s}(V_e - WV)\right\} + \frac{\sqrt{\alpha}}{W_e}\frac{\partial^2 V}{\partial \eta^2}, \tag{5.15.5}$$

$$-U\frac{\partial W}{\partial \eta} + V\frac{\partial W}{\partial \psi} + \frac{W}{J}\frac{\partial W}{\partial s} = \frac{\alpha\sin\psi}{J}(VW - V_e)$$

$$+ \frac{1}{W_e}\left\{\frac{\partial W_e}{\partial \psi}(V_e - VW) + \frac{1}{J}\frac{\partial W_e}{\partial s}(1-W^2)\right\} + \frac{\sqrt{\alpha}}{W_e}\frac{\partial^2 W}{\partial \eta^2} \tag{5.15.6}$$

and the continuity equation

$$-\frac{\partial U}{\partial \eta} + \frac{\partial V}{\partial \psi} + \frac{1}{J}\frac{\partial W}{\partial s} - \frac{\alpha V\sin\psi}{J} + \frac{1}{W_e}\left\{V\frac{\partial W_e}{\partial \psi} + \frac{W}{J}\frac{\partial W_e}{\partial s}\right\} = 0, \tag{5.15.7}$$

where J is defined by

$$J \equiv 1 + \alpha\cos\psi. \tag{5.15.8}$$

In deriving Eqs. (5.15.5)–(5.15.7), the radial coordinate r has been approximated by the radius of the pipe, a, since the boundary layer is assumed thin compared with the pipe radius. The boundary conditions are

$$U = V = W = 0 \quad \text{at } \eta = 0,$$

$$U = U_e, \quad V = V_e, \quad W = 1, \quad \frac{\partial U}{\partial \eta} = \frac{\partial V}{\partial \eta} = \frac{\partial W}{\partial \eta} = 0, \quad \eta \to \infty \quad (5.15.9)$$

and U, V, W given at $s = s_i$, an initial station.

We shall now derive the basic integral relations from Eqs. (5.15.5)–(5.15.7). Introducing a complete set of linearly independent functions $\{g_k(W)\}$ the elements of which satisfy the condition

$$\lim_{W \to 1} g_k(W) = 0, \quad \text{for all } k \tag{5.15.10}$$

one obtains the first integral by multiplying (5.15.7) and (5.15.6) by $g_k(W)$ and $g'_k(W)$, respectively, adding and integrating the result from $\eta = 0$ to $\eta \to \infty$ and finally changing the variable of integration to W by defining Z to be

$$Z = \left(\frac{\partial W}{\partial \eta} \right)^{-1}. \tag{5.15.11}$$

The result is

$$\frac{\partial}{\partial \psi} \int_0^1 V Z g_k \, dW + \frac{1}{1 + \alpha \cos \psi} \frac{\partial}{\partial s} \int_0^1 W Z g_k \, dW$$

$$= \left(\frac{\alpha \sin \psi}{1 + \alpha \cos \psi} - \frac{1}{W_e} \frac{\partial W_e}{\partial \psi} \right) \int_0^1 [(VW - V_e) g'_k + V g_k] Z \, dW$$

$$+ \frac{1}{1 + \alpha \cos \psi} \frac{1}{W_e} \frac{\partial W_e}{\partial s} \int_0^1 [(1 - W^2) g'_k - W g_k] Z \, dW$$

$$- \frac{\sqrt{\alpha}}{W_e} \frac{g'_k(0)}{Z_0} - \frac{\sqrt{\alpha}}{W_e} \int_0^1 \frac{g''_k}{Z} \, dW, \tag{5.15.12}$$

where

$$g'_k(W) \equiv \frac{dg_k(W)}{dW},$$

where Z_0 is the value of Z at the wall.

The second integral relation is derived by first transforming (η, ψ, s) to (W, ψ, s) in Eq. (5.15.5), then multiplying the transformed equation by a weighting function $h_k(W)$, an element of another complete linear set $\{h_k(W)\}$. The result is integrated from $W=0$ to $W=1$, and we have

$$\frac{\partial}{\partial \psi} \int_0^1 \frac{V^2}{2} h_k \, dW + \frac{1}{1+\alpha \cos \psi} \frac{\partial}{\partial s} \int_0^1 W V h_k \, dW$$

$$= \frac{1}{W_e} \frac{\partial W_e}{\partial \psi} \times \int_0^1 \left\{ (V_e^2 - V^2) - \frac{\partial V}{\partial W} (V_e - V W) \right\} h_k(W) \, dW$$

$$+ \frac{1}{1+\alpha \cos \psi} \frac{1}{W_e} \frac{\partial W_e}{\partial s} \int_0^1 \left\{ V_e - V W - \frac{\partial V}{\partial W} (1 - W^2) \right\} h_k \, dW$$

$$+ \frac{\alpha \sin \psi}{1+\alpha \cos \psi} \int_0^1 \left\{ (1 - W^2) - \frac{\partial V}{\partial W} (V W - V_e) \right\} h_k(W) \, dW$$

$$+ \left(V_e \frac{\partial V_e}{\partial \psi} + \frac{1}{1+\alpha \cos \psi} \frac{\partial V_e}{\partial s} \right) \int_0^1 h_k(W) \, dW$$

$$+ \frac{\sqrt{\alpha}}{W_e} \int_0^1 \frac{\partial^2 V}{\partial W^2} \frac{1}{Z^2} h_k(W) \, dW. \tag{5.15.13}$$

In the plane of symmetry we obtain two auxiliary integral relations from Eqs. (5.15.12) and (5.15.13) for W and a new variable S given by

$$S = \frac{dV}{d\psi}. \tag{5.15.14}$$

The unknowns Z, V and S are represented by

$$Z = \left\{ b_{01} + \sum_{j=1}^{N-1} b_{j1} g_j(W) \right\} / (1 - W). \tag{5.15.15}$$

$$V = \left\{ V_e + \sum_{j=1}^{N} b_{j2} h_j(W) \right\} W, \tag{5.15.16}$$

$$S = \left[S_e + \sum_{j=1}^{N} e_j h_j(W) \right] W. \tag{5.15.17}$$

The functions $\{g_k(W)\}$ and $\{h_k(W)\}$ are chosen as orthonormal sets derived from the functions $(1-W)^k$, $k=1, 2, \ldots$ as follows:

$$g_k(W) = \sum_{j=1}^{k} a_{kj}(1-W)^j, \tag{5.15.18}$$

$$h_k(W) = \sum_{j=1}^{k} c_{kj}(1-W)^j \tag{5.15.19}$$

such that

$$\int_0^1 g_i(W)g_j(W)\frac{W}{1-W}dW = \delta_{ij} \tag{5.15.20}$$

and

$$\int_0^1 W^2 h_i(W)h_j(W)dW = \delta_{ij}, \tag{5.15.21}$$

where δ_{ij} is the Kronecker delta. The existence and uniqueness of g_k and h_k are the content of the Gram-Schmidt process [ISAACSON and KELLER (1966)].

Substitution of (5.15.15)–(5.15.17) into the basic integral relations yield a system of first-order partial differential equations of hyperbolic type. In the N-th approximation, there are $2N$ such equations in $2N$ unknowns generated from Eqs. (5.15.12)–(5.15.13).

The boundary layer flow is matched with flow in the inviscid core of the pipe by an iterative process. The core flow is firstly represented very simply by a two-dimensional point vortex profile and the resulting velocity and pressure distributions are used to calculate the displacement thickness of the pipe boundary layer. The inviscid flow is then recalculated in a pipe of thickness reduced by this thickness and the boundary layer calculation is modified accordingly. The sequence is continued until the two flow fields match.

The flow field was calculated for a pipe with curvature ratio $\alpha = a/R = 0.1$, Reynolds number $\mathrm{Re} = 10^4$. Both the first and second approximations were used with 40 iterations required in the latter case. Figure 5.33 shows the distributions of $\bar{V} = \int_0^1 V dW$, the average azimuthal velocity, in successive cross flow planes. Typical profiles of axial and azimuthal velocity across the boundary layer are shown in Fig. 5.34.

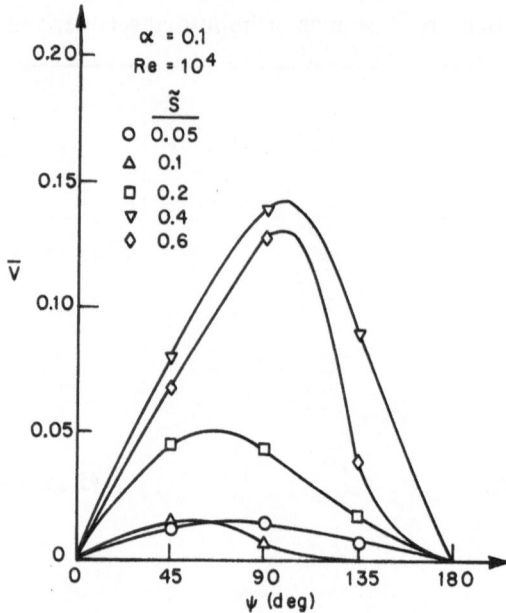

Fig. 5.33   Variation of $\overline{V}$ with ψ at different streamwise stations

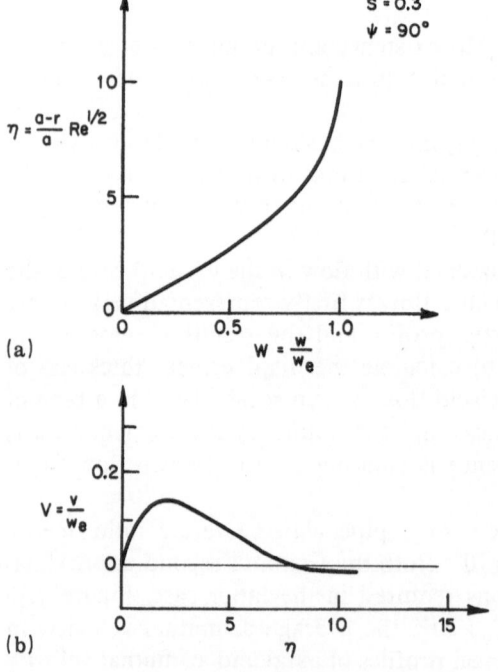

Fig. 5.34   Velocity profiles across the boundary layer

Turbulent Boundary Layers: YEUNG and YANG (1981) extended the modified Method of Integral Relations to apply to turbulent boundary layers, using an eddy viscosity model applicable to regions of favorable, zero and unfavorable pressure gradient. The model uses an expression near the wall derived by SPALDING (1961) and KLEINSTEIN (1967) and, in the outer boundary layer, uses a wake flow formula due to CLAUSER (1956). In both regions the eddy viscosity is expressed in terms of streamwise mean velocity, rather than distance from the wall.

The two dimensional turbulent boundary layer equations in incompressible flow may be written

$$u \frac{\partial u}{\partial x} + v \frac{\partial u}{\partial y} = -\frac{1}{\varrho} \frac{\partial p}{\partial x} + \frac{\partial}{\partial y}\left[\left(v + \frac{\varepsilon}{\varrho}\right)\frac{\partial u}{\partial y}\right], \tag{5.15.22}$$

$$\frac{\partial u}{\partial x} + \frac{\partial v}{\partial y} = 0, \tag{5.15.23}$$

where u and v are the mean velocity components, in the x and y directions, respectively, and p, ϱ, v are pressure, density and kinematic viscosity, respectively. The eddy viscosity, ε, is defined by

$$-\varrho\overline{u'v'} = \varepsilon \frac{\partial u}{\partial y}, \tag{5.15.24}$$

where $-\varrho\overline{u'v'}$ is the Reynolds stress. We now change to the following dimensionless variables

$$U = \frac{u}{u_e}, \quad V = \frac{v \, \mathrm{Re}^{1/2}}{u_e}, \quad \bar{x} = \frac{x}{L}, \quad \bar{y} = \frac{y \, \mathrm{Re}^{1/2}}{L}, \quad U_e = \frac{u_e}{u_\infty}, \tag{5.15.25}$$

where L and u_∞ are the characteristic length and velocity scales, respectively, u_e is the velocity at the outer boundary-layer edge, and the Reynolds number Re is defined as

$$\mathrm{Re} = u_\infty L/v. \tag{5.15.26}$$

The pressure gradient is related to the outer edge velocity by Bernoulli's equation

$$-\frac{1}{\varrho} \frac{\partial p}{\partial x} = u_e \frac{du_e}{dx}. \tag{5.15.27}$$

The transformed equations of motion are

$$U\frac{\partial U}{\partial \bar{x}} + V\frac{\partial U}{\partial \bar{y}} = \frac{1}{U_e}\frac{dU_e}{d\bar{x}}(1-U^2) + \frac{1}{U_e}\frac{\partial}{\partial \bar{y}}\left[\left(1+\frac{\varepsilon}{\mu}\right)\frac{\partial U}{\partial \bar{y}}\right], \qquad (5.15.28)$$

$$\frac{\partial U}{\partial \bar{x}} + \frac{\partial V}{\partial \bar{y}} = -\frac{U}{U_e}\frac{dU_e}{d\bar{x}}. \qquad (5.15.29)$$

The basic integral relation derived from Eqs. (5.15.28) and (5.15.29) after factoring by weighting functions f_i' and f_i, respectively, adding and integrating across the boundary layer is

$$\frac{\partial}{\partial \bar{x}} \int_0^1 f_i UZdU = \frac{1}{U_e}\frac{dU_e}{d\bar{x}}\int_0^1 [(1-U^2)f_i' - Uf_i]ZdU$$

$$-\frac{1}{U_e}f_i'(0)\frac{1}{Z_0} - \frac{1}{U_e}\int_0^1\left(1+\frac{\varepsilon}{\mu}\right)\frac{f_i''}{Z}dU, \qquad (5.15.30)$$

where

$$Z = \left(\frac{\partial U}{\partial \bar{y}}\right)^{-1}. \qquad (5.15.31)$$

The functions f_i are a complete set of orthonormal functions formed from combinations of $(1-U)^k$, $k=1, 2, 3, \ldots$ and satisfying

$$\int_0^1 f_k f_j \frac{U}{1-U}dU = \delta_{kj}, \qquad (5.15.32)$$

where δ_{kj} is the Kronecker delta. The unknown Z is represented by

$$Z = \frac{b_0 + \sum_{j}^{N-1} b_j f_j(U)}{1-U}, \qquad (5.15.33)$$

where the coefficients $b_0, b_1, b_2, \ldots$ are to be determined.

Introducing f_i, $i=1, \ldots, N$ into Eq. (5.15.30) we obtain the following ordinary differential equations for $b_0, b_1, \ldots, b_{N-1}$:

$$\frac{db_0}{d\bar{x}}\int_0^1\frac{f_1 U}{1-U}dU + \frac{db_i}{d\bar{x}} = \frac{1}{U_e}\frac{dU_e}{d\bar{x}}\int_0^1[(1-U^2)f_i' - Uf_i]ZdU$$

$$-\frac{1}{U_e}\frac{f_i'(0)}{Z_0} - \frac{1}{U_e}\int_0^1\left(1+\frac{\varepsilon}{\mu}\right)\frac{f_i''}{Z}dU,$$

$$i = 1, 2, \ldots, N-1 \qquad (5.15.34)$$

and

$$\frac{db_0}{d\bar{x}} \int\limits_0^1 \frac{f_N U}{1-U} dU = \frac{1}{U_e} \frac{dU_e}{d\bar{x}} \int\limits_0^1 [(1-U^2)f'_N - Uf_N]ZdU$$

$$- \frac{1}{U_e} f'_N(0) \frac{1}{Z_0} - \frac{1}{U_e} \int\limits_0^1 \left(1+\frac{\varepsilon}{\mu}\right) \frac{1}{Z} f''_N dU, \qquad (5.15.35)$$

which are integrated subject to given values at some initial station.

To represent ε/μ in Eqs. (5.15.34) and (5.15.35) we use the following model in the wall region due to SPALDING (1961) and KLEINSTEIN (1967):

$$\frac{\varepsilon}{\mu} = 0.04432\{e^{0.4u^+} - 1 - 0.4u^+ - 0.08u^{+2}\}. \qquad (5.15.36)$$

In the outer boundary layer we use the CLAUSER (1956) wake model

$$\frac{\varepsilon}{\mu} = 0.0168 \, \mathrm{Re}_{\delta^*}, \qquad (5.15.37)$$

where $u^+ = u/u_\tau$ and $u_\tau = (\tau_w/\varrho)^{1/2}$, the wall frictional velocity. The Reynolds number is based on the displacement thickness δ^*

$$\mathrm{Re}_{\delta^*} = \frac{u_e \delta^*}{\nu}. \qquad (5.15.38)$$

In terms of U and Z, the eddy viscosity can be written as

$$\frac{\varepsilon}{\mu} = 0.04432 \, [e^{0.4U \sqrt{U_e Z_0 \mathrm{Re}^{1/2}}}]$$

$$- 1 - 0.4 U\sqrt{U_e Z_0 \mathrm{Re}^{1/2}} - 0.08 \, U^2 \, U_e \, \mathrm{Re}^{1/2} Z_0$$

$$\text{for } 0 \le U < U_m \qquad (5.15.39)$$

and

$$\frac{\varepsilon}{\mu} = 0.0168 \, U_e \, \mathrm{Re}^{1/2} \int\limits_0^1 (1-U)ZdU \quad \text{for } U_m \le U \le 1, \qquad (5.15.40)$$

where U_m is calculated so that the wall and wake models match continuously with continuous slopes.

This formulation of the Method of Integral Relations is applied to simple turbulent boundary layer flows which have been investigated experimentally with (i) zero pressure gradient, (ii) adverse pressure gradient, (iii) favorable pressure gradient. The measured velocity profiles and derived characteristic flow properties such as displacement thickness and skin friction coefficient are due to COLES and HIRST (1968). In case (i) the development of the velocity profiles is shown in Fig. 5.35 while Figs. 5.36 and 5.37

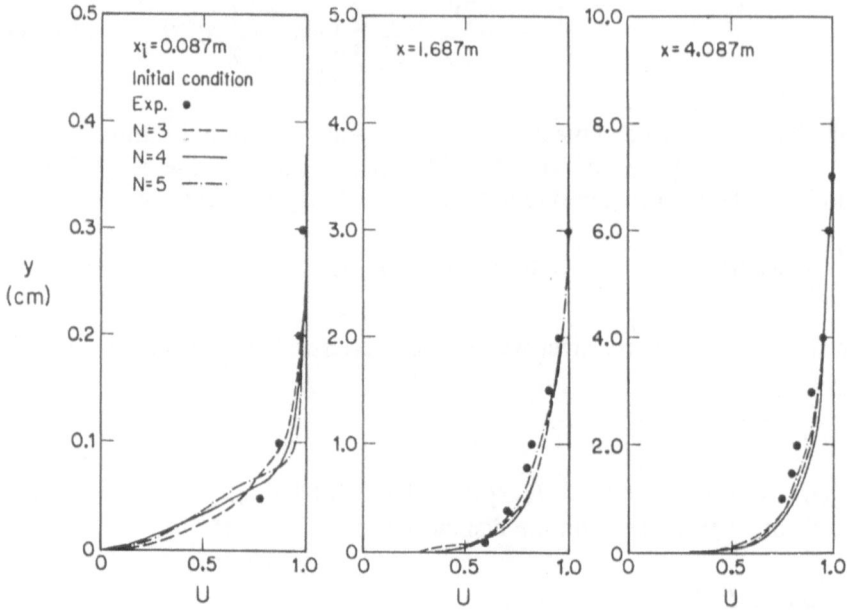

Fig. 5.35   Initial velocity profile and its development (ID 1400). (This ID number is from HIRST and COLES (1968))

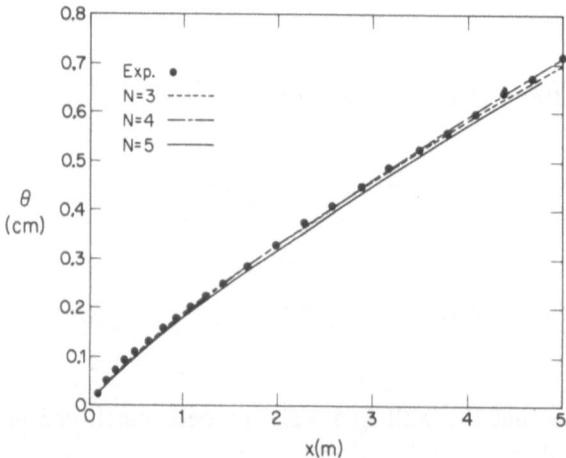

Fig. 5.36   Momentum thickness (ID 1400)

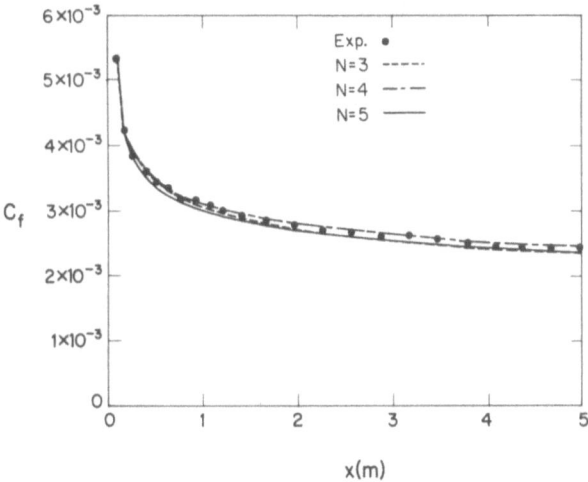

Fig. 5.37   Skin friction coefficient (ID 1400)

show the growth of displacement thickness and decrease in skin friction coefficient in the main stream direction. Agreement between calculated and measured results is excellent. The numerical method was applied in the approximations $N = 3, 4$ and 5 and results show that no higher order approximations are needed. Similarly good agreement between computation and experiment was obtained in case (iii). Results for positive pressure gradient, case (ii), are good for $N = 5$ but indicate a need to improve on the models (5.15.36) and (5.15.37) as the separation region in a turbulent boundary layer is approached.

Boundary Layer Control by Means of Strong Injection: The products of coal gasification (sulfides and oxides) contain corrosive species which can attack the walls of pipes in the heat exchange system. To deal with this problem the effect of normal and parallel injection on boundary layer flow has been investigated, in both laminar and turbulent regimes. In the laminar case the mixing between injectant and main stream, and the merged boundary layer flow are calculated by the modified Method of Integral Relations, extended to account for two or more species reactions, as described in Sect. 5.12. For parallel injection the flow is at constant pressure. For normal injection, the dividing streamline between the injectant and main stream is determined by an inviscid solution corresponding to a line source. This procedures a negative pressure gradient which further delays the effect of corrosion on the pipe walls.

In the turbulent case only parallel injection has been investigated, using the YEUNG-YANG (1982) formulation of the Method of Integral Relations. It is found that injection reduces corrosion below the critical level for a distance along the tube wall approximately 100 times the height of the parallel jet.

Model Problem (C. K. CHU and K. GONG, 1975)

Use the method of integral relations to solve the problem of laminar mixing of two uniform supersonic streams shown

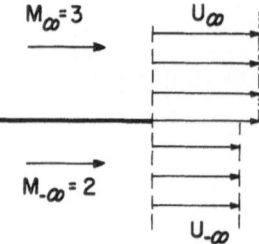

Fig. 5.38   Mixing of two parallel supersonic streams

The equations of motion are (for $Pr = 1$)

$$\frac{\partial(\varrho u)}{\partial x} + \frac{\partial(\varrho v)}{\partial y} = 0$$

$$\varrho u \frac{\partial u}{\partial x} + \varrho v \frac{\partial u}{\partial y} = \frac{\partial}{\partial y}\left(\mu \frac{\partial u}{\partial y}\right)$$

$$\varrho u \frac{\partial T}{\partial x} + \varrho v \frac{\partial T}{\partial y} = \frac{\partial}{\partial y}\left(\mu \frac{\partial T}{\partial y}\right) + (\gamma - 1) M_\infty^2 \mu \left(\frac{\partial u}{\partial y}\right)^2$$

$$\varrho T = 1 .$$

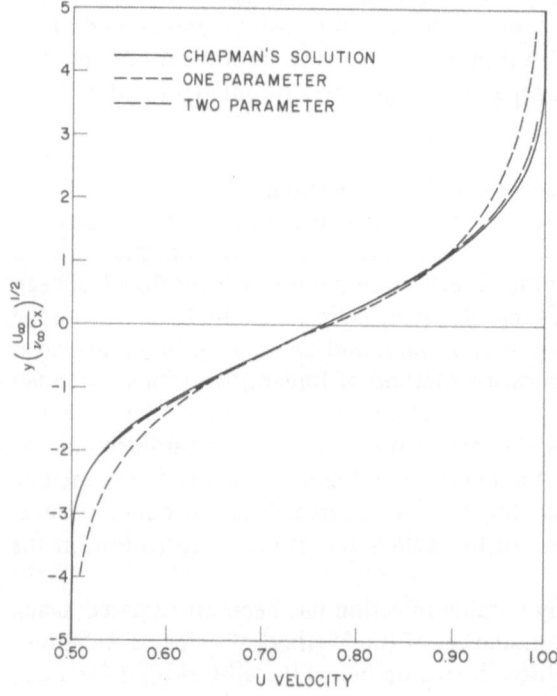

Fig. 5.39   Velocity profiles for two stream mixing problems

The boundary conditions are

$$y \to \infty, \qquad T \to 1, \qquad u \to 1$$

$$y \to -\infty, \qquad T \to T_0, \qquad u \to T_0^{1/2} \frac{M_{-\infty}}{M_\infty} = u_0.$$

Take $u_0 = 1/2$, $T_0 = (225)/400$.

The method of solution is given in the listing attached.

The results are shown in Fig. 5.39.

RTRAN COMPILER VERSION 2.3 B.3   27 MAY 75

```
        PROGRAM MIR(INPUT,OUTPUT,PUNCH)
C       ..... FIRST APPROX .....
        U1 = .7787
        A = 6.5545
        F= -1.8*F2(U1)+3.575*F1(U1)-.775*F0(U1)
        G = 1.8*G2(U1)-3.575*G1(U1)+.775*G0(U1)
        U = 1.
        DO 100 I = 1,22
        U = U-.01
        Y = A*(1.-U1)*(-1.8*F2(U)+3.575*F1(U)-.775*F0(U)-F)
        PUNCH 10,Y,U
100     PRINT 10,Y,U
        DO 200 I = 1,27
        U = U-.01
        Y = -A*(U1-.5)*(1.8*G2(U)-3.575*G1(U)+.775*G0(U)-G)
        PUNCH 10,Y,U
200     PRINT 10,Y,U
        PRINT 20
C       ..... SECOND APPROX .....
        U1 = .7672
        A = 8.0038
        F = 1.2*F3(U1)+(.6*U1-4.1833)*F2(U1)+(4.0917-1.1917*U1)*F1(U1)
       1+(.2583*U1-.775)*F0(U1)
        G=1.2*G3(U1)+(.6*U1-3.2833)*G2(U1)+(2.3042-1.1917*U1)*G1(U1)
       1+(.2583*U1-.3875)*G0(U1)
        U=1.
        DO 300 I = 1,23
        U = U-.01
        Y=A*(1.2*F3(U)+(.6*U1-4.1833)*F2(U)+(4.0917-1.1917*U1)*F1(U)+(.258
       13*U1-.775)*F0(U)-F)
        PUNCH 10,Y,U
300     PRINT 10,Y,U
        DO 400 I = 1,26
        U = U-.01
        Y=-A*(1.2*G3(U)+(.6*U1-3.2833)*G2(U)+(2.3042-1.1917*U1)*G1(U)+(.25
       183*U1-.3875)*G0(U)-G)
        PUNCH 10,Y,U
400     PRINT 10,Y,U
10      FORMAT (5X,2F20.6)
20      FORMAT (//)
        STOP
        END


        FUNCTION F3(U)
        V = 1.-U
        F3 = V*V*V/3.-1.5*V*V+3.*V-ALOG(V)
        RETURN
        END
```

```
FUNCTION F2(U)
V = 1.-U
F2 = -.5*V*V+2.*V-ALOG(V)
RETURN
END


FUNCTION F1(U)
V = 1.-U
F1 = V-ALOG(V)
RETURN
END


FUNCTION F0(U)
V = 1.-U
F0 = -ALOG(V)
RETURN
END


FUNCTION G3(U)
V = U-.5
G3 = V*V*V/3.+.75*V*V+.75*V+.125*ALOG(V)
RETURN
END


FUNCTION G2(U)
V = U-.5
G2 = .5*V*V+V+.25*ALOG(V)
RETURN
END


FUNCTION G1(U)
V = U-.5
G1 = V+.5*ALOG(V)
RETURN
END


FUNCTION G0(U)
V = U-.5
G0 = ALOG(V)
RETURN
END
```

CHU and GONG used the original form of MIR to solve this problem and found that the second approximation gave close agreement with the exact solution. In 1976 C. O. AJAGU and U. S. CHOI solved the same problem using the modified form of MIR and again found the second approximation to be accurate.

References

Abbott, D. E., Bethel, H. E.: Ingenieur Archiv 37, 110–124 (1968).
Alikhashkin, Y. I., Favorskii, A. P., Chushkin, P. I.: Zh. Vych. Mat. Mat. Fiz. 3, 1136–1140 (1963).
Bachalo, W. D.: J. App. Mech. 42, 289–294 (1975).
Bachalo, W. D., Holt, M.: AGARD Symposium on Flow Separation, Preprint No. 168, Göttingen 1975.

Bashkin, V. A.: Zh. Vych. Mat. Mat. Fiz. **8**, 1280–1290 (1968).

Bashkin, V. A.: Zh. Vych. Mat. Mat. Fiz. **10**, 1491–1502 (1970).

Belotserkovskii, O. M.: Prik. Mat. Mekh. **22**, 206–219 (1958).

Belotserkovskii, O. M.: Prik. Mat. Mekh. **24**, 511–517 (1960).

Belotserkovskii, O. M.: Zh. Vych. Mat. Mat. Fiz. **2**, 1062–1085 (1962).

Belotserkovskii, O. M., Chushkin, P. I.: Zh. Vych. Mat. Mat. Fiz. **2**, 731–759 (1962).

Brown, S. N., Stewartson, K.: Laminar Separation. *Ann. Rev. Fluid Mechanics*, Vol. 1, 1969.

Catherall, D., Mangler, K. W.: J. Fluid Mechanics **26**, 163–182 (1966).

Chapman, D. R., Kuehn, D. M., Larson, K. L.: NACA TR 1356, 1958.

Cherry, T. M.: J. Australian Math. Soc. **1**, 357 (1960).

Chushkin, P. I.: Sb. "Vych. Mat." **2**, 20–44 (1957).

Chushkin, P. I.: Sb. "Vych. Mat." **3**, 99–110 (1958).

Chushkin, P. I.: Dok. Akad. Nauk SSSR **125**, 748–751 (1959).

Clauser, F. H.: The Turbulent Boundary Layer. *Advances in Applied Mechanics, Vol. 4*, p. 1–51. Academic Press, New York (1956).

Cohen, C. B., Reshotko, E.: NACA TR 1293, 1956.

Coles, P., Hirst, E. (eds.): Computation of Turbulent Boundary Layers. *Proc. AFOSR-IFP Stanford Conference, Vol. 2*, Compiled Data. Stanford (1968).

Der, J. Jr., Raetz, G. S.: Inst. Aerospace Sciences Preprint No. 62–70, 1962.

Dorodnitsyn, A. A.: Solution of mathematical and logical problems on high-speed digital computers. *Proc. Conf. Develop. Soviet Mach. Machines and Devices*, Part 1, 44–52, VINITI Moscow 1956.

Dorodnitsyn, A. A.: *Advances in Aeronautical Sciences*, Vol. II, p. 832–844 m. New York: Pergamon Press 1959.

Dorodnitsyn, A. A.: Advances in Aeronautical Sciences, Vol. 3. New York: Pergamon Press 1960.

Dorodnitsyn, A. A.: Arch. Mech. Stosowanej **14**, 343–357 (1962).

Dwyer, H. A.: AIAA J. **6**, 1336–1342 (1968).

Dwyer, H. A., Sanders, B. R.: Proc. 4th Int. Conf. on Numerical Methods in Fluid Dynamics. *Lecture Notes in Physics* **19**, p. 144–150. Berlin-Heidelberg-New York: Springer 1975.

Emmons, H. W.: NACA TN 932, 1944.

Falkner, V. M., Skan, S. W.: British Aeronautical Research Council R & M 1314, 1930.

Finlayson, B. A.: The Method of Weighted Residuals and Variational Principles. Academic Press, New York (1972).

Fletcher, C. A. J.: AIAA Journal, **13**, 1073–1078 (1975).

Fletcher, C. A. J., Computational Galerkin Methods. *Springer Series in Computational Physics.* Berlin-Heidelberg-New York-Tokyo: Springer (to appear in 1984).

Fletcher, C. A. J., Holt, M.: J. Computational Physics **18**, 154–164 (1975).

Fletcher, C. A. J., Holt, M.: Supersonic Flow over Cones at Large Angles of Attack, J. Fluid Mechanics **74**, p. 561–591, 1976.

Holt, M.: *Symposium Transsonicum* (Ed. K. Oswatitsch), p. 310–324. Berlin-Göttingen-Heidelberg-New York: Springer 1964.

Holt, M. (Ed.): *Basic Developments in Fluid Dynamics, Part I.* New York: Academic Press 1965.

Holt, M.: *Proc. AGARD Conference on Separated Flows*, p. 69–87. Rhode St. Genèse 1966.

Holt, M.: *Proc. XVII Int. Astronautical Congress*, Polish Scientific Publishers, p. 383–401, 1967.

Holt, M.: *Proc. XVIII Int. Astronautical Congress*, Polish Scientific Publishers, p. 385–397, 1968.

Holt, M.: A review of numerical techniques for calculating supercritical airfoil flows. *Symposium Transsonicum II*, Göttingen 1975, 362–368. Berlin-Heidelberg-New York: Springer 1976.

Holt, M., Chan, W. K.: An Integral Method for Unsteady Laminar Boundary Layers. *Symposium on Unsteady Aerodynamics*, Univ. Arizona, March 1975.

Holt, M., Chan, W.-K.: Computing Methods in Applied Sciences and Engineering, 1977 II. 3rd Int. Symp. *Lecture Notes in Physics* **91**, p. 75–89. Berlin-Heidelberg-New York: Springer 1979.

Holt, M., Lu, T. A.: Acta Astronautica **2**, 409–429 (1975).

Holt, M., Masson, B. S.: Proc. 2nd Int. Conf. on Numerical Methods in Fluid Dynamics. *Lecture Notes in Physics* **8** (Ed. M. Holt), p. 207–214. Berlin-Heidelberg-New York: Springer 1971.

Holt, M., Meng, J. C. S.: *Proc. XIX Int. Astronautical Congress.* New York: Pergamon Press 1970; Polish Scientific Publishers 1970.

Holt, M., Meng, J. C. S.: Astronautica Acta **18**, 91–98 (1973).

Holt, M., Modarress, D.: Proc. Roy. Soc. A **353**, 319–347 (1977).

Holt, M., Ndefo, D. E.: J. Comp. Phys., **5**, 463–486 (1970).

Howarth, L.: Proc. Roy. Soc. A **194**, 1036 (1948).

Isaacson, E., Keller, H. B.: *Analysis of Numerical Methods*, p. 199. New York: Wiley 1966.

Kleinstein, G.: AIAA Journal **5**, 1402–1407 (1967).

Krause, E.: AIAA J. **7**, 575–576 (1969).

Krause, E.: Numerical treatment of boundary layer problems. AGARD Lecture Series No. 64. *Advances in Numerical Fluid Dynamics* **4** (1973).

Liu, S. T.: Zh. Vych. Mat. Mat. Fiz. **2**, 666–683 (1962a).

Liu, S. T.: Zh. Vych. Mat. Mat. Fiz. **2**, 868–883 (1962b).

Loos, H. G.: J. Aero. Sci. **22**, 35–40 (1954).

Lubard, S., Helliwell, W. S.: AIAA J. **12**, 965–974 (1974).

Melnik, R. E., Ives, D. C.: Proc. 2nd Int. Conf. on Numerical Methods in Fluid Dynamics, *Lecture Notes in Physics* **8** (Ed. M. Holt), p. 243–251. Berlin-Heidelberg-New York: Springer 1970.

Modarress, D., Holt, M.: Laminar Boundary Layer Separations in Three Dimensions. Proceedings Fifth International Conference on Numerical Methods in Fluid Dynamics, Twente University, Holland, June-July 1976 (to be published at Springer-Verlag).

Moore, F. K.: NACA TN 2279, 1951.

Nielsen, J. N., Goodwin, F. K., Kuhn, G. D.: Review of the method of integral relations applied to viscous interaction problems including separation. *Symposium on Viscous Interaction Phenomena in Supersonic and Hypersonic Flow*, Hypersonic Research Lab., Aeronautical Research Labs., Wright Patterson Air Force Base, 1969.

Pavlovskii, Y. N.: Zh. Vych. Mat. Mat. Fiz. **2**, 884–901 (1962).

Rosenhead, L. (Ed.): *Laminar Boundary Layers*, p. 457–461. Oxford: Oxford University Press 1966.

Schlichting, H.: *Boundary Layer Theory* (translated by J. Kestin), 6th Ed. New York: McGraw-Hill 1968.

Schuh, H.: Z. Flugwiss. **5**, 122–131 (1953).

Sfeir, A. A.: Rev. Roum. Sci. Tech. Mec. Appl. **15**, 1375–1391 (1970).

Sowerby, L.: J. Fluid Mechanics **22**, 587–598 (1955).

Spalding, D. B.: ASME J. Appl. Mech. **28**, 455–457 (1961).

Stewartson, K.: Proc. Roy. Soc. A **200**, 84–100 (1949).

Stewartson, K., Williams, P. G.: Proc. Roy. Soc. A **312**, 81–206 (1969).

Tracy, R. R.: Calif. Inst. Tech. Memo. No. 69, 1963.

Tai, T. C.: *AIAA 13th Aerospace Sciences Meeting*, Preprint No. 75–78, Pasadena 1975.

Wang, K. C.: J. Fluid Mechanics **43**, 187–209 (1970).

Yang, R.-J., Holt, M.: Boundary Layer Control by Means of Strong Injection (to appear in J. Appl. Mech.).

Yeung, W. C.-C., Holt, M.: ZAMM **62**, 391–399 (1982).

Yeung, W.-S.: J. Appl. Mech. **47**, 697–702 (1980).

Yeung, W.-S., Holt, M.: ASME J. Appl. Mech. **50**, 239–243 (1983).

Yeung, W.-S., Yang, R.-J.: ASME J. Appl. Mech. **48**, 701–706 (1981).

CHAPTER 6

Telenin's Method and the Method of Lines

6.1 Introduction

The Method of Integral Relations, described in Chapt. 5, is one technique for reducing the amount of finite difference computation in the numerical solution of partial differential equations. The reduction is achieved by integrating the governing equations in one or more coordinate directions and representing unknowns in integrands by polynomials or trigonometrical expansions in the respective coordinates. We then solve ordinary or partial differential equations (of lower order) for the unknown coefficients in these expansions.

Telenin's method and the Method of Lines are alternative techniques for eliminating finite difference calculations in selected coordinate directions. Polynomials or other smooth interpolating functions are again used to represent the variations of the unknowns in one or more of the independent variables. The governing partial differential equations are now solved in their original form without first recasting them in integral form and the interpolating functions are simply used to calculate derivatives transverse to the direction (or directions) of finite difference calculations. These methods avoid the algebra required in the Method of Integral Relations but are less easily applied to problems in which discontinuities, such as shock waves, arise in the interior of the range of integration.

Telenin's method was first applied to the blunt body problem (GILINSKII and TELENIN, 1964) and provided rapid solutions of this both for symmetrical flow and for flow at angles of attack (TINYAKOV, 1965). It was subsequently applied to calculate supersonic flow past circular cones at moderate angles of attack (HOLT and NDEFO, 1970). It also forms part of the solution for flow past circular cones at high angles of attack, when the shock is still attached but the cross flow is partially supersonic (FLETCHER, 1975). The method has been used to calculate supercritical flow past ellipses, in both the symmetrical case (GROSS and HOLT, 1975) and in the unsymmetrical case (LI and HOLT, 1982). It has also been applied to transonic flow in symmetrical channels and nozzles (KLOPFER and HOLT, 1975). The method has not been exploited sufficiently in viscous flow problems; the only application known to the writer is to laminar base flow on a re-entry shuttle vehicle (MENG, 1973).

The Method of Lines has been applied to most of the problems just listed and is similar to Telenin's method in representing transverse derivatives by analytical formulae rather than by finite difference approximations. Its principal difference from Telenin's method is in the use of local interpolation over 3 to 5 points rather than interpolation extending over a whole range in one coordinate direction.

In common with the Method of Integral Relations both the methods described in this chapter are applicable to partial differential equations of elliptic, mixed elliptic-hyperbolic, or parabolic type.

The chapter begins with the formulation of Telenin's method for a Dirichlet problem associated with Laplace's equation over a rectangle (transforming this to a Cauchy problem to be solved iteratively). The method is then applied to a model linear mixed elliptic-hyperbolic equation. This basic formulation is derived from GILINSKII, TELENIN and TINYAKOV (1964). The application of the mixed model by TELENIN and his co-authors to the blunt body problem (both symmetrical and unsymmetrical cases) is then described. Later sections deal with the applications carried out here to the supersonic cone problem, at moderate and large angles of yaw.

The Method of Lines is next introduced and compared with Telenin's method. Subsequent sections deal with the applications of both methods to the supersonic cone problem (over a whole range of angles of yaw), the problem of supercritical flow past an ellipse, and transonic nozzle flow.

6.2 Solution of Laplace's Equation by Telenin's Method

We first consider the solution of a boundary value problem, in which the unknown satisfies Laplace's equation in the interior of a closed domain with values of the unknown specified on the enclosing boundary. This is the well known Dirichlet problem. There are two principal ways to solve the problem by a finite difference method. Firstly we may iterate successively on estimates of the unknown at interior network points, that is, use a relaxation method. Secondly we may introduce time dependency deliberately into the problem, converting the governing equation into a two dimensional wave equation or diffusion equation, solve the problem as an initial value problem by a time marching technique, and continue until a steady state solution is attained.

In contrast to the Dirichlet problem, the Cauchy problem in two dimensions is much easier to solve. In this the unknown satisfies the wave equation subject to specified values of the unknowns and its normal derivative on an open segment. Its solution by finite differences simply requires advance along characteristics proceeding away from points on the initial segment into its domain of dependence.

Telenin's method, as applied to Laplace's equation, consists essentially in solving a Dirichlet problem as a Cauchy problem. Taking the closed boundary as a rectangle, Cauchy data (i. e., values of the unknown and its normal derivative) are prescribed on one side and used to integrate Laplace's equation step by step up to the opposite side. The calculated values of the unknown on this side are then compared with the prescribed values. If they disagree the normal derivatives on the first side are changed and the integration repeated. This is continued until all the prescribed boundary values are satisfied.

As pointed out by GILINSKII, TELENIN and TINYAKOV (1964), the general Cauchy problem for an elliptic problem is incorrectly posed since solutions in a region adjacent to an open initial segment are unstable with respect to data prescribed on that segment—this behavior can be called Hadamard instability (PETROVSKII, 1954). If, however, the elliptic problem is restricted it can be posed correctly; in particular it is correct if the initial data and the corresponding solutions are analytic functions.

Referring to Fig. 6.1 GILINSKII, TELENIN and TINYAKOV consider the Cauchy problem associated with Laplace's equation

$$\frac{\partial^2 \phi}{\partial x^2} + \frac{\partial^2 \phi}{\partial y^2} = 0 \tag{6.2.1}$$

in the rectangle shown, $-1 \le x \le 1$, $0 \le y \le b$, with initial data

$$\phi(x,0) = \sigma(x), \qquad \phi_y(x,0) = \tau(x) \tag{6.2.2}$$

It is assumed that, in a slightly larger rectangle enclosing that shown, ϕ is a harmonic function with $|\text{grad } \phi| \le M$ (M finite) while, on AB, $|\sigma(x)| \le (1/2)m$, $|\tau(x)| \le (1/2)m$ (m finite).

If $\sigma(x)$ and $\tau(x)$ are analytic functions of the real variable x then we can define analytic continuations of these functions denoted by $\sigma(\zeta)$ and $\tau(\zeta)$ into a complex plane $\zeta = x + i\eta$. If we denote $T(x) = \int_0^x \tau(\xi) d\xi$ and its analytic continuation by $T(\zeta)$, then the solution of the Cauchy problem (6.2.1), (6.2.2) can be shown to be

$$\phi(x,y) = \text{Re}\left[\sigma(z) - i\,T(z)\right] \tag{6.2.3}$$

$$= \text{Re}\left[\sigma(z) - i\int_0^z \tau(\xi) d\xi\right] \tag{6.2.3a}$$

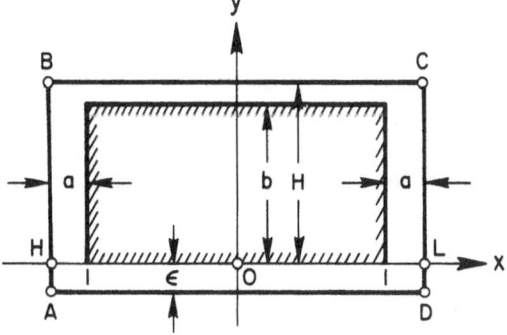

Fig. 6.1 Telenin's method applied to Dirichlet problem in a rectangle

Now divide the initial segment $(-1,1)$ into $2n$ intervals at $2n+1$ nodes and approximate the functions $\sigma(x)$, $\tau(x)$ by Lagrange polynomials with coefficients determined from values of $\sigma(x)$ and $\tau(x)$ at nodal points x_j ($j=0,...,2n$). Then in the complex region ζ we write

$$\sigma(\zeta) \simeq \sigma_n^*(\zeta) = \sum_{j=0}^{2n} a_j \zeta^j$$

$$\tau(\zeta) \simeq \tau_n^*(\zeta) = \sum_{j=0}^{2n} b_j \zeta^j \tag{6.2.4}$$

where a_j, b_j are linear combinations of the nodal values $\sigma(x_j)$ and $\tau(x_j)$. If $\sigma(x)$ and $\tau(x)$ are entire functions then each sequence of polynomials (6.2.4) converges smoothly in any finite region of the complex domain.

If we substitute (6.2.4) into (6.2.3a) we obtain the following approximate solution to the Cauchy problem,

$$\phi(x,y) \simeq \phi_n^*(x,y) = \text{Re} \left\{ \sum_{j=0}^{2n} a_j z^j - i \sum_{j=0}^{2n} \frac{b_j}{j+1} z^{j+1} \right\} \tag{6.2.5}$$

satisfying Laplace's equation and satisfying the initial data

$$\phi_n^*(x,0) = \sigma_n^*(x), \qquad \phi_{ny}^*(x,0) = \tau_n^*(x).$$

Within and on the rectangle $ABCD$ it can be shown that the sequence ϕ_n^* converges smoothly to the exact solution $\phi(x,y)$ of the Cauchy problem (6.2.1), (6.2.2).

The modified Cauchy problem (that is, the problem with initial data represented by Lagrange polynomials) is now solved by a numerical technique. For this purpose the unknown ϕ is approximated by a Lagrange polynomial in x

$$\phi(x,y) \simeq \phi_n^*(x,y) = \sum_{j=0}^{2n} \phi_{jn}^\circ(y) x^j \tag{6.2.6}$$

where $\phi_{jn}^\circ(y)$ is a linear combination of the values $\phi_{jn}^*(y) = \phi_n^*(x_j,y)$ of ϕ evaluated at nodal points (x_j,y), $j=0,...,2n$.

Substituting (6.2. in (6.2.1) and noting that the latter is satisfied identically for all x in $-1 \leq x \leq 1$ we obtain the following system of ordinary differential equations

$$\frac{d^2 \phi_{kn}^*}{dy^2} + \sum_{j=2}^{2n} j(j-1) \phi_{jn}^\circ(y) x_k^{j-2} = 0 \quad (k=0,1,2,...,2n) \tag{6.2.7}$$

with boundary conditions (at $y=0$)

$$\phi_{kn}^{*}(0) = \sigma(x_k), \qquad \phi_{kny}^{*}(0) = \tau(x_k) \tag{6.2.8}$$

It can be verified that any solution of (6.2.1) which is a polynomial in x of degree $2n$, for fixed y, satisfies (6.2.7) on each ray $x=x_k$. It follows that, using (6.2.5)

$$\phi_{kn}^{*}(y) = \phi_{n}^{*}(x_k, y) = \mathrm{Re} \left\{ \sum_{j=0}^{2n} a_j z_k^j - \mathrm{i} \sum_{j=0}^{2n} \frac{b_j}{j+1} z_k^{j+1} \right\} \qquad (z_k = x_k + \mathrm{i}y) \tag{6.2.9}$$

is the solution of the approximate system (6.2.7) with boundary values (6.2.8).

The numerical solution derived from the system (6.2.7) with boundary values (6.2.8) therefore converges to the exact solution of (6.2.1), (6.2.2) as $n \to \infty$.

In the model problem just considered solutions are found algebraically both in the exact and approximate formulations. In a more general elliptic problem, when the governing equation corresponding to (6.2.1) is non-linear, the equations corresponding to system (6.2.7) can only be solved numerically and the influence of errors must be considered. GILINSKII, TELENIN and TINYAKOV point out that errors are of three types, two resulting from approximations in the x and y directions $\delta^x \phi$, $\delta^y \phi$, and the third from round-off, $\delta^r \phi$. It is shown that the $\delta^x \phi$ errors decrease exponentially with increase in number of nodes. The $\delta^y \phi$ and $\delta^r \phi$ errors, on the other hand, arising from actual finite difference calculations, increase exponentially with the number of nodes. The $\delta^y \phi$ errors can be controlled, and kept within the order of the $\delta^x \phi$ errors by using a stable difference scheme. On the other hand, the round-off errors depend on the number of digits carried by the computer used and are not subject to reduction by change in difference scheme. To avoid inaccuracies in Telenin schemes it is therefore desirable to keep round-off errors to a minimum (by use of high precision arithmetic) and to use the minimum number of nodes possible in representing the unknowns, both on the initial line and at each subsequent stage of the integration. In many of the applications to be described remarkably few nodes suffice to represent the physical behavior investigated, and in consequence the method is frequently effective.

6.3 Solution of a Model Mixed Type Equation by Telenin's Method

The blunt body problem, namely, the calculation of uniform supersonic flow past a cylindrical body or body of revolution with a rounded nose, is of mixed elliptic-hyperbolic type. It is illustrated in Fig. 6.2, in the case of symmetrical plane flow with ED representing the given body contour, ABG a detached shock wave (the shape of which is determined as part of the solution), a sonic line DB, and a limiting characteristic DK. We consider a high Mach number flow where the sonic line is bent upstream.

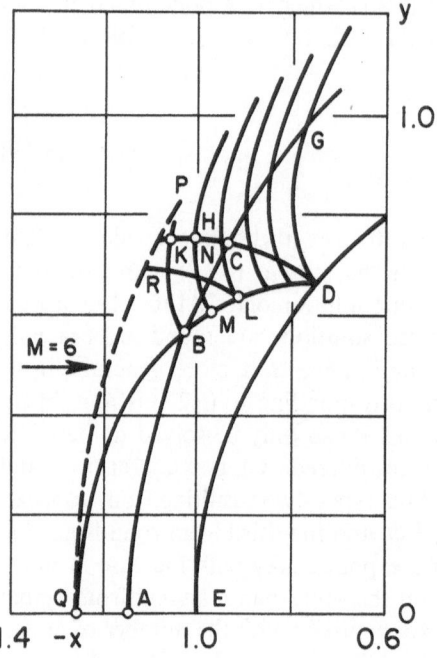

Fig. 6.2   Boundaries in blunt body problem

It can be shown that the governing equations of motion (which are elliptic in $AEDB$ and hyperbolic in DKB), with solid boundary conditions satisfied on ED, have a unique solution in the region $EDKBAE$, simultaneously fixing the positions of the boundaries ABC, DK and DB. (See GILINSKII, TELENIN and TINYAKOV, 1964).

GILINSKII, TELENIN and TINYAKOV consider a model of this problem, governed by the linear equation

$$\frac{\partial^2 \phi}{\partial x^2} + (1 - y^2)\frac{\partial^2 \phi}{\partial y^2} = 0 \qquad (6.3.1)$$

applied to the region shown in Fig. 6.3. Here x represents the dimensionless polar coordinate and y the angular coordinate in the blunt body problem. The stagnation point on the body is at $x = 1/2$. The y axis corresponds to the shock wave and EDF the body boundary. Evidently the sonic line corresponds to $y = 1$ (where (6.3.1) changes type) and DK is a characteristic of the first family through $x = 1/2$. The characteristic curves of (6.3.1) are

$$y = \cosh(x - c) \qquad (6.3.2)$$

which are tangential to $y=1$ at $x=c$. The branch $x<c$ belongs to the first family and the branch $x>c$ to the second family.

We seek solutions of (6.3.1) which satisfy certain boundary conditions on the open boundary $KBAED$. If the solutions are to have physical meaning they must be continuous and their derivatives must be everywhere finite.

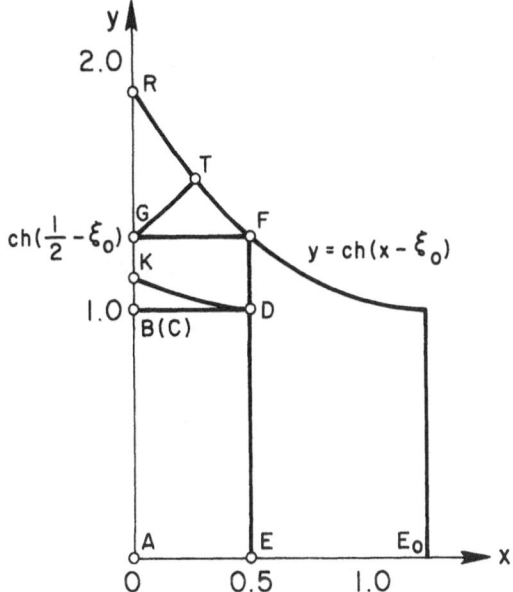

Fig. 6.3 Telenin's method applied to model blunt body problem

Since (6.3.1) is linear, we may seek solutions by separation of variables of the form

$$\phi = F(x)f(y).$$

Then from (6.3.1)

$$F_k''(x) - (\pm\lambda_k^2)F_k(x) = 0 \tag{6.3.3}$$

$$f_k''(y) + \frac{(\pm\lambda_k^2)f_k(y)}{(1-y^2)} = 0 \tag{6.3.4}$$

where λ_k^2 are real positive eigen values. For continuous solutions with bounded derivatives we must satisfy the conditions

$$\phi_{xx} = 0, \qquad y = 1$$
$$\phi_y = 0, \qquad y = 0.$$

Hence, we seek solutions of (6.3.4) satisfying

$$f_k(1) = 0, \qquad f_k'(0) = 0 \tag{6.3.5}$$

To find this solution we solve the problem

$$f_k(0) = \varepsilon_k, \qquad f_k'(0) = 0 \tag{6.3.6}$$

and iterate on ε_k until condition (6.3.5) is satisfied. If we choose the minus sign in the coefficient $\pm \lambda_k^2$ in (6.3.4), then

$$f_k''(y) = \frac{\lambda_k^2 f_k(y)}{1 - y^2}$$

so that f_k'' and f_k have the same sign in $0 \le y \le 1$. It follows that $|f_k| \ge \varepsilon_k$ for all points in the interval and the only solution satisfying (6.3.5) is the trivial one

$$f_k(y) \equiv 0 \quad \text{everywhere.}$$

To obtain non-trivial solutions we must therefore take the positive sign in $\pm \lambda_k^2$. Putting $f_k = (1 - y^2) g_k$ in (6.3.4) we find that g_k satisfies the equation

$$(1 - y^2) g_k'' - 4 y g_k' + (\lambda_k^2 - 1) g_k = 0. \tag{6.3.7}$$

It can be shown that the eigen values satisfying (6.3.5) are

$$\lambda_k^2 = k(k+1) \tag{6.3.8}$$

and (6.3.7) then defines the derivatives of the Legendre function

$$g_k = P_k'(y).$$

The non-trivial solutions of (6.3.1) are therefore of the form

$$\phi_k(x, y) = (1 - y^2) P_k'(y) \left[A_k \sinh \lambda_k x + B_k \cosh \lambda_k x \right]. \tag{6.3.9}$$

Now consider the solution of the model equation (6.3.1) satisfying the boundary conditions

$$\phi(0, y) = \sigma_1(y), \qquad \phi(\tfrac{1}{2}, y) = \sigma_2(y), \qquad \phi_y(x, 0) = 0. \tag{6.3.10}$$

Suppose firstly that σ_1, σ_2 are prescribed only on $0 \le y \le 1$. Then we can show that the solution satisfying (6.3.1), (6.3.10), and the condition of smooth transition through $y = 1$ must be unique.

Suppose that ϕ_1, ϕ_2 are two such solutions. Let

$$u = \phi_1 - \phi_2 .$$

Then u satisfies (6.3.1) and the boundary conditions

$$u(0,y) = 0, \qquad u(\tfrac{1}{2},y) = 0, \qquad u_y(x,0) = 0. \qquad (6.3.11)$$

To find this solution, simultaneously to have regular behavior at $y=1$, we solve the Cauchy problem

$$u(x,0) = \varepsilon(x), \qquad u_y = 0. \qquad (6.3.12)$$

Eq. (6.3.11) shows that the eigen values λ_k^2 are given by

$$\lambda_k = 2k\pi$$

and the minus sign must be taken in (6.3.3) to give

$$F_k(x) = D_k \sin 2k\pi x .$$

(Note $F_k = 0$, $x=0$, $x=1/2$.)

We therefore write

$$\varepsilon(x) = \sum_{k=1}^{\infty} \varepsilon_k \sin 2k\pi x .$$

Substituting in (6.3.4) we must again find the solution for f_k satisfying $f_k(0) = \varepsilon_k$, $f_k'(0) = 0$. As before the null solution is the only solution satisfying $f_k(1) = 0$. Hence $\varepsilon_k = 0$ for all k so that

$$u(x,y) \equiv 0, \qquad 0 \le y \le 1, \qquad 0 \le x \le \tfrac{1}{2},$$

proving uniqueness of the solution of (6.3.1) and (6.3.10) in $0 \le y \le 1$.

Since the derivatives of ϕ are finite across $y=1$ the solution can be continued into the hyperbolic region $y>1$. If the boundary values σ_1, σ_2 are given on EF and AG, the solution is uniquely determined in the rectangle $AEFGA$.

In solving the boundary value problem defined by (6.3.1) and (6.3.10) we may use two formulations of a Cauchy problem. In the first we prescribe $\phi(x,0) = b(x)$ in addition to $\phi_y(x,0) = 0$ and iterate on $b(x)$ until the condition $\phi_{xx} = 0$ at $y=1$ is satisfied for all $x, 0 \le x \le 1/2$. Now, if b differs by any amount, $\varepsilon(x)$, no matter how small, from its correct (and unique) value at one or more points in $0 \le x \le 1/2$ the solution of the Cauchy problem will have an infinite second derivative ϕ_{yy} at $y=1$. In other words, if we solve the problem defined by (6.3.1) and (6.3.10) by using interpolation functions in the x direction and reducing the problem to one governed by a system of ordinary differential equations in the y direction, then these equations will all have saddle points at $y=1$ and the task of finding regular solutions passing through the saddle points is compounded as the refinement of the x interpolation is increased. This difficulty has already been brought out in the application of the Method of Integral Relations to the Blunt Body Problem (see Chapt. 5).

In the second formulation of a Cauchy problem we prescribe the following initial data on $x=0$

$$\phi(0,y) = \sigma_1(y), \qquad \phi_x(0,y) = \tau_1(y)$$

and iterate on the function $\tau_1(y)$ until the condition $\phi(1/2,y)=\sigma_2(y)$ (of conditions (6.3.10)) is satisfied.

Now even powers of y, y^{2k}, can be expressed in terms of the functions $P'_r(y)(1-y^2)$ $r=1,\ldots,2k-1$, appearing in (6.3.9). Therefore, if we solve this Cauchy problem approximately using even polynomial representations for $\sigma_1(y)$ and $\tau_1(y)$ and combinations (6.3.9) for ϕ in $x\geq0$, each component of the solution will automatically satisfy the symmetry condition $\phi_y=0$, $y=0$ and the regularity condition $\phi_{xx}=0$, $y=1$. Provided a convergent iteration process can be found to determine $\tau_1(y)$ to give $\phi(1/2,y)=\sigma_2(y)$ the second formulation of the Cauchy problem will therefore lead to a unique and bounded solution of the boundary value problem defined by (6.3.1) and (6.3.10).

To solve this problem approximately we consider the second Cauchy problem. We divide the y interval on which σ_1 and σ_2 are given into m equal intervals $y=y_j$ $(j=0,\ldots,m)$ and represent ϕ by the even powered Lagrange polynomial

$$\phi(x,y) = \phi_m^*(x,y) = \sum_{j=0}^{m} \phi_{jm}^\circ(x)y^{2j}. \tag{6.3.13}$$

The coefficients ϕ_{jm}° are linear combinations of the values of ϕ, $\phi_{jm}^* = \phi_m^*(x,y_j)$ at nodal points $y=y_j$ and, by substitution in (6.3.1) and equating corresponding powers of y, evidently satisfy the system of ordinary differential equations

$$\frac{d^2\phi_{km}^*}{dx^2} + (1-y^2)\sum_{j=1}^{m} 2j(2j-1)\phi_{jm}^\circ(x)y^{2j-2} = 0 \quad (k=0,1,\ldots,m). \tag{6.3.14}$$

As an example, consider the boundary value problem (6.3.1) and (6.3.10) with the conditions

$$\phi(0,y) = \sigma_1(y) = a_0 + a_2 y^2 + a_4 y^4$$
$$\phi(\tfrac{1}{2},y) = \sigma_2(y) = b_0 + b_2 y^2 + b_4 y^4 \tag{6.3.15}$$
$$\phi_y(x,0) = 0$$

Then if we use three rays in $y>0$ ($y=0$, $y=y_1$, $y=y_2$) and reflected rays in $y=0$, the solution (6.3.13), with coefficients found by solving (6.3.14), is

$$\phi(x,y) = M + N x + P_1'(y)(y^2-1)\{C_1 e^{\sqrt{2}x}+C_2 e^{-\sqrt{2}x}\}$$
$$+\tfrac{2}{15} P_3'(y)(y^2-1)\{A_1 e^{\sqrt{12}x}+A_2 e^{-\sqrt{12}x}\} \qquad (6.3.16)$$

where M, N, C_1, C_2, A_1, A_2 are given in terms of $a_0, a_2, a_4, b_0, b_2, b_4$. Eq. (6.3.16) is also the solution of the exact problem defined by (6.3.1) and (6.3.10), since the order of the approximation corresponds to the degree of the polynomial defining the boundary data ($m=2$ in (6.3.13)). In other words, if the number of rays used in the polynomial (6.3.13) and in (6.3.14) is N, and $2n$ is the degree of polynomial in (6.3.15), then $N=2m+1=2n+1$ and $m=n$. If $m<n$ and $N<2m+1$ the solution (6.3.13) is only an approximation to the exact solution of (6.3.1) and (6.3.10). It can be shown that the error in the approximate solution is of the order $|d\phi^*/dh|$ where h is the position of the extreme ray.

As with elliptic equations, when we apply the approximation scheme to the non-linear mixed type equations in fluid dynamics, equations corresponding to (6.3.14) must be integrated numerically and round-off errors must be carefully controlled by using the lowest order of approximation consistent with a faithful representation of physical phenomena in the problem.

6.4 Application of Telenin's Method to the·Symmetrical Blunt Body Problem

The axially symmetrical blunt body problem is illustrated in Fig. 6.4. EF defines the body contour and AG the position of the detached shock wave ahead of the body. The equations of motion are referred to polar coordinates (r, θ) based on an origin 0 on the axis of symmetry inside the body. The radial coordinate is then replaced by a dimensionless coordinate ξ, defined by

$$\xi = \frac{r - r_B(\theta)}{\varepsilon(\theta)} \qquad (6.4.1)$$

where

$$\varepsilon = r_S(\theta) - r_B(\theta) \qquad (6.4.2)$$

is the distance from the body to the shock (an unknown), r_B is the body radius and r_S the shock radius. Note that $\xi=0$ on the body and $\xi=1$ on the shock.

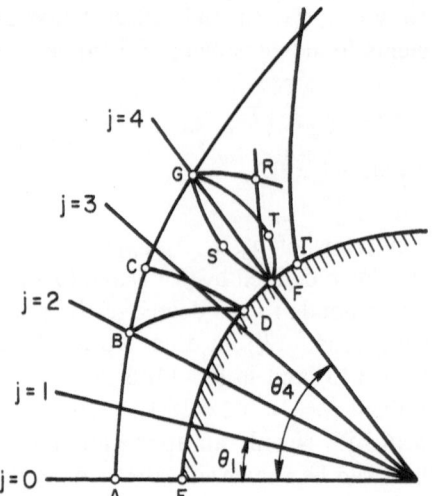

Fig. 6.4 Sector boundaries in blunt body problem

The physical variables are taken as (u, v), the velocity components in the (ξ, θ) directions, the pressure p (expressed as a ratio of $\varrho_\infty q_{max}^2$), the density ϱ (referred to ϱ_∞), and the stream function ψ (referred to $\varrho_\infty q_{max} L$ where L is a typical length). The stream function is, of course, a universal function of entropy. The equations are as follows

$$\left[\gamma p^{(\gamma-1)/\gamma}\phi(\psi)-u^2\right]\frac{\partial u}{\partial r} - uv\frac{\partial v}{\partial r} + \frac{1}{r}\left\{\left[\gamma p^{(\gamma-1)/\gamma}\phi(\psi)-v^2\right]\frac{\partial v}{\partial r} - uv\frac{\partial u}{\partial \theta}\right\}$$

$$+ \frac{\gamma\phi(\psi)}{r}p^{(\gamma-1)/\gamma}(2u+v\,ctg\,\theta)=0$$

$$\frac{\partial v}{\partial r} = \frac{1}{r}\frac{\partial u}{\partial \theta} - \frac{v}{r} - \frac{\gamma}{\gamma-1}r\sin\theta\frac{p}{\phi(\psi)}\frac{d\phi(\psi)}{d\psi} \qquad (6.4.3)$$

$$\frac{\partial p}{\partial r} = \frac{1}{\phi(\psi)}p^{1/\gamma}\left(\frac{v^2}{r} - \frac{v}{r}\frac{\partial u}{\partial \theta} - u\frac{\partial u}{\partial r}\right)$$

$$\frac{\partial \psi}{\partial r} = \frac{1}{\phi(\psi)}p^{1/\gamma}rv\sin\theta, \qquad \left(\phi(\psi)=\frac{1}{\varrho}p^{1/\gamma}\right)$$

The boundary and shock conditions are

$$u = vr_B'/r_B \qquad (6.4.4)$$

$$u = v_y\sin\theta - v_x\cos\theta, \qquad v = v_y\cos\theta + v_x\sin\theta$$

$$v_x = \frac{V_\infty}{V_{max}}\left[1 - \frac{2}{\gamma+1}\sin^2\sigma\left(1 - \frac{1}{M_\infty^2\sin^2\sigma}\right)\right], \qquad v_y = \left(\frac{V_\infty}{V_{max}} - v_x\right)ctg\,\sigma$$

$$p = \frac{2}{\gamma+1}\left[\left(\frac{V_\infty}{V_{max}}\right)^2 \sin^2\sigma - \left(1 - \frac{V_\infty^2}{V_{max}^2}\right)\frac{(\gamma-1)^2}{4\gamma}\right] \tag{6.4.5}$$

$$\varrho = \frac{\gamma+1}{\gamma-1}\frac{V_\infty^2}{V_{max}^2}\frac{\sin^2\sigma}{1-(V_\infty^2/V_{max}^2)\cos^2\sigma}$$

$$\psi = \frac{1}{2}\frac{V_\infty}{V_{max}}r_S^2\sin^2\theta,$$

where v_x, v_y are velocity components parallel and normal to the undisturbed stream directions.

To solve the governing equations, together with the surface boundary condition, shock conditions and symmetry conditions, by Telenin's method, we divide the disturbed region $AGFEA$ into a number of equiangular sectors. The upper boundary FG is chosen to be entirely within the supersonic region, shortly downstream of the sonic lin (GILINSKII et al. note that the region of validity of the mixed elliptic-hyperbolic problem may be somewhat larger than the region corresponding to Tricomi's problem since analyticity conditions are imposed in Telenin's method.) The sectors are reflected in the axis of symmetry.

We then represent the unknowns by even or odd powered polynomials in θ

$$u \simeq \sum_{j=0}^{m} u_j^\circ(\xi)\theta^{2j}$$

$$v \simeq \sum_{j=0}^{m} v_j^\circ(\xi)\theta^{2j+1} \tag{6.4.6}$$

$$r_S \simeq \sum_{j=0}^{m} r_j^\circ\theta^{2j}$$

where $2m+1$ is the total number of rays (including those in the lower half plane). The θ derivatives of the unknowns are found by differentiating the polynomials in (6.4.6).

The equations of motion are satisfied on each ray and are expressed as relations between first order derivatives in u, v, p and ψ. [The equation of state reduces to an algebraic relation between p, ϱ and ψ.] In the first of these equations the ξ derivatives of p, ψ and v are eliminated to yield an equation for u only. We then obtain the following system of ordinary differential equations for u, v, p, ψ and ϱ on the k-th ray.

$$\frac{du_k}{d\xi} = \frac{\varepsilon_k(r_{Bk}+\xi\varepsilon_k)}{\Delta_k}\left\{c_k\left[u_k v_k(r_{Bk}+\xi\varepsilon_k)^2 + b_k(r_{Bk}'+\xi\varepsilon_k')\cdot(r_{Bk}+\xi\varepsilon_k)\right]\right.$$

$$\left. - d_k(2u_k+v_k\,\mathrm{ctg}\,\theta_k) - b_k v_k' + (u_k'-v_k')\frac{b_k(r_{Bk}'+\xi\varepsilon_k')}{r_{Bk}+\xi\varepsilon_k} + (2u_k'-v_k)u_k v_k\right\}$$

$$\frac{dv_k}{d\xi} = \varepsilon_k \left[\frac{1}{r_{Bk} + \xi\varepsilon_k} \left(u_k' - \frac{r_{Bk}' + \xi\varepsilon_k'}{\varepsilon_k} \frac{du_k}{d\xi} \right) - \frac{v_k}{r_{Bk} + \xi\varepsilon_k} + c_k(r_{Bk} + \xi\varepsilon_k) \right] \quad (6.4.7)$$

$$\frac{dp_k}{d\xi} = \frac{\varepsilon_k p^{1/\gamma}}{\phi(\psi_k)} \left\{ \frac{v_k^2}{r_{Bk} + \xi\varepsilon_k} - \frac{v_k}{r_{Bk} + \xi\varepsilon_k} \left[u_k' - \frac{r_{Bk}' + \xi\varepsilon_k'}{\varepsilon_k} \frac{du_k}{d\xi} \right] - \frac{u_k}{\varepsilon_k} \frac{du_k}{d\xi} \right\}$$

$$\frac{d\psi_k}{d\xi} = \varepsilon_k(r_{Bk} + \xi\varepsilon_k) \frac{1}{\phi(\psi_k)} p_k^{1/\gamma} v_k \sin\theta_k$$

$$\varrho_k = \left[\frac{p_k^{1/\gamma}}{\phi(\psi_k)} \right] \quad (k = 0, 1, \dots, m)$$

where

$$a_k = \gamma p_k^{(\gamma-1)/\gamma} \phi(\psi_k) - u_k^2, \qquad\qquad b_k = \gamma p_k^{(\gamma-1)/\gamma} \phi(\psi_k) - v_k^2$$

$$c_k = -\sin\theta_k \frac{\gamma}{\gamma-1} \frac{p_k}{\phi(\psi_k)} \frac{d\phi(\psi)}{d\psi}\Bigg|_{\psi=\psi_k}, \qquad d_k = \gamma p_k^{(\gamma-1)/\gamma} \phi(\psi_k) \quad (6.4.8)$$

$$\Delta_k = a_k(r_{Bk} + \xi\varepsilon_k)^2 + 2(r_{Bk}' + \xi\varepsilon_k')(r_{Bk} + \xi\varepsilon_k)u_k v_k + b_k(r_{Bk}' + \xi\varepsilon_k')^2.$$

The θ derivatives on the right of (6.4.7) are determined from (6.4.6). The unknown shock detachment distance is determined from the geometrical relation

$$\frac{d\varepsilon_k}{d\theta} = -(r_{Bk} + \varepsilon_k)\cot(\sigma_k + \theta_k) - \frac{dr_{Bk}}{d\theta} \quad (6.4.9)$$

where σ_k is the shock angle (measured from the undisturbed stream direction).

The procedure in carrying out Telenin's method is to start at the shock $\xi = 1$ and estimate values of r_S at the nodal points on the shock. We then use (6.4.6) to fit a shock shape, determine the shock angle distributions and calculate the values of all unknowns behind the shock from the shock relations. We fit these into (6.4.6) to determine all the θ derivatives at shock nodal points. In this way we determine all the terms on the right of (6.4.7) and hence can start the step-by-step integration of these equations in the ξ direction (moving towards $\xi = 0$). At the end of the first ξ step we then know values of all the dependent variables on each ray. We use these to determine the coefficients in the expansions (6.4.6), recalculate the θ derivatives and perform the next integration step in (6.4.7). The integration is continued up to the body $\xi = 0$ and a test is made to determine if the condition $q_{normal} = 0$ is satisfied at each body nodal point. If the conditions are not satisfied the whole process is repeated, starting from revised estimates of r_S at shock nodal points. The changes in r_S can be related to the errors in q_{normal} after each cycle of integration.

The solution satisfying $q_{normal} = 0$ is unique and automatically satisfies smooth transition across the sonic line, by analogy with results for the model problem.

TELENIN and his collaborators successfully applied the method to a wide variety of body shapes, including ellipses, Cassini ovals (some of which have concave sections near the stagnation point), bodies with discontinuous curvature points and bodies with sharp corners (for the latter the sonic point coincides with the sharp shoulder).

Figure 6.5 shows the shapes of shock wave and sonic line for flow past an oblate ellipsoid (axis ratio 2:1) at four free stream Mach numbers. Figure 6.6

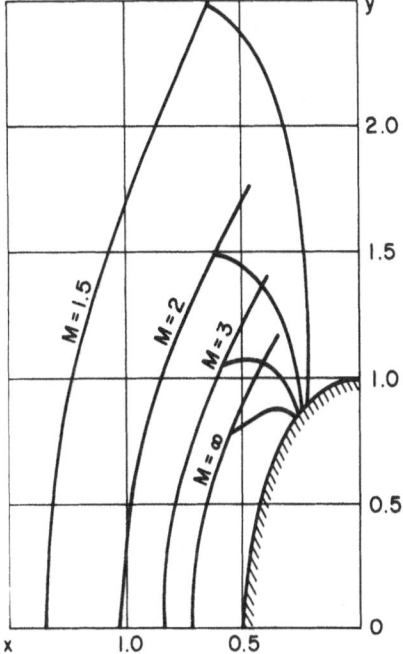

Fig. 6.5 Supersonic flow past an oblate ellipsoid

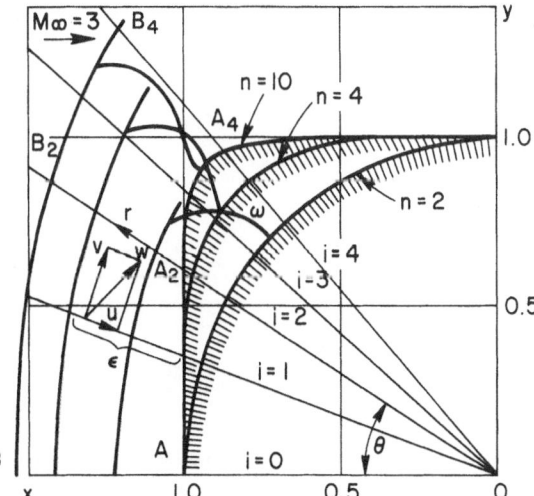

Fig. 6.6 Supersonic flow past bodies $x^n + y^n = 1$

illustrates flow at $M_\infty = 3$ past a family of bodies with Cartesian equation $x^n + y^n = 1$, $n = 2, 4, 10$. For $n > 2$ the nose radius of curvature is infinite and as $n \to \infty$ the shape approaches that of a flat-headed cylinder. The flow patterns generated by Cassini ovals, with equation

$$(x^2 + y^2)^2 + 2c^2(x^2 - y^2) = a^4 - c^4$$

are given in Fig. 6.7 for $M_\infty = 3$. The maximum height of the body $D/2 = (a^2 + c^2)^{1/2}$ is taken as unit distance and ovals shown correspond to values of c of 0.6, 0.678, 0.702. The latter two shapes have reentrant sections near the stagnation point but these have remarkably little effect on the shock shapes. The final shape is a circular segment joined to a straight section at a sharp shoulder shown in Fig. 6.8. The equations of motion have a singular

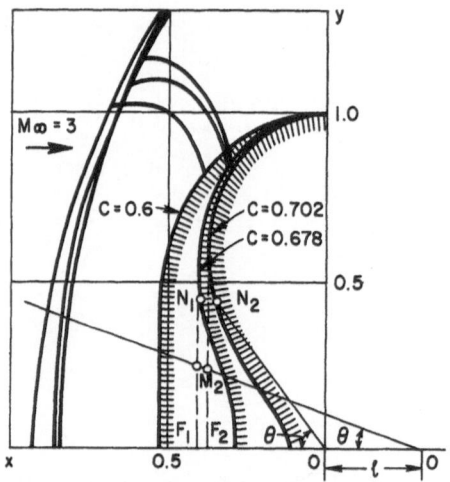

Fig. 6.7   Supersonic flow past Cassini ovals

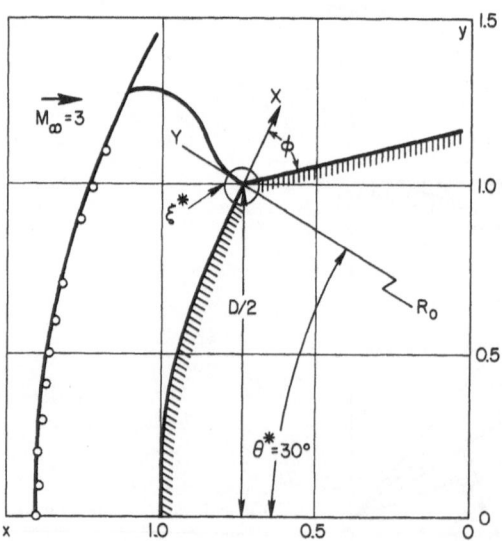

Fig. 6.8   Supersonic flow past bodies with sharp shoulders

point at the shoulder where the flow becomes sonic on the body and expands through a centered expansion round the corner. In applying Telenin's method the corner is surrounded by a small circle within which the flow is represented by the similarity solutions developed by FRIEDMAN (1962). The curvature of the sonic line changes sign since it is normal to the circular arc section at the shoulder. The open circles are experimentally measured shock points obtained (but not published) by Yu. Ya. KARPESKII. The many blunt body calculations are discussed at length in GILINSKII and TELENIN (1964).

6.5 Extension to Unsymmetrical Blunt Body Flows

TINYAKOV (1965) extended the Telenin method to apply to flow past bodies without axial symmetry and illustrates this with calculations of flow past ellipsoids of revolution at angles of attack.

The flow field for a general body is referred to spherical coordinates r, θ, ψ shown in Fig. 6.9 (it is assumed that the body and flow field have a plane of symmetry).

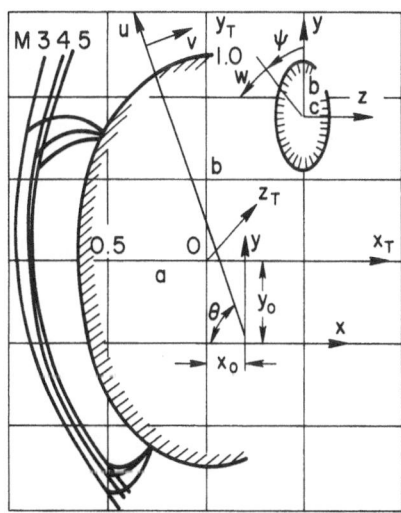

Fig. 6.9 Supersonic flow past blunt bodies at angles of attack

The equations of the body and shock contours are written

$$r = r_B(\theta, \psi) \quad \text{and} \quad r = r_S(\theta, \psi), \quad \text{respectively}$$

and the radial coordinate is replaced by a normalized coordinate ξ,

$$\xi = \frac{r - r_B(\theta, \psi)}{r_S(\theta, \psi) - r_B(\theta, \psi)} \tag{6.5.1}$$

Then the equations of motion are

$$A\frac{\partial u}{\partial \xi} + \frac{1}{\varrho}\frac{\partial p}{\partial \xi} + a\left(v\frac{\partial u}{\partial \theta} + \frac{w}{\sin\theta}\frac{\partial u}{\partial \psi} - v^2 - w^2\right) = 0$$

$$A\frac{\partial v}{\partial \xi} - \frac{B_\theta}{\varrho}\frac{\partial p}{\partial \xi} + a\left(v\frac{\partial v}{\partial \theta} + \frac{w}{\sin\theta}\frac{\partial v}{\partial \psi} + \frac{1}{\varrho}\frac{\partial p}{\partial \theta} + uv - w^2\,\text{ctg}\,\theta\right) = 0$$

$$A\frac{\partial w}{\partial \xi} - \frac{B_\psi}{\varrho}\frac{\partial p}{\partial \xi} + a\left(v\frac{\partial w}{\partial \theta} + \frac{w}{\sin\theta}\frac{\partial w}{\partial \psi} + uw + vw\,\text{ctg}\,\theta + \frac{1}{\varrho\sin\theta}\frac{\partial p}{\partial \psi}\right) = 0$$

$$\frac{\partial u}{\partial \xi} + \frac{A}{\varrho}\frac{\partial \varrho}{\partial \xi} - B_\theta\frac{\partial v}{\partial \xi} - B_\psi\frac{\partial w}{\partial \xi}$$

$$+ a\left(\frac{\partial v}{\partial \theta} + \frac{1}{\sin\theta}\frac{\partial w}{\partial \psi} + 2u + v\,\text{ctg}\,\theta + \frac{v}{\varrho}\frac{\partial \varrho}{\partial \theta} + \frac{w}{\varrho\sin\theta}\frac{\partial \varrho}{\partial \psi}\right) = 0 \tag{6.5.2}$$

$$A\left(\frac{\partial p}{\partial \xi} - \frac{\gamma p}{\varrho}\frac{\partial \varrho}{\partial \xi}\right) + a\left[v\left(\frac{\partial p}{\partial \theta} - \frac{\gamma p}{\varrho}\frac{\partial \varrho}{\partial \theta}\right) + \frac{w}{\sin\theta}\left(\frac{\partial p}{\partial \psi} - \frac{\gamma p}{\varrho}\frac{\partial \varrho}{\partial \psi}\right)\right] = 0$$

$$A = u - B_\theta v - B_\psi w, \qquad a = \frac{\varepsilon}{r_B + \xi\varepsilon}, \qquad \varepsilon = r_s - r_B$$

$$B_\theta = \frac{\partial r_B/\partial \theta + \xi\partial\varepsilon/\partial\theta}{r_B + \xi\varepsilon}, \qquad B_\psi = \frac{1}{\sin\theta}\frac{\partial r_B/\partial\psi + \xi\partial\varepsilon/\partial\psi}{r_B + \xi\varepsilon}$$

where u, v, w are velocity components in the spherical coordinate system. The unknowns must satisfy the surface boundary condition

$$u - \frac{v}{r_B}\frac{\partial r_B}{\partial\theta} - \frac{w}{r_B\sin\theta}\frac{\partial r_B}{\partial\psi} = 0 \quad \text{for} \quad \xi = 0 \tag{6.5.3}$$

and the shock conditions

$$\varrho = \frac{\gamma+1}{\gamma-1}\frac{V_{n\infty}^2}{1 - V_{\tau\infty}^2}, \qquad p = \frac{\gamma-1}{2\gamma}(1 - V_\infty^2) - V_{n\infty}^2\left(\frac{1}{\varrho} - 1\right) \tag{6.5.4}$$

$$u = u_\infty + F, \qquad v = v_\infty - \frac{1}{r_s}\frac{\partial r_s}{\partial\theta}F, \qquad w = w_\infty - \frac{1}{r_s\sin\theta}\frac{\partial r_s}{\partial\psi}F \quad \text{for} \quad \xi = 1.$$

Here V_∞ is the speed in the undisturbed flow, and

$$F = V_{n\infty}^2 \left(\frac{1}{\varrho} - 1\right) \left(u_\infty - \frac{v_\infty}{r_s}\frac{\partial r_s}{\partial \theta} - \frac{w_\infty}{r_s\sin\theta}\frac{\partial r_s}{\partial \psi}\right)^{-1}$$

$$V_{n\infty}^2 = \left(u_\infty - \frac{v_\infty}{r_s}\frac{\partial r_s}{\partial \theta} - \frac{w_\infty}{r_s\sin\theta}\frac{\partial r_s}{\partial \psi}\right)^2 \left[1 + \left(\frac{1}{r_s}\frac{\partial r_s}{\partial \theta}\right)^2 + \left(\frac{1}{r_s\sin\theta}\frac{\partial r_s}{\partial \psi}\right)^2\right]^{-1}$$

$$V_{\tau\infty}^2 = V_\infty^2 - V_{n\infty}^2$$

$$u_\infty = V_\infty(-\cos\alpha\cos\theta + \sin\alpha\sin\theta\cos\psi)$$

$$v_\infty = V_\infty(\cos\alpha\sin\theta + \sin\alpha\cos\theta\cos\psi)$$

$$w_\infty = -V_\infty\sin\alpha\sin\psi .$$

In (6.5.2)–(6.5.4) quantities are made dimensionless in terms of pressure $\varrho_\infty V_{max}^2$, density ϱ_∞ and velocity V_{max}.

To solve (6.5.2) we draw $4k+4$ half planes $\psi = \pm\psi_0,\ldots,\pm\psi_k$, $\pm\pi\mp\psi_k,\ldots,\pm\pi\mp\psi_0$ $(0\le\psi<\pi/2)$ and n conical surfaces $\theta = \theta_1,\ldots,\theta_n$. Their intersections generate $n(4k+4)$ rays, including the axis $\theta = 0$.

The flow variables for $\xi = $ constant will be approximated from their values at $n(4k+4)+1$ points of intersection with surfaces $\xi = $ constant. In each plane $\psi = \psi_i$, $(-\pi+\psi_i)$ parameters are approximated by Lagrange polynomials, determined from values at $(2n+1)$ points

$$f = \sum_{i=0}^{2n} f_i^\circ\, \theta^{\circ i}, \qquad \theta^\circ = \begin{cases} \theta & \text{for } |\psi| < \tfrac{1}{2}\pi \\ -\theta & \text{for } |\psi| > \tfrac{1}{2}\pi \end{cases} \tag{6.5.5}$$

Using symmetry conditions we obtain a unique polynomial for the half planes $\psi = \psi_i$, and $\psi = \pi - \psi_i$ with coefficients which are linear functions of the values of the unknowns on $2n+1$ rays in the half space $0\le\psi\le\pi$. For fixed θ the unknowns are approximated by trigonometric polynomials in cosines for even functions and sines for odd functions

$$f = \sum_{j=0}^{2k+1} f_j^{\circ\circ}\cos j\psi \qquad (f = u,v,p,\varrho) \tag{6.5.6}$$

$$w = \sum_{j=1}^{2k+2} w_j^{\circ\circ}\sin j\psi \tag{6.5.7}$$

The coefficients in (6.5.6), (6.5.7) are linear combinations of the values of the unknowns on rays lying in the half plane $0\le\psi\le\pi$. If $\psi_0 = 0$ the upper index in (6.5.7) is equal to $2k$. Substituting these expressions in (6.5.2) and requiring that they be satisfied on all $n(2k+2)+1$ rays we arrive at the following system of ordinary differential equations (i refers to rays):

$$\frac{dp_i}{d\xi} = \frac{-1}{\Delta_i}\left[\gamma p_i(R_{1i}+B_{\theta i}R_{2i}+B_{\psi i}R_{3i}+A_iR_{4i})+A_iR_{5i}\right]$$

$$\frac{d\varrho_i}{d\xi} = \frac{\varrho_i}{\gamma p_i}\left(\frac{dp_i}{d\xi}+\frac{R_{5i}}{A_i}\right), \qquad \frac{du_i}{d\xi} = \frac{1}{A_i}\left(R_{1i}-\frac{1}{\varrho_i}\frac{dp_i}{d\xi}\right)$$

$$\frac{dv_i}{d\xi} = \frac{1}{A_i}\left(\frac{B_{\theta i}}{\varrho_i}\frac{dp_i}{d\xi}-R_{2i}\right), \qquad \frac{dw_i}{d\xi} = \frac{1}{A_i}\left(\frac{B_{\psi i}}{\varrho_i}\frac{dp_i}{d\xi}-R_{3i}\right)$$

$$\Delta_i = \frac{\gamma p_i}{\varrho_i}(1+B_{\theta i}^2+B_{\psi i}^2)-A_i^2$$

$$R_{1i} = a_i\left[v_i\left(v_i-\frac{\partial u_i}{\partial \theta}\right)+w_i\left(w_i-\frac{1}{\sin\theta_i}\frac{\partial u_i}{\partial \psi}\right)\right] \qquad (6.5.8)$$

$$R_{2i} = a_i\left(v_i\frac{\partial v_i}{\partial \theta}+\frac{w_i}{\sin\theta_i}\frac{\partial v_i}{\partial \psi}+u_iv_i-w_i^2\,\text{ctg}\,\theta_i+\frac{1}{\varrho_i}\frac{\partial p_i}{\partial \theta}\right)$$

$$R_{3i} = a_i\left(v_i\frac{\partial w_i}{\partial \theta}+\frac{w_i}{\sin\theta_i}\frac{\partial w_i}{\partial \psi}+u_iw_i+v_iw_i\,\text{ctg}\,\theta_i+\frac{1}{\varrho_i\sin\theta_i}\frac{\partial p_i}{\partial \psi}\right)$$

$$R_{4i} = a_i\left(\frac{\partial v_i}{\partial \theta}+\frac{1}{\sin\theta_i}\frac{\partial w_i}{\partial \psi}+v_i\,\text{ctg}\,\theta_i+2u_i+\frac{v_i}{\varrho_i}\frac{\partial \varrho_i}{\partial \theta}+\frac{w_i}{\varrho_i\sin\theta_i}\frac{\partial \varrho_i}{\partial \psi}\right)$$

$$R_{5i} = a_i\left[v_i\left(\frac{\partial p_i}{\partial \theta}-\frac{\gamma p_i}{\varrho_i}\frac{\partial \varrho_i}{\partial \theta}\right)+\frac{w_i}{\sin\theta_i}\left(\frac{\partial p_i}{\partial \psi}-\frac{\gamma p_i}{\varrho_i}\frac{\partial \varrho_i}{\partial \psi}\right)\right]$$

$$(i=1,2,\ldots,n(2k+2)+1)$$

Boundary conditions must be satisfied at points of intersection of rays with the body surface and the shock wave, using the forms already given in (6.5.3) and (6.5.4).

The unknown surface of the shock front $r=r_S(\theta,\psi)$ is approximated by the same scheme and contains $n(2k+2)+1$ arbitrary parameters. These are selected iteratively so that the results of integrating (6.5.8) satisfy conditions on $\xi=0$.

Calculations are carried out with $\psi_0=0$, $k=1$, $n=3$ for a variety of ellipsoids of revolution. The resulting trigonometric series go up to $\cos 3\psi$ for even functions and up to $\sin 2\psi$ for odd functions. Integration with respect to ξ is carried out by a Runge-Kutta method of third order accuracy, usually with step $k=0.1$.

Figure 6.10 shows the traces of shock shapes and sonic lines in the plane of symmetry for three ellipsoids of revolution with axis ratios $\delta=0.5, 2, 3.07$. The full lines correspond to $M_\infty=3$, while the dashed line shows contours for $\delta=2$ at $M_\infty=\infty$. Angles of attack range from $\alpha=5°$ to $\alpha=15°$. Figure 6.11

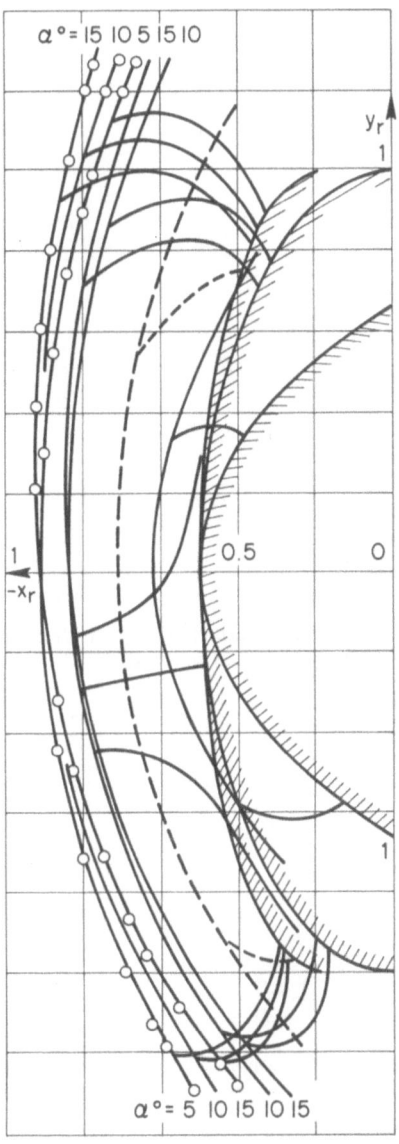

Fig. 6.10 Shock lines and sonic lines on supersonic ellipsoids

compares the theoretical shock shape for an ellipsoid with $\delta = 3.07$, $\alpha = 10°$, $M_\infty = 3$ with the experimental shape determined by Yu. Ya. KARPEISKII (see TELENIN and TINYAKOV, 1964). Finally, Fig. 6.12 gives the surface pressure distributions in the plane of symmetry for (1) $\delta = 3.07$, (2) $\delta = 2$, (3) $\delta = 1.5$, (4) $\delta = 0.5$, and angles of attack 1. $\alpha = 0°$, 2. $\alpha = 5°$, 3. $\alpha = 10°$, 4. $\alpha = 15°$.

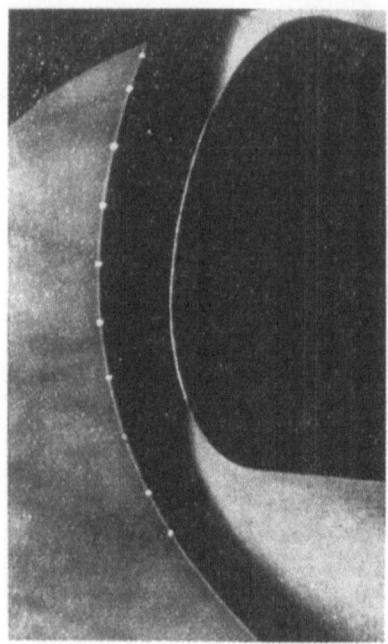

Fig. 6.11 Theoretical and experimental shock shapes on an ellipsoid

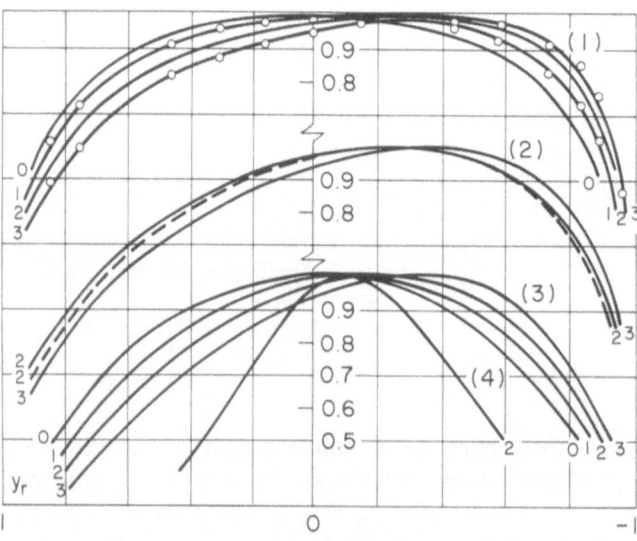

Fig. 6.12 Pressure distributions on ellipsoids

6.6 Application of Telenin's Method to the Supersonic Yawed Cone Problem

The problem of calculating uniform supersonic flow past cones at angles of attack has received considerable attention over the past thirty years. Conical flows always play an important role in the aerodynamics of supersonic vehicles both because sections of such vehicles are geometrically conical (for example, swept back delta wings) and because the flow around supersonic pointed bodies is dominated by a conical flow at the nose. Conical flows are instructive and rewarding to investigate since they exhibit the principal characteristics of general three dimensional flows while, as a result of similarity properties, they can be described in terms of only two reduced (angular) coordinates.

Since the wing components of supersonic configurations are usually of small thickness their aerodynamic characteristics can be adequately described by small disturbance theory. This is developed as the linearized theory of supersonic conical fields (see GOLDSTEIN and WARD, 1950). On the other hand, the geometrical surfaces bounding the body components of supersonic vehicles are normally inclined at moderate or large angles to the undisturbed flow and the aerodynamics of conical bodies must be formulated in a non-linear manner. Theoretical investigation of flow past conical bodies therefore requires numerical methods.

A whole series of numerical methods have been applied to conical flows. The earliest methods, developed by STONE (1948, 1952) and KOPAL (1949), before large scale computers became available, depended on trigonometrical expansions in the angle of attack. These were followed by several finite difference approaches culminating in the BVLR method described in Chapt. 3. The Method of Integral Relations was used by CHUSHKIN and SHCHENNIKOV (1962) and NDEFO (1968) but was found to be difficult to extend beyond the first approximation. A survey of these methods as well as the asymptotic expansion techniques of MELNIK (1967) and others, is given in HOLT and NDEFO (1970).

Both Telenin's method and the Method of Lines have proved to be particularly effective in calculating conical flows. Conical flows are referred to a spherical polar coordinate system centered on the apex of the conical body in question. The flow properties are independent of the radius measured along a conical ray and can be completely determined by solving the governing equations in a cross flow plane (more precisely, the trace of the flow pattern on a sphere centered on the cone apex). The outer boundary of this flow is a conical shock and its section in the cross flow plane is a closed curve completely enclosing the solid cone boundary.

At small and moderate angles of attack the cross flow is entirely subsonic (that is, the resultant transverse velocity component is subsonic). It is analogous to two dimensional flow past an airfoil but is simpler than this in one respect, since the disturbance does not extend to infinity but is confined to the interior of the shock contour.

At larger angles of attack the cross flow is subsonic on the windward side but accelerates to supersonic speeds near the central shoulder of the body, becoming subsonic again as the leeward side is approached. In this case .the flow is therefore analogous to two dimensional supersonic flow past a blunt body.

Telenin's method, as seen in Sect. 6.4, has been very successfully applied to the blunt body problem for a wide range of geometrical configurations. It should therefore be very suitable for the calculation of flow past highly yawed cones, at least on the windward side up to the shoulder. Telenin's method should also yield the solution to the yawed cone problem at moderate angles of attack, since this is an elliptic problem which, in principle, can be treated by the technique developed by GILINSKII et al. (1964) for Laplace's equation.

The calculation at low and moderate angles of attack was carried out for circular cones by HOLT and NDEFO (1970). This appears to be the first application of Telenin's method to an elliptic problem in Fluid Dynamics. FLETCHER (1974, 1975) obtained the solution for highly yawed circular cones with the aid of Telenin's method. The cross flow in the highly yawed case has some elements not present in the usual blunt body problem since the flow field must be closed on the leeward side. An interior shock must be fitted to permit the supersonic cross flow near the shoulder to decelerate to subsonic conditions on the leeward plane of symmetry and to satisfy symmetry conditions there. FLETCHER develops an ingenious form of Telenin's method and the Method of Lines (integrating in the circumferential direction) to fit this shock and determine the flow field on the leeward side.

The equations of motion for supersonic flow past a cone are expressed in terms of spherical polar coordinates, shown in Fig. 6.13. The radius r is measured from the apex of the cone, θ is the angle between a ray and the axis of the cone,

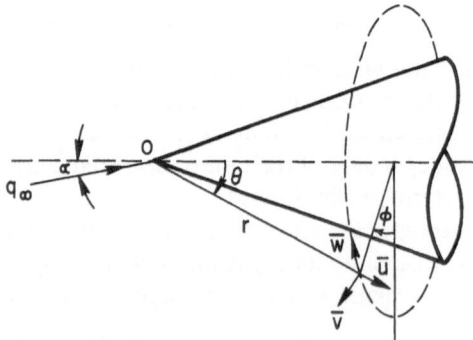

Fig. 6.13   Conical flow coordinate system

while ϕ is the angle between a general meridian plane and the plane of symmetry. The velocity component (u, v, w) in the directions (r, θ, ϕ), respectively, are expressed as dimensionless ratios to the critical velocity, a^*, the density ϱ is given in terms of the undisturbed density ϱ_∞ and the pressure p is in terms of $\varrho_\infty a^{*2}$.

Since the flow is conical the physical variables u, v, w, p, ϱ are independent of r and the equations of motion are

Conservation of Mass:

$$v_\theta + \frac{1}{\sin\theta}\cdot w_\phi + \frac{v}{\varrho}\cdot\varrho_\theta + \frac{w}{\varrho\sin\theta}\cdot\varrho_\phi + 2u + v\cot\theta = 0, \tag{6.6.1}$$

Conservation of 'r' Momentum:

$$v\cdot u_\theta + \frac{w}{\sin\theta}\cdot u_\phi - (v^2 + w^2) = 0, \tag{6.6.2}$$

Conservation of 'θ' Momentum:

$$v\cdot v_\theta + \frac{w}{\sin\theta}\cdot v_\phi + \frac{1}{\varrho}\cdot p_\theta + u\cdot v - w^2\cdot\cot\theta = 0, \tag{6.6.3}$$

Conservation of 'ϕ' Momentum:

$$v\cdot w_\theta + \frac{w}{\sin\theta}\cdot w_\phi + \frac{1}{\varrho\sin\theta}\cdot p_\phi + uw + vw\cot\theta = 0, \tag{6.6.4}$$

Conservation of Energy:

$$v\cdot p_\theta + \frac{w}{\sin\theta}\cdot p_\phi - a^2\left(v\cdot\varrho_\theta + \frac{w}{\sin\theta}\cdot\varrho_\phi\right) = 0, \tag{6.6.5}$$

where the sound speed, $a^2 = \gamma p/\varrho$.
An entropy function S is determined from the equation of state (here we assume perfect gas behavior)

$$S = p/\varrho^\gamma \tag{6.6.6}$$

We also use Bernoulli's equation, derived from (6.6.1)–(6.6.5), in place of one of the latter equations:

$$\frac{\gamma p}{(\gamma-1)\varrho} + \frac{1}{2}(u^2 + v^2 + w^2) = \frac{(\gamma+1)}{2(\gamma-1)} \tag{6.6.7}$$

Eqs. (6.6.1)–(6.6.7) are solved in conjunction with boundary conditions satisfied at the solid cone and shock conditions behind the conical shock.
For a circular cone, defined by $\theta = \theta_c$, we must satisfy

$$v = 0, \quad \theta = \theta_c \tag{6.6.8}$$

In the two planes of symmetry

$$\left.\begin{array}{r} w=0 \\ d\theta_S/d\phi=0 \end{array}\right\} \phi=0, \quad \phi=\pi \qquad\qquad (6.6.9)$$

u, v, p, ϱ are even functions of ϕ.

On the shock surface $\theta=\theta_S$. (θ_S is a function of ϕ).

$$\begin{aligned} u_S &= q_\infty n_1, \\ v_S &= q_\infty n_3 \sin\varepsilon\cos\varepsilon - n_4, \\ w_S &= q_\infty n_3 \cos^2\varepsilon + n_4\tan\varepsilon, \\ \varrho_S &= q_\infty n_2 \cos^2\varepsilon/n_4, \\ p_S &= \left(\frac{2}{\gamma+1}\right) q_\infty^2 n_2^2 \cos^2\varepsilon - \left(\frac{\gamma-1}{2\gamma}\right)(1-\mu^2 q_\infty^2), \end{aligned} \qquad (6.6.10)$$

where

$$\begin{aligned} n_1 &= \cos\alpha\cos\theta_S - \sin\alpha\sin\theta_S\cos\phi, \\ n_2 &= \cos\alpha\sin\theta_S + \sin\alpha\cos\theta_S\cos\phi + \sin\alpha\sin\phi\tan\varepsilon, \\ n_3 &= \frac{\sin\alpha\sin\phi - n_2\sin\varepsilon\cos\varepsilon}{\cos^2\varepsilon}, \\ n_4 &= \frac{1-\mu^2 q_\infty^2(1-n_2^2\cos^2\varepsilon)}{q_\infty n_2}, \\ \mu^2 &= \frac{\gamma-1}{\gamma+1}, \\ \varepsilon &= \tan^{-1}\left(\frac{1}{\sin\theta_S}\frac{d\theta_S}{d\phi}\right), \\ q_\infty^2 &= \frac{M_\infty^2}{\mu^2 M_\infty^2 + 2/(\gamma+1)}. \end{aligned} \qquad (6.6.11)$$

ε is the shock angle in the cross flow plane, namely, it is the angle between the normal to the shock and the meridian plane $\phi=$ constant.

It is convenient to use angular coordinates which vary between 0 and 1 in the disturbed region. We thus define

$$\begin{aligned} \eta &= \phi/\pi \\ \xi &= \frac{\theta_S(\eta)-\theta}{\sigma(\eta)} \end{aligned} \qquad (6.6.12)$$

where $\sigma(\eta)=\theta_S(\eta)-\theta_c$.

It is sufficient to work in the half space, $0 \le \phi \le \pi$, because of symmetry. The boundaries of the cross flow plane, in both θ, ϕ and ξ, η systems are then as shown in Fig. 6.14.

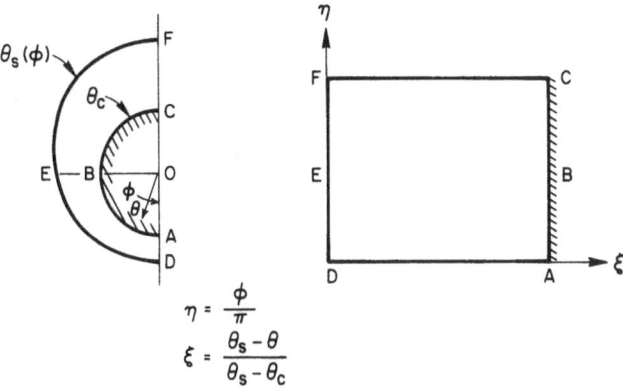

$$\eta = \frac{\phi}{\pi}$$

$$\xi = \frac{\theta_s - \theta}{\theta_s - \theta_c}$$

Fig. 6.14   Transformed coordinates in conical flow problem

In terms of the new independent variables ξ, η the equations of motion transform to the following

$$-\frac{1}{\sigma} v_\xi + \frac{B}{\pi \sin \theta} w_\xi + \frac{E}{\varrho} \varrho_\xi + \frac{1}{\pi \sin \theta} \left(w_\eta + \frac{w}{\varrho} \varrho_\eta \right) + 2u + v \cot \theta = 0,$$

$$E u_\xi + \frac{w}{\pi \sin \theta} u_\eta - (v^2 + w^2) = 0,$$

$$E v_\xi - \frac{1}{\varrho \sigma} p_\xi + \frac{w}{\pi \sin \theta} v_\eta + uv - w^2 \cot \theta = 0, \qquad (6.6.13)$$

$$E w_\xi + \frac{B}{\varrho \pi \sin \theta} p_\xi + \frac{1}{\pi \sin \theta} \left(w w_\eta + \frac{1}{\varrho} p_\eta \right) + uw + vw \cot \theta = 0,$$

$$E(p_\xi - a^2 \varrho_\xi) + \frac{w}{\pi \sin \theta} (p_\eta - a^2 \varrho_\eta) = 0,$$

where

$$E = \frac{Bw}{\pi \sin \theta} - \frac{v}{\sigma}$$

$$B = \pi \left(\frac{1 - \xi}{\sigma} \right) \frac{d\theta_s}{d\phi}.$$

The analysis and discussion up to this point apply equally to both moderate and high angle of attack cases. However, the methods of interpolation used to represent the unknowns in the circumferential (η) direction are different in the two cases, partly because the governing equations are elliptic in the first case and of mixed type in the second, and partly because the highly yawed cross flow on the windward side is independent of that on the leeward side, while in the first case the whole cross flow must be considered at the same time. Our further discussion therefore treats the two cases separately.

Low and Moderate Yaw Angles—Elliptic Case: At small and moderate angles of yaw the cross flow is subsonic and must be considered as a whole, from the windward to the leeward planes of symmetry. Since symmetry conditions are satisfied in $\phi=0$, $\phi=\pi$, the unknowns are periodic in ϕ with period π. In applying Telenin's method we therefore use trigonometric interpolation expressions, with cosine terms for the even functions and sine terms for the odd functions. We divide the region $0\le\phi\le2\pi$ into $2m+2$ equal sectors and write

$$
\begin{pmatrix} u(\xi,\eta) \\ v(\xi,\eta) \\ p(\xi,\eta) \\ \varrho(\xi,\eta) \end{pmatrix} \approx \sum_{k=0}^{m+1} \begin{pmatrix} u_k^\circ(\xi) \\ u_k^\circ(\xi) \\ p_k^\circ(\xi) \\ \varrho_k^\circ(\xi) \end{pmatrix} \cos(k\pi\eta) \tag{6.6.14}
$$

$$
w(\xi,\eta) \approx \sum_{k=0}^{m+1} w_k^\circ(\xi) \sin(k\pi\eta) \tag{6.6.15}
$$

The shock shape is approximated by the equation

$$
\theta_s(\eta) \approx \sum_{k=0}^{m+1} \theta_{sk}^\circ \cos(k\pi\eta) \tag{6.6.16}
$$

Working in $0\le\eta\le1$ we integrate (6.6.13) along $m+2$ rays $\eta=j/(m+1)$ $j=0,\dots,m+1$. On each ray we regard (6.6.13) as a system of ordinary differential equations for u, v, w, p, ϱ with ξ as independent variable, calculating the η derivatives from expressions (6.6.14), (6.6.15).

The integration is started at the shock $\xi=0$. Estimates are made of the $m+2$ angular shock detachment distances $\theta_s(\eta)$, $\eta=j/(m+1)$, $j=0,\dots,m+1$. Eq. (6.6.16) is then solved for θ_{sk}°. Then (6.6.16) represents the shock shape for all $\eta, 0\le\eta\le1$ and the cross flow angle ε can be calculated. The values of θ_s and ε are then substituted in the shock equation (6.6.10) to determine the values of $u_s, v_s, w_s, p_s, \varrho_s$ at each shock nodal point. These values in turn are used to calculate the coefficients in the trigonometric expansions (6.6.14), (6.6.15). The unknowns and their η derivatives can then be calculated at all shock nodal points and (6.6.13) can be integrated across the first ξ step. The integration gives the values of u, v, w, p, ϱ at the end of the step. These are used to determine new coefficients $u_k^\circ, \ldots,$ in the trigonometric series (6.6.14), (6.6.15). The η derivatives of the unknowns can then be determined from (6.6.14), (6.6.15) in readiness for performing the next step in the integration of (6.6.13) along the rays $\eta = j/(m+1)$, $j = 0, \ldots, m+1$. The alternating step-by-step integration of (6.6.13) and inversion of the coefficient matrix on the right of (6.6.14), (6.6.15) is continued up to the solid cone $\xi = 1$. Here, we test whether the surface boundary conditions

$$v(\eta) = 0, \quad \xi = 1 \quad \text{for} \quad \eta = \frac{j}{m+1}, \quad j = 0, \ldots, m+1$$

are satisfied. If not, the residual values of $v_{\xi=1}$ are determined and used to revise the estimates of the shock angles θ_s. The whole numerical procedure is then repeated and iteration between the shock and body conditions is continued until all the values $v_{\xi=1}$ equal zero.

The technique described was applied to calculate the flow field around a circular cone of half angle $\theta_c = 20^\circ$ at $M_\infty = 3.53$ for a range of angles of attack $\alpha = 5^\circ, 10^\circ, 15^\circ, 20^\circ$. Figures 6.15 and 6.16 show the comparison between surface pressure distributions for $\alpha = 10^\circ, 20^\circ$ calculated by the method and

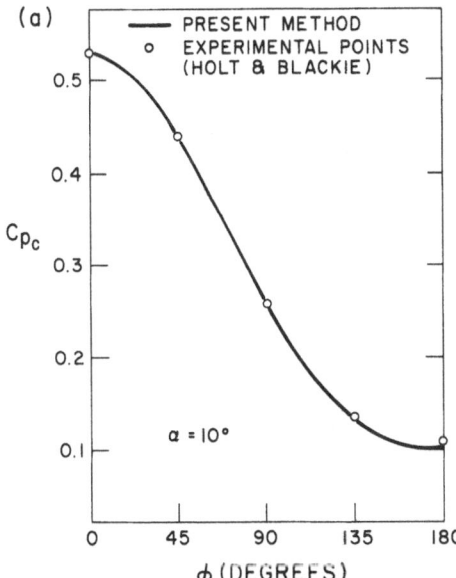

Fig. 6.15 Surface pressure distributions on 10° cone

measured distributions taken from HOLT and BLACKIE (1952). The agreement is excellent. The only other methods giving comparable accuracy are the BVLR method (which requires more numerical calculations) and the Method of Lines (the latter application was developed independently at the same time—see JONES (1968)). Figures 6.17 and 6.18 show cross flow streamlines at $\alpha = 10^\circ$,

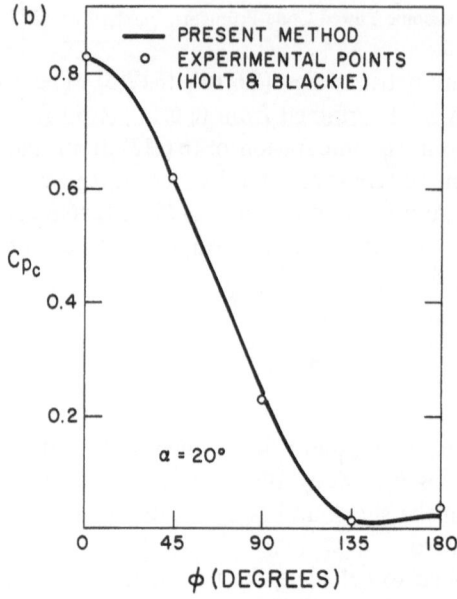

(b)

─── PRESENT METHOD
○ EXPERIMENTAL POINTS
(HOLT & BLACKIE)

C_{p_c}

$\alpha = 20°$

ϕ (DEGREES)

Fig. 6.16 Surface pressure distributions on 20° cone

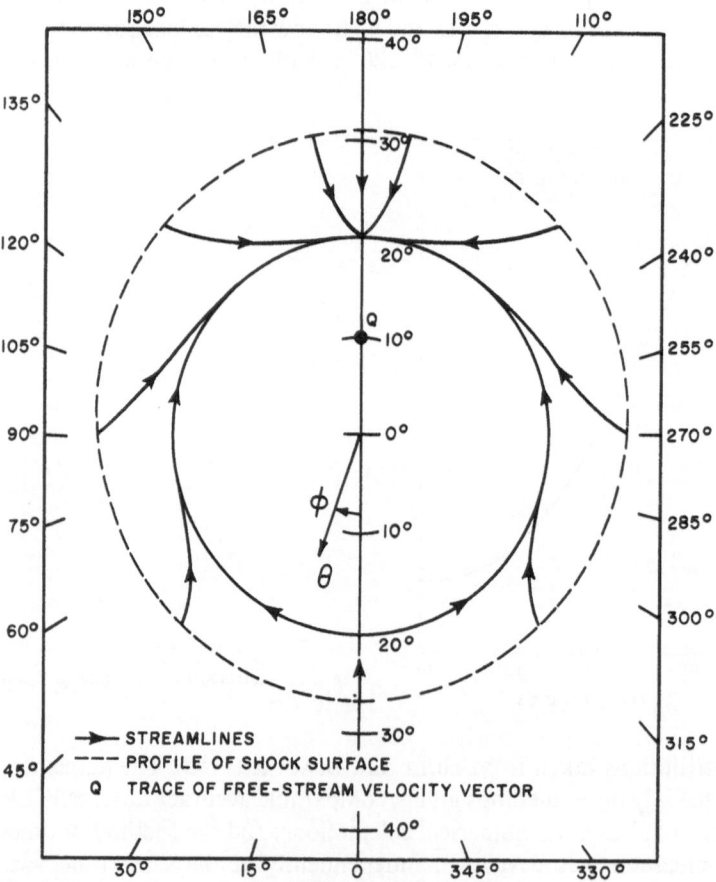

➤ STREAMLINES
--- PROFILE OF SHOCK SURFACE
Q TRACE OF FREE-STREAM VELOCITY VECTOR

Fig. 6.17 Cross flow streamline pattern for 20° cone $\alpha = 10°$

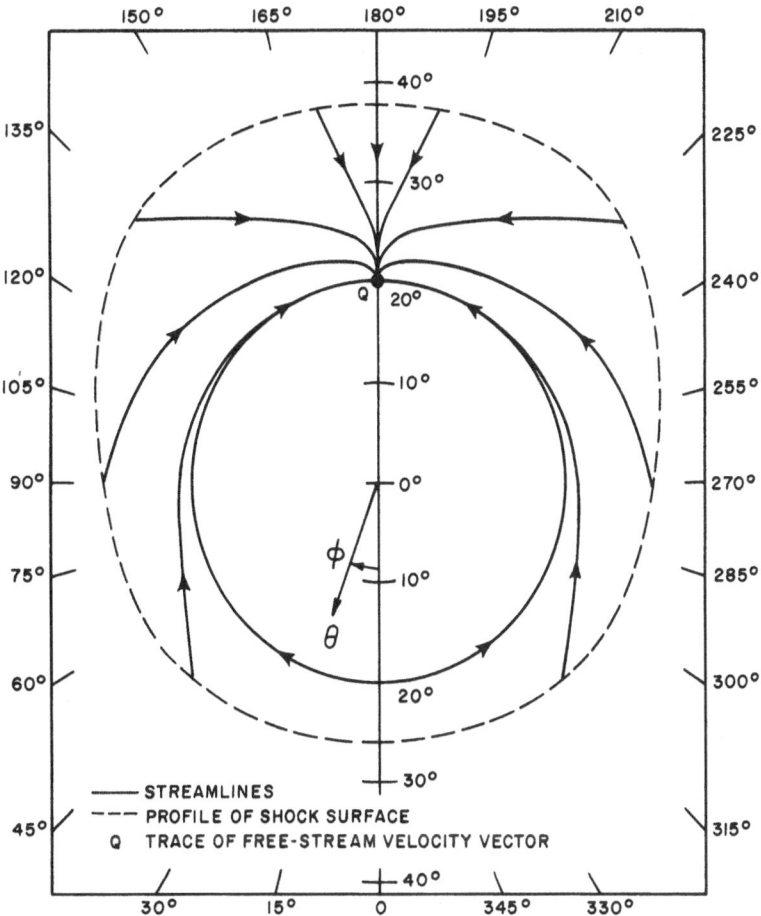

Fig. 6.18   Cross flow streamline pattern for 20° cone $\alpha = 20°$

$\alpha = 20°$, together with shock shapes. The streamline pattern near the vortical singularity at $\theta = \theta_1$, $\phi = \pi$ is of interest. At the lower yaw angle most streamlines follow close to the cone surface before reaching the singularity although those entering on the far leeward side approach the singularity over a range of angles. At the limiting yaw angle for elliptic cross flow $\alpha = 20°$, all streamlines except those crossing the shock on the forward windward side approach the singularity tangential to the leeward plane of symmetry.

These results were obtained with the bare minimum number of rays ($m = 3$, in fact) taken along $\eta = 0$, $\pi/4$, $\pi/2$, $3\pi/4$, π. This low order harmonic representation was sufficient to describe all features of the flow field with precision. The calculations were thus very economical in computer time and also met the requirement for minimizing round-off errors discussed in Sect. 6.9.

High Yaw Angles. Mixed Elliptic-Hyperbolic Case: When the component of the free stream Mach number normal to the cone axis, $M_\infty \sin\alpha$, is larger than unity, the cross flow is supersonic in the shoulder region, while remaining subsonic near the windward and leeward planes of symmetry. The principal boundaries are shown in Fig. 6.19 for a half space $0 \leq \phi \leq \pi$. A conical shock

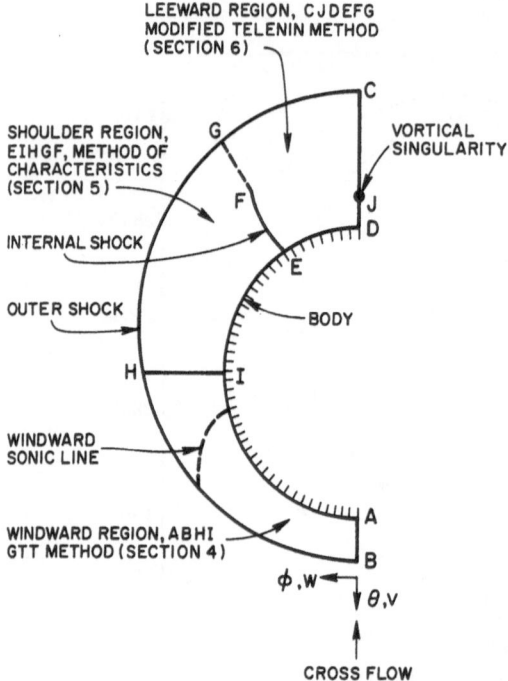

Fig. 6.19   Classification of the different flow regions on the basis of the numerical method used

$BMHGC$ encloses the whole disturbed flow field. The cross flow on the windward side accelerates smoothly to supersonic speeds across a sonic line LM. A supersonic region extends to a region close to the leeward plane of symmetry DC. It cannot extend all the way to this plane because the flow must be partially subsonic there. A cross flow stagnation point ($v=0$, $w=0$) always exists at the point D. In addition, at high angles of attack, a vortical singularity (a sink like point through which all cross flow streamlines pass) lifts away from the position D which it occupies at low yaw to a position J. This point is also a cross flow stagnation point $v=0$, $w=0$, besides being a point where u and S are many valued. The cross flow must therefore be subsonic on $\phi = \pi$ over the segment DJ and, in practice, it is also subsonic over the remaining segment JC.

To reduce the supersonic cross flow near the shoulder to subsonic conditions at the leeward plane of symmetry an internal shock EF is required. This does not extend all the way to the main enclosing shock since the outer cross flow adjacent to this shock can decelerate continuously across a leeward sonic line FG (this situation is analogous to the shock-sonic line pattern on a transonic airfoil; see CHATTOT and HOLT (1972)).

The complete inviscid solution in the highly yawed case has recently been determined by FLETCHER (1975). He uses an extension of Telenin's technique, as applied to the blunt body problem, to determine the mixed flow on the windward side. He then calculates the flow field in the shoulder region by a method of characteristics. This calculation is continued to the leeward plane of symmetry by applying a Method of Lines combined with fitting of an interior shock—this phase is discussed later in Sect. 6.8.

The equations of motion, (6.6.13), are the same as in the low yaw case, referred to coordinates ξ and η. The region of integration, however, is now reduced and the symmetry conditions imposed on the leeward plane of symmetry in the low yaw case are dropped in favor of regularity conditions satisfied near the sonic line, where the governing equations change from elliptic to hyperbolic type.

FLETC. R proposes two ways of interpolating for u, v, w, p, ϱ in the η direction. In the first, the trigonometric expansions (6.6.14), (6.6.15) and (6.6.16) are used with five equally spaced rays extending from $\eta=0$ to $\eta=1/2$. These expressions build in the symmetry requirements on $\eta=0$ and guarantee smooth transition across the sonic line. The Method of Telenin is applied to the equations in $0\le\xi\le1, 0\le\eta\le1/2$, and an iterative technique is again used to adjust the shock nodal point positions to the conditions $v=0$ at the five body nodal points, as in the low yaw case. In his more recent calculation the iteration process is speeded up by use of Powell's method, which adjusts the nodal values of the shock angles systematically so as to minimize the sum of the squares of the normal velocity residuals at the body nodal points.

Fletcher's second treatment of the interpolation in the η direction is suggested by the Method of Lines. Using the same number of rays and nodal points on the shock and body, he represents the η derivatives at a nodal point by expressions of the type

$$(u_\eta)_i = \sum_{k=0}^{m-1} a_{0k} u_k \tag{6.6.17}$$

Eq. (6.6.17), in effect, represents u_η at nodal points by a difference formula of fifth order accuracy. Representatior (6.6.17) are equivalent to using fifth degree polynomials in the variables u, v, w, p, ϱ themselves with coefficients determined from values at nodal points. The second treatment is therefore the same as that of the blunt body problem by TELENIN et al. In this case the order of accuracy of the difference formula used to represent derivatives is the same as the number of rays and Telenin's method is exactly equivalent to the Method of Lines.

Both methods of interpolation are equally good. Convergence of the iteration process depends on the closeness of the initial estimates of the shock angles at nodal points to their true values. A series of calculations was carried out at $M_\infty=7$ for a cone with semi-angle $\theta_b=30°$. These started from the known solution for $\alpha=30°$ (a limiting case of elliptic flow which can be found by the Holt-Ndefo technique) and solutions up to $\alpha = 43°$ were generated by increasing α in 1° steps using, at each stage, the last found solution to make shock angle estimates in the next. The whole range of solutions obtained in this manner required a total of 6.2 seconds on a CDC 7600 computer.

Figures 6.20 and 6.21 show results for a 20° cone (semi-apex angle) at 30° angle of incidence and free stream Mach number of 7.00. Figure 6.20 shows

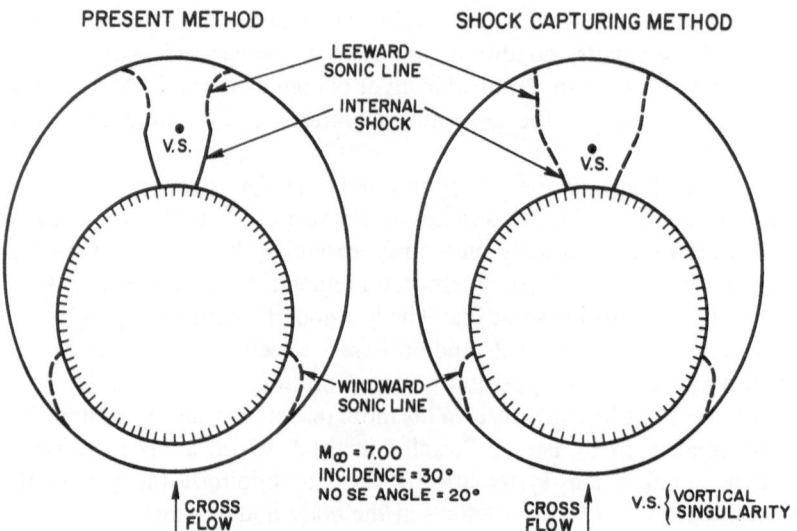

Fig. 6.20   Comparison with shock capturing method-shock wave and sonic line locations

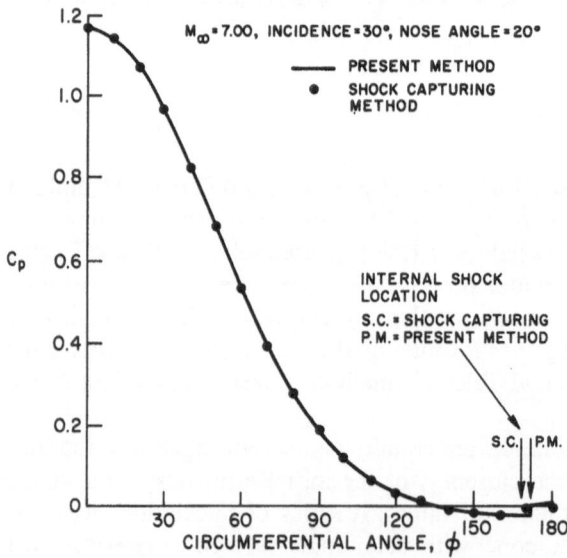

Fig. 6.21   Comparison with shock capturing method-pressure distribution

external and internal shock shapes, together with sonic line positions, while Fig. 6.21 gives the surface pressure distribution. In both figures comparisons are made with calculations for the same configuration obtained by the shock capturing method of KUTLER and LOMAX (1970). Very close agreement is obtained apart from small differences in the results near the leeward plane of symmetry. In this region, of course, viscous effects are important, affecting both calculations.

6.7 The Method of Lines. General Description

The Method of Lines (MOL) is essentially a technique for replacing a system of partial differential equations in two independent variables by an approximate system of ordinary differential equations in one of these variables. If the independent variables are x, y the region of integration is divided into strips parallel to one of the coordinates (say y) by lines $y =$ constant and the partial derivatives with respect to y on each of these lines are represented by finite difference expressions using values of the unknowns on the line and on neighboring lines to either side of it. There results a system of ordinary differential equations determining the unknowns along each strip boundary in terms of the other independent variable, x.

The method has been studied and used extensively in the Soviet Union for over thirty years, mostly in applications to linear partial differential equations; it is discussed at length in LISKOVETS (1965). The method was recently formulated independently by JONES (1968) to solve the Yawed Cone Problem (low and moderate yaw angle case). The connection between Jones' formulation and the earlier Soviet work on linear equations is clarified in a further study of the Unsymmetrical Supersonic Cone Problem by KLUNKER, SOUTH and DAVIS (1971). An up to date presentation of the method, illustrated with applications to both linear and non-linear problems is given by JONES, SOUTH and KLUNKER (1972). Messrs. JONES and SOUTH (1979) have recently completed a report on the method and current applications.

As with Telenin's method, the Method of Lines is suitable for the numerical solution of partial differential equations of elliptic, mixed elliptic-hyperbolic, and parabolic type. The boundary within which the method is applied is taken to be a rectangle (boundaries of more general shape can be reduced to this form by suitable coordinate transformations). Boundary conditions are prescribed on all or some sides of the rectangle. For elliptic equations, values of the unknowns, their normal derivatives or combinations of these are prescribed on all four sides. For mixed type equations boundary conditions on one of these sides are left open and replaced by regularity conditions within the rectangle. For parabolic equations boundary conditions are only needed on three sides.

Following JONES, SOUTH and KLUNKER (1972), we suppose that $\psi(x, y)$ is an unknown satisfying a second order partial differential equation of elliptic or mixed type with independent variables x, y (the discussion can easily be extended to parabolic equations). We seek a solution of the governing equation for ψ within a rectangle bounded by $x=x_0$, $x=x_1$, $y=y_0$, $y=y_1$ satisfying certain boundary conditions on the sides. Without loss of generality let us consider a Dirichlet problem associated with an elliptic equation, with

$$
\begin{aligned}
\psi = a(y) \quad &\text{on} \quad x=x_0 \\
\psi = b(y) \quad &\text{on} \quad x=x_1
\end{aligned} \qquad y_0 \le y \le y_1
$$

$$
\begin{aligned}
\psi = c(x) \quad &\text{on} \quad y=y_0 \\
\psi = d(x) \quad &\text{on} \quad y=y_1
\end{aligned} \qquad x_0 \le x \le x_1
$$
(6.7.1)

Divide the rectangle into n strips of equal width δy, $y_0+j\delta y$ $(j=0,\ldots,n)$. Next, let us solve the Dirichlet problem as an iteration on the following Cauchy problem with

$$
\left.
\begin{aligned}
\psi &= a(y) \\
\frac{\partial \psi}{\partial x} &= F(y)
\end{aligned}
\right\} \quad \text{on} \quad x=x_0, \qquad y_0 \le y \le y_1
$$

$$
\begin{aligned}
\psi &= c(x) \quad \text{on} \quad y=y_0 \\
\psi &= d(x) \quad \text{on} \quad y=y_1
\end{aligned} \qquad x_0 \le x \le x_1
$$
(6.7.2)

The function $F(y)$ is initially unknown and must be so chosen that the solution of the Cauchy problem is compatible with the second condition in (6.7.1)

$$
\psi = b(y) \quad \text{on} \quad x=x_1 .
$$

To solve the Cauchy problem defined by (6.7.2) we consider the nodal points $y=y_0+j\delta y$ on $x=x_0$ and calculate the y derivatives of ψ at each of the interior points using, either the second order finite difference formulae

$$
\left(\frac{\partial \psi}{\partial y}\right)_{x,y} = \frac{\psi(x, y+\delta y)-\psi(x, y-\delta y)}{2\delta y} + O(\delta y^2 \, \psi^{\text{III}}),
$$

$$
\left(\frac{\partial^2 \psi}{\partial y^2}\right)_{x,y} = \frac{\psi(x, y+\delta y)-2\psi(x, y)+\psi(x, y-\delta y)}{\delta y^2} + O(\delta y^2 \, \psi^{\text{IV}}),
$$
(6.7.3)

or the fourth order formulae

$$\left(\frac{\partial \psi}{\partial y}\right)_{x,y} = \frac{4}{3}\left[\frac{\psi(x,y+\delta y)-\psi(x,y-\delta y)}{2\delta y}\right.$$

$$\left.-\frac{1}{3}\left[\frac{\psi(x,y+2\delta y)-\psi(x,y-2\delta y)}{4\delta y}\right]+O(\delta y^4 \psi^{V})\right.$$

$$\left(\frac{\partial^2 \psi}{\partial y^2}\right)_{x,y} = \frac{4}{3}\left[\frac{\psi(x,y+\delta y)+\psi(x,y-\delta y)-2\psi(x,y)}{\delta y^2}\right]$$

$$-\frac{1}{3}\left[\frac{\psi(x,y+2\delta y)+\psi(x,y-2\delta y)-2\psi(x,y)}{4\delta y^2}\right]+O(\delta y^4 \psi^{VI}).$$

(6.7.4)

These are substituted in the governing equation for ψ. Then at each nodal point this equation reduces to a relation between the x derivatives in ψ, in other words, to a second order ordinary differential equation for ψ in x. Assuming values of $F(y)$ at each nodal point we can thus, with the aid of condition (6.7.2), integrate these equations numerically through a short x step (Δx, say) to determine the values of ψ at all nodal points on $x=x_0+\Delta x$. At each nodal point of $x=x_0+\Delta x$ we can then calculate finite difference approximations to $\partial \psi/\partial y$ and $\partial^2 \psi/\partial y^2$ and continue the integration of the ordinary differential equations in x up to $x_0+2\Delta x$. This procedure is continued up to the right boundary $x=x_1$ at which stage we evaluate the values $\psi(x_1,y_0+j\Delta y)$ $j=0,1,\ldots,n$. In general these values will not agree with the required boundary values

$$\psi(x_1,y)=b(y)$$

of (6.7.1) and we record the residuals

$$\varepsilon(x,y+j\Delta y)=\psi(x,y)-b(y) \tag{6.7.5}$$

We determine $F(y)$ at nodal points on $x=x_0$ by minimizing the sum

$$E=\sum_{k=0}^{n}\varepsilon^2(y_0+k\delta y) \tag{6.7.6}$$

with respect to $F(y_0+k\delta y)$, $k=0,1,\ldots,n$. This is achieved by Powell's minimization technique for functions of several variables (POWELL, 1964), described in Sect. 6.9. The technique provides a systematic way of calculating improved estimates of $F(y_0+k\delta y)$ at the end of each iteration, before proceeding with the following iteration.

The stability and convergence of the method are analyzed by JONES et al. in relation to the Dirichlet problem:

Determine the solution ψ of Laplace's equation

$$\psi_{xx}+\psi_{yy}=0 \tag{6.7.7}$$

within the rectangle $0 \le x \le 1$, $-b \le y \le b$, satisfying the boundary conditions

$$\psi(0, y) = \psi(1, y) = 0$$
$$\psi(x, b) = \psi(x, -b) = \sin \pi x.$$

(6.7.8)

The exact solution to this problem obtained by separation of variables is

$$\psi(x, y) = \frac{\cosh \pi y \sin \pi x}{\cosh \pi b}$$

(6.7.9)

The same problem is now solved approximately by the Method of Lines. Because of symmetry about $x = 1/2$ and $y = 0$ it is sufficient to solve in the quadrant $0 \le x \le 1/2, 0 \le y \le b$. This rectangle is divided into N strips parallel to the x axis by the lines

$$y = y_n = nh = \frac{nb}{N}, \quad n = 0, ..., N.$$

The right hand boundary condition is now

$$\psi_x(\tfrac{1}{2}, y) = 0 \quad \text{(replacing } \psi(1, y) = 0)$$

In addition ψ is even in y so that

$$\psi(x, y) = \psi(x, -y).$$

In applying MOL we replace ψ_{yy} in (6.7.7) by the three point difference formula (6.7.3) and obtain the following system of ordinary differential equations

$$\psi_n'' + (\psi_{n+1} - 2\psi_n + \psi_{n-1})/h^2 = 0, \quad n = 0, 1, ..., N-1.$$

(6.7.10)

The boundary and symmetry conditions associated with these equations are

$$\psi_n(0) = 0$$
$$\psi_n'(\tfrac{1}{2}) = 0$$
$$\psi_N = \sin \pi x$$
$$\psi_{-n}(x) = \psi_n(x).$$

(6.7.11)

The solution of (6.7.10) with conditions (6.7.11) can be found analytically in the following form

$$\psi_n(x) = \sum_{m-1,3,...}^{2N-1} T_n(\theta_m)(A_m e^{\mu m x} + B_m e^{-\mu m x}) + \frac{\cosh nz}{\cosh Nz} \sin \pi x$$

(6.7.12)

where

$$\mu_m = \left(\frac{2N}{b}\right) \sin\frac{m\pi}{4N}, \qquad \theta_m = \cos\frac{m\pi}{2N}$$

$$z = \cosh^{-1}\left\{1 + \frac{1}{2}\left(\frac{\pi b}{N}\right)^2\right\}$$

(6.7.13)

and $T_n(\theta_m)$ is the Chebyshev polynomial of order n. To satisfy the first two of conditions (6.7.11) we must have

$$A_m = B_m = 0, \qquad m = 1, 3, \ldots, 2N-1.$$

The MOL solution to the problem therefore reduces the last term in (6.7.12), for each $y = y_n$. If we expand (6.7.13) in $(\pi b/N)$ the MOL solution becomes, to order (N^{-4})

$$\psi_n(x) = \frac{\cosh\pi y_n}{\cosh\pi b}\sin\pi x\left[1 + \left(\tanh\pi b - \frac{y_n}{b}\tanh\pi y_n\right)\left(\frac{\pi^3 b^3}{24N^2} + O(N^{-4})\right)\right]$$

(6.7.14)

Comparing (6.7.9) with (6.7.14) we see that the MOL solution converges to the exact solution as $h \to 0$ with error $O(1/N^2)$.

It should be noted that the coefficients A_m, B_m in the MOL solution (6.7.12) only vanish for the particular boundary conditions $\psi(0, y) = \psi(1, y) = 0$. If these boundary values are different from zero the MOL solution $\psi_n(x)$ grows exponentially with x for $x > 0$. Therefore when we apply MOL to a non-linear elliptic equation with similar boundary conditions to (6.7.8) we face the problem of exponential growth of round-off error inherent in Telenin's method (see Sect. 6.2). These are again contolled by combining high precision arithmetic with use of the smallest possible number of y strips.

6.8 Applications of the Method of Lines

JONES (1968) applied his own formulations of the Method of Lines to calculate surface pressure distributions, shock shapes and velocity profiles on circular and elliptic cones at small and moderate angles of attack in a uniform supersonic stream. He used the reduced independent variables ξ, η defined so that the governing equations of motion are (6.7.13). In applying MOL he represents the shock shape by a finite cosine series, using as many harmonics as needed depending on the angle of incidence (only two are used at very small yaw angles). He then initiates the numerical integration of (6.7.13) in the ξ direction along strip boundaries $\eta = \eta_j = j/m+1, j = 1, \ldots, m-1$. At the end of the first ξ step the values of the unknowns are then determined at all nodal points and finite difference formulae (6.7.3) or (6.7.4) are used to determine η derivatives. Eq. (6.7.13) can then be integrated a further step in ξ and this process is continued until the conical body is reached. The integration between $\xi=0$ and $\xi=1$ is combined with Powell's method of adjusting the shock position to the normal velocity components on the body until a converged solution is obtained. JONES (1969) publishes results in the form of tables for circular cones of semi angles between 5°–25°, at angles of attack from 0° to 20° and a range of Mach numbers from 1.8 to 7 (JONES, SOUTH and KLUNKER, 1972). He also gives results for a series of elliptic cones.

These calculations were extended by KLUNKER, SOUTH and DAVIS (1971) to higher angles of yaw, and to cones with thin elliptic cross sections. Their results include many cases where regions of supersonic cross flow are present. However, in all cases the Method of Lines is evidently applied over the whole cross flow region from the windward to the leeward planes of symmetry. Their results can therefore be expected to be good for thin elliptic cones approaching delta wing conditions.

Figure 6.22 shows Jones' calculated pressure distribution for a circular cone $\theta_c=12.5°$ at $M_\infty=4.25$ and $\alpha=8.24°$, $\alpha=12.5°$. These are compared

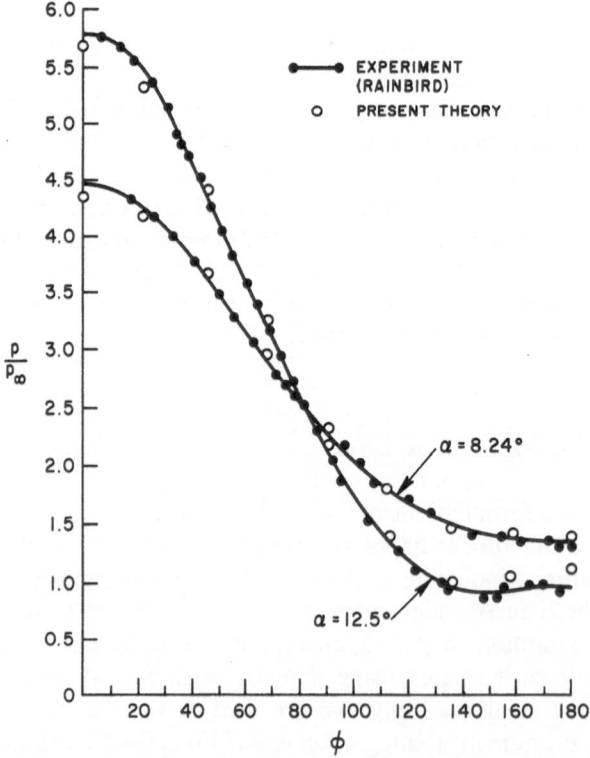

Fig. 6.22   Circular Cone. Comparison of surface pressure between present theory and experiment; $M_\infty=4.25$, $\theta_c=12.5°$, $\alpha=8.24°$ and 12.5°

with measured distributions under the same conditions obtained by RAINBIRD (1968). Figure 6.23 shows the pressure distributions on a thin parabolic section delta wing calculated by KLUNKER et al. and compared with calculations by the Method of Characteristics and the BVLR method.

GROSS and HOLT (1975) applied both Telenin's method and the Method of Lines to calculate supercritical flow past ellipses. So far the calculations have been limited to flows with two axes of symmetry, when the flow field

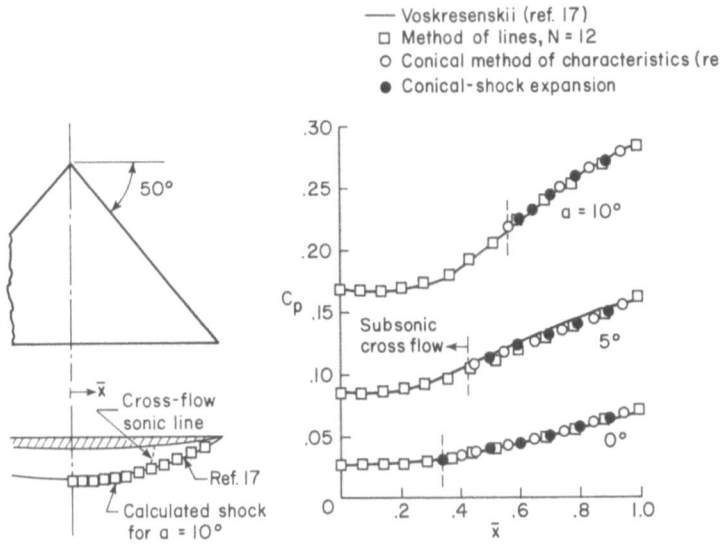

Fig. 6.23 Comparison of computations for shock shape and spanwise pressure distribution for a conical delta wing. Parabolic-arc cross section; $M_\infty = 4°$, $\theta_0 = 3°$, $\Lambda = 50°$, $N = 12$

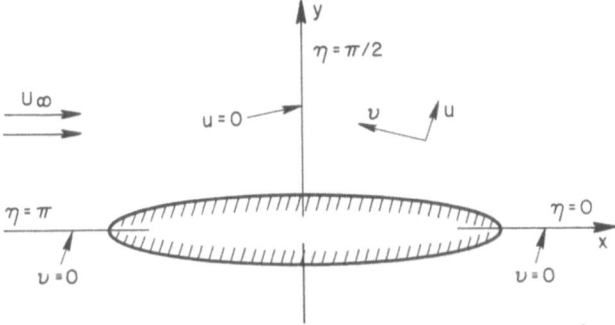

Fig. 6.24 Symmetrical flow field with boundary conditions

is shock free. The flow can be assumed to be irrotational and isentropic, and is referred to an elliptic coordinate system ξ, η shown in Fig. 6.24. The velocity components (u, v) in the (ξ, η) direction are taken as the principal dependent variables and using dimensionless forms for the unknowns (the same as in conical flow) the governing equations are

$$\frac{\partial u}{\partial \xi} = \frac{2uv\frac{\partial v}{\partial \xi} - (\gamma-1)(1-u^2-v^2)\left(\frac{1}{\varrho h}\frac{\partial \varrho vh}{\partial \eta} + \frac{u\sinh(2\xi)}{2h^2}\right)}{(\gamma-1)(1-v^2)-(\gamma+1)u^2} \tag{6.8.1}$$

$$\frac{\partial v}{\partial \xi} = \frac{1}{h}\frac{\partial uh}{\partial \eta} - \frac{v\sinh(2\xi)}{2h^2} \tag{6.8.2}$$

where

$$\varrho = (1-u^2-v^2)^{1/(\gamma-1)} \tag{6.8.3}$$

$$h = (\sinh^2\xi + \sin^2\eta)^{1/2} \tag{6.8.4}$$

The symmetry and surface boundary conditions are

$$u = 0 \qquad \eta = \pi/2 \tag{6.8.5}$$

$$v = 0 \qquad \eta = \pi \tag{6.8.6}$$

$$u = 0 \quad \text{on} \quad \xi = \xi_0, \quad \text{the body} \tag{6.8.7}$$

The velocity approaches the free stream value at O so that

$$u \to U_\infty \cos\eta, \qquad v = -U_\infty \sin\eta, \qquad \eta \to \infty \tag{6.8.8}$$

It is sufficient to solve the problem in the quadrant $0 \le \eta \le \pi/2$. To apply Telenin's method we divide the quadrant by N equally spaced hyperbolic rays in η and calculate derivatives in the η direction occurring in (6.8.1), (6.8.2) from expressions of the type

$$hu = \sum_{i=1}^{N} a_i(\xi)\eta^{(i-1)} \tag{6.8.9}$$

or

$$hu = \sum_{i=1}^{N} b_i(\xi)\cos(2i-1)\eta. \tag{6.8.10}$$

When applying MOL we use the expression

$$\frac{\partial(hu)}{\partial \eta}\bigg|_i = \sum_j F_{ij}(hu)_j \tag{6.8.11}$$

to calculate derivatives; j need not be summed from 1 to N but over 3 or 5 points at and on either side of $\eta = \eta_i$ (usually 5 points were used in the MOL calculations). Values of v are assumed at nodal points on the body $\xi = \xi_0$, η derivatives can then be calculated from equations of type (6.8.9), (6.8.10) or (6.8.11) and (6.8.1), (6.8.2) can be integrated in the ξ direction. The values of u, v at large distances from the ellipse must approach the values (6.8.8) on each

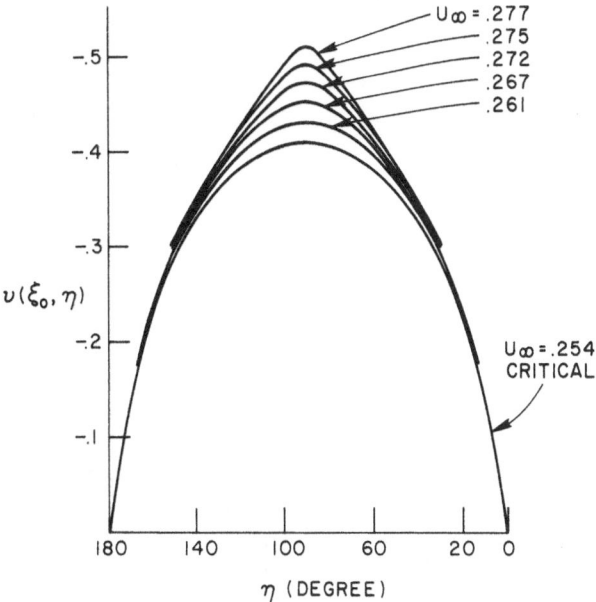

Fig. 6.25 Critical and supercritical velocity profiles on the ellipse with $\delta = 0.4$

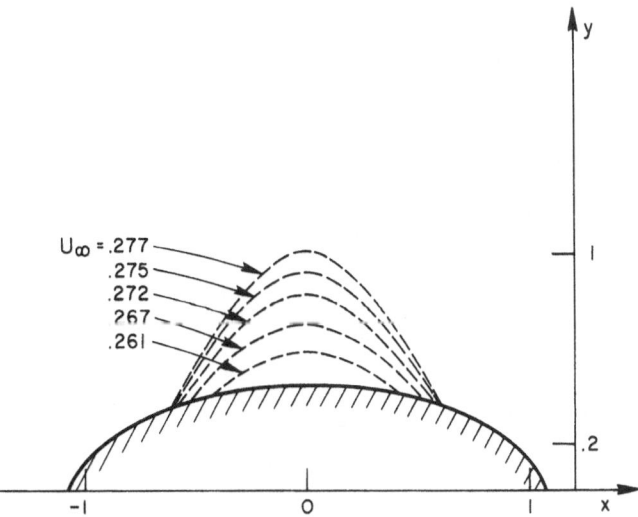

Fig. 6.26 Sonic lines for supercritical flows

ray. To adjust the values $(v)_{\xi_0}$ to satisfy these conditions we can either use the trigonometric series solution of the linearized Prandtl-Glauert equation developed by CHUSHKIN (see Sect. 5.3) or a dipole solution representing the disturbance of the ellipse when $\xi \to \infty$. Results for an ellipse with axis ratio 0.4 are shown in Figs. 6.25 and 6.26 for free stream velocities ranging from the critical value $U_\infty = 0.254$ up to $U_\infty = 0.277$.

CHATTOT (1977) applied the method of Telenin to calculate supercritical flow past a symmetrical double wedge. The physical plane is shown in Fig. 6.27.

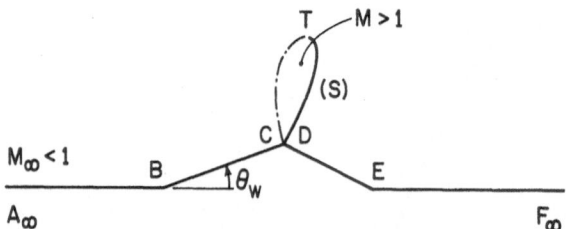

Fig. 6.27   Transsonic flow past a symmetrical double wedge

The flow is subsonic everywhere except in a small supersonic region attached to the shoulder of the wedge CTD where CT is the sonic line; this region is terminated on the downstream side by the shock DT.

To simplify the solution of the problem we transform it into the hodograph plane. In this plane the straight lines bounding the edge transform into straight lines in the same directions. The point $C-D$ is singular and represents a Prandtl-Meyer expansion which transforms into a characteristic CD_1 (see Fig. 6.28). The shock wave (S) transforms into curves $D_2 T(S_2)$ and

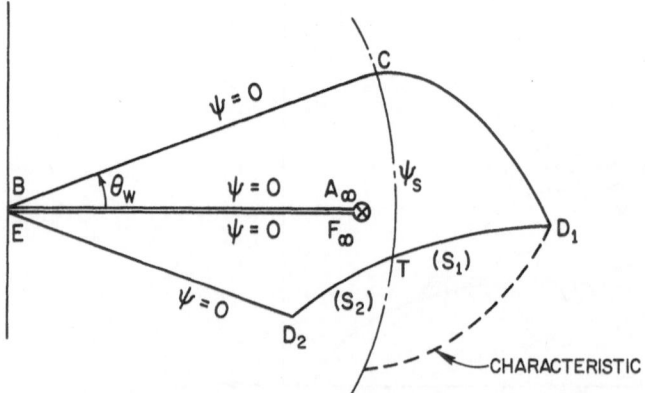

Fig. 6.28   Hodograph for the symmetrical flow past a double wedge

$TD_1(S_1)$, the shapes of which are unknown at the start of the calculation. The unknown in the hodograph plane is the stream function ψ which satisfies Chaplygin's equation. The point at infinity in the physical plane transforms into the point A_∞, F_∞. This is a singular point in the neighborhood of which the behavior of ψ is known.

Telenin's method is applied along three directions, BC, BA_∞, ED_2, each extending up to the characteristic CD (and its continuation). The boundary conditions are $\psi = 0$ along the three lines shown and along the characteristic through C. To the right of the curve $D_2 D_1$ the transformation of this solution on the physical plane is triple valued and a shock wave $D_2 D_1$ is fitted step by step to exclude this region. A full account of the work is given in CHATTOT (1977).

In a recent paper (KLOPFER and HOLT, 1975) both Telenin's method and MOL are applied to calculate a series of steady transonic nozzle flows. Emphasis is placed on flow characteristics in the neighborhood of the nozzle throat and sonic line. Transonic nozzle problems are of two types, the direct and the inverse (or design). In the direct problem the shape of the nozzle contour is prescribed and the flow field through the nozzle, for given conditions upstream and downstream of the throat, must be calculated. In the inverse problem the shape of nozzle boundary producing a prescribed axial velocity or pressure distribution is to be determined. Both types of problems are solved and both plane and axially symmetric flows are considered. Mathematically, the problems are of mixed elliptic hyperbolic type with the governing equations changing type as the sonic line is crossed in the streamwise direction. In conformity with the findings of Sect. 6.3 related to the model equation (6.3.1) we therefore interpolate for the unknowns in a streamwise coordinate and integrate with respect to a transverse coordinate, when applying Telenin's method or MOL. The governing equations are the steady Euler equations of motion, the continuity equation, and a geometric relation defining the flow direction of a streamline. We consider isentropic flow of an ideal gas.

Normalizing the velocities by the maximum velocity, the pressure and density by their stagnation values, the governing equations become, after applying the von Mises transformation,

$$\frac{\partial p}{\partial \psi} = -\frac{2\gamma}{\gamma-1} \frac{1}{y^j} \frac{\partial}{\partial x} \left\{ \left(\frac{v}{u}\right) \left[\frac{1-p^{(\gamma-1)/\gamma}}{1+(v/u)^2}\right]^{1/2} \right\} \tag{6.8.12}$$

$$\frac{\partial y}{\partial \psi} = \frac{1}{y^j p^{1/\gamma}} \left[\frac{1+(v/u)^2}{1-p^{(\gamma-1)/\gamma}}\right]^{1/2} \tag{6.8.13}$$

$$\frac{\partial(v/u)}{\partial \psi} = \frac{\partial}{\partial x} \left\{ \frac{1}{y^j p^{1/\gamma}} \left[\frac{1+(v/u)^2}{1-p^{(\gamma-1)/\gamma}}\right]^{1/2} \right\} \tag{6.8.14}$$

where $j = 0, 1$ for the plane and axisymmetric cases, respectively, and γ is the specific heat ratio.

The choice of the dependent variables $p, y, (v/u)$ is dictated by two requirements of the present numerical method: 1) the transformed equations should be divergence free in the plane case, and 2) x derivatives should be of the lowest order possible.

Boundary Condition: The domain for the inverse problem, namely, the determination of nozzle shape corresponding to a prescribed axial velocity distribution is (Fig. 6.29)

$$x_a \leq x \leq x_b \quad \text{and} \quad \psi_a \leq \psi < \infty$$

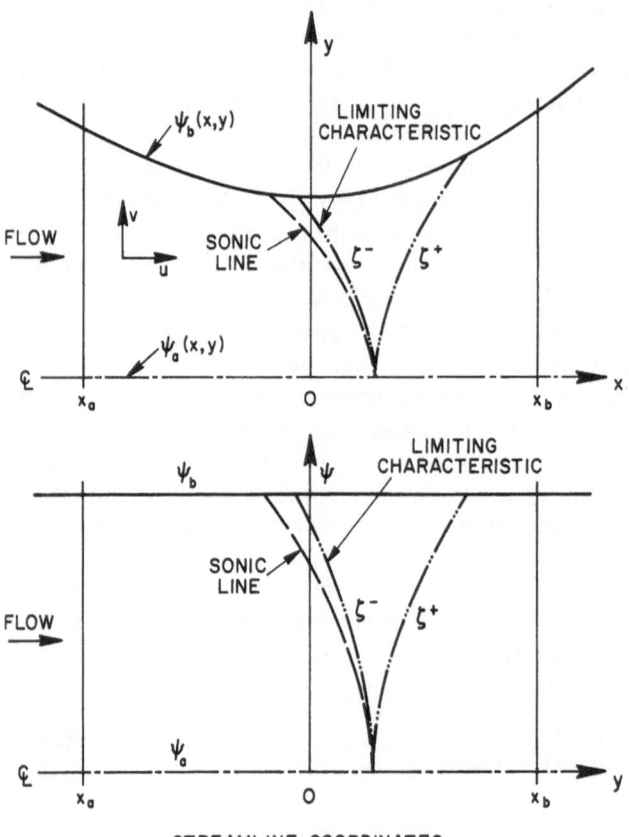

STREAMLINE COORDINATES

Fig. 6.29   Physical and streamline coordinates for the plane transonic nozzle ($\zeta^+, \zeta^- =$ left and right running characteristics)

Cauchy data are required on ψ_a, which can consist of the (x, y) coordinates of ψ_a and the x distribution of the pressure p. However, according to the Cauchy-Kowalewsky theorem (PETROVSKII, 1954), the data must be sufficiently smooth and analytic if a unique analytic solution is to exist in the neighborhood of ψ_a.

No precise boundary conditions need to be prescribed on x_a and x_b provided that the solution upstream of x_a and downstream of x_b is consistent with the initial data on ψ_a.

The direct problem has the domain (Fig. 6.29)

$$x_a \leq x \leq x_b \quad \text{and} \quad \psi_a \leq \psi \leq \psi_b.$$

Dirichlet or Neumann data must be given along the entire part of the boundary which may influence the elliptic region. The (x, y) coordinates of ψ_a, ψ_b are required up to and including the limiting characteristics and Dirichlet or Neumann conditions on the dependent variables are required on x_a. It should be recognized that a shock-free solution may not exist for arbitrary streamlines; they must be sufficiently smooth and analytic.

The numerical method used to solve the various flow problems is based on both Telenin's method and MOL. The equations of motion are integrated simultaneously in the ψ direction along N rays which are spaced parallel to the ψ axis. The x derivatives are represented by a three point Lagrangian difference scheme with variable spacing of the rays for the direct problem. The resulting set of $3N$ ordinary differential equations is integrated simultaneously by a predictor-corrector scheme.

The direct problem is solved, as by HOPKINS and HILL (1966), as a sequence of inverse problems. To start the integration, estimates are made of the value of ψ_b, $\bar{\psi}_b$ say, and the x distribution of p along ψ_a. The estimates are iterated by Powell's method (1964) until the differences between the y coordinates of ψ_b and $\bar{\psi}_b$ are minimized. The major difficulty with the above formulation, as in other applications already described, arises from the phenomenon we have called Hadamard instability. Any errors in the initial data (or subsequent integration) grow like $\exp(N\psi)$, where $N =$ the number of rays of constant x and ψ is the integration direction. The truncation errors in the approximation for the x derivatives are $O(1/(N-1))^{p-1}$ for a 'p' point Lagrangian difference scheme ($p \leq N$) and $O(1/(N-1))^4$ for the cubic spline approximation. The choice of N is a compromise between a large truncation error and a large subsequent growth of the truncation error. Also p should be as low as possible for maximum stabilization.

The choice of the dependent variable in (6.8.12), (6.8.13) and (6.8.14) was dictated by the Hadamard instability. One source of errors in the ψ derivatives comes from truncation errors in the approximate evaluation of the x derivatives. To make these as small as possible we require that the transformed equations be divergence free, if possible, and also that only first order derivatives in x appear. Uncertainty analysis was used to determine the errors of the ψ derivatives from these sources.

For the axisymmetric case the governing equations become singular for $y = 0$. A series expansion is required to remove this singularity. Following

PIRUMOV (1967), the dependent variables are expanded in a power series of ψ about $y=0$ $(\psi=0)$.

$$f = \sum_{k=0}^{\infty} f_k \psi^k,$$

where f is any of the variables $p, \varrho, u, (v/\psi^a)$ and (y/ψ^a), $a=1/2$ for the axisymmetric case and $a=1$ for the plane case. The coefficients f_k are obtained by substitution of the series into the governing equation and equating like powers of ψ.

Results are presented for two inverse plane nozzle problems and one direct axisymmetric nozzle.

The accuracy of the present method depends on the degree of control achieved over the Hadamard instability. The accuracy is monitored by separate x momentum balances around control volumes consisting of the entry and exit section, the centerline, and each of the streamlines. The influence of the

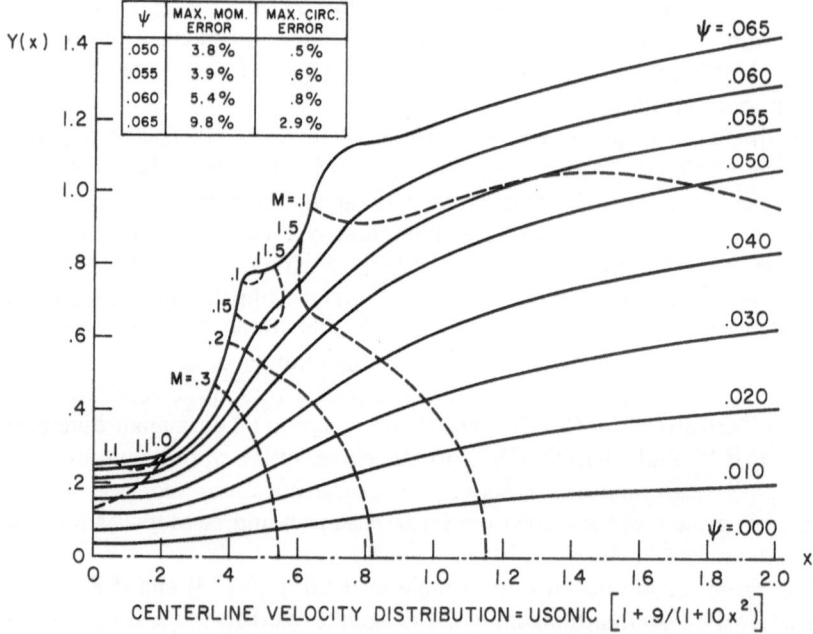

Fig. 6.30   Pirumov's rectilinear sonic line nozzle ($q^2 - v/u$ simplified formulation; 3 point Lagrange; 20 rays)

Hadamard instability is obvious from Fig. 6.30, where momentum errors of several percentage points result (referred to the momentum influx at the entry section). The momentum errors are only fractional percentage points for the plane hyperbolic nozzle (Fig. 6.31) and axisymmetric nozzle (Fig. 6.32). The

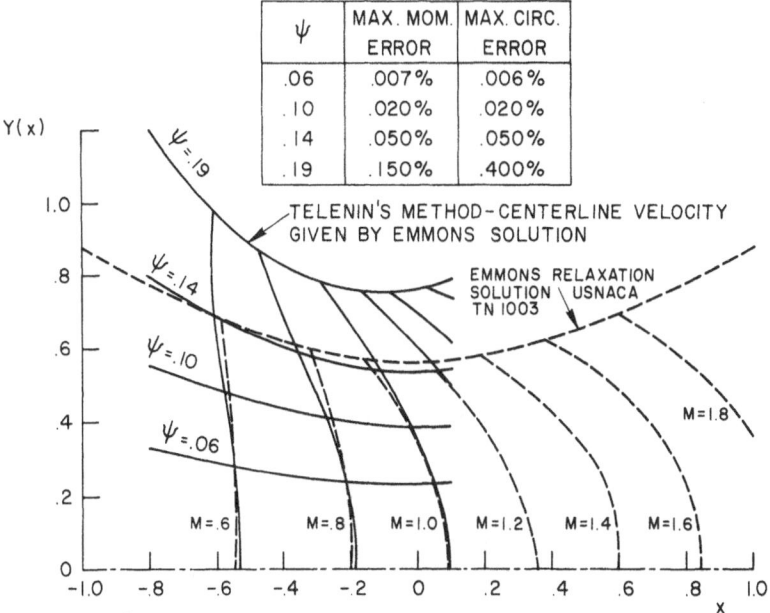

ψ	MAX. MOM. ERROR	MAX. CIRC. ERROR
.06	.007%	.006%
.10	.020%	.020%
.14	.050%	.050%
.19	.150%	.400%

Fig. 6.31 Inverse problem-comparison of Telenin's method ($q^2 - v/u$ simplified; 3-point Lagrange; 9 rays) with Emmons' relaxation solution

truncation errors in representing the x derivatives by a three point Lagrange difference approximation are of order $\Delta x^2 \cong 0.05$ for the inverse problem, Figs. 6.30 and 6.31, and by a cubic spline approximation are of order Δx^4 where $\Delta x = 0.1$ for the direct problem, Figs. 6.32, 6.33, and 6.34. The accuracy specified for the predictor-corrector integration in the ψ direction is 10^{-5}.

The execution times varied from 0.5 to 1 second for the inverse problem and 30 to 60 seconds for the direct problem, using a CDC 7600 computer. The direct problem usually required approximately 500 iterations before the computed streamline and specified nozzle wall matched within the specified tolerance (10^{-3}).

The inverse problem for a plane nozzle with a rectilinear transition surface is shown in Fig. 6.30. The results are similar to those obtained by PIRUMOV (1967) for the same centerline velocity distribution. The plane hyperbolic nozzle solved by EMMONS (1946) by a relaxation method was solved inversely by the present method using Emmons' solution for the centerline velocity distribution. This problem has also been treated by MIR (see Sect. 5.5 and HOLT (1964) in References, Chapt. 5).

The third example is an axisymmetric nozzle with conical entry and exit sections joined by a circular arc of varying radius of curvature. Figure 6.32 shows the isovels for the case of a 45° entry and 15° exit and radius curvature ratio $R_C/R_T = 0.625$ calculated by the present method. These are compared with the experimental data of CUFFEL, BACK, and MASSIER (1969), and with the series solution of KLIEGEL and LEVINE (1966). The results (Figs. 6.32 and 6.33)

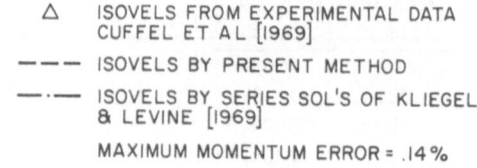

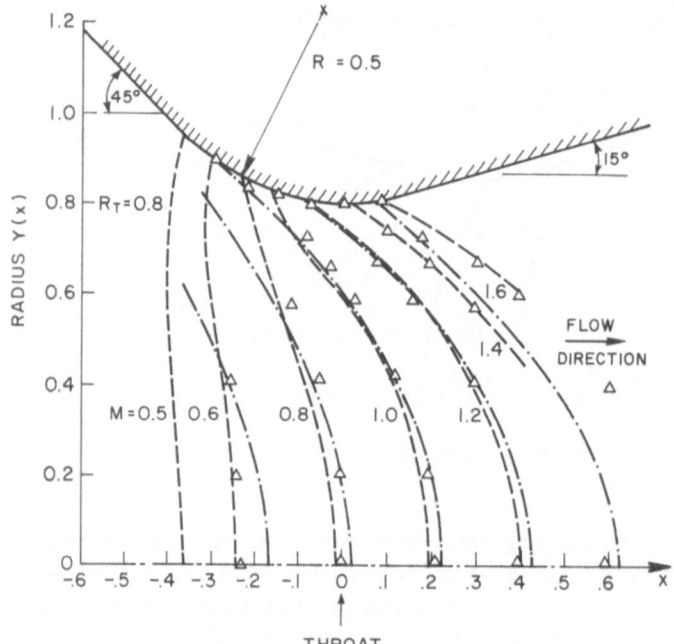

Fig. 6.32   Isovels for $45° - 15° - 0.625$ axisymmetric nozzle (spline; 9 rays; 7 opt. variables)

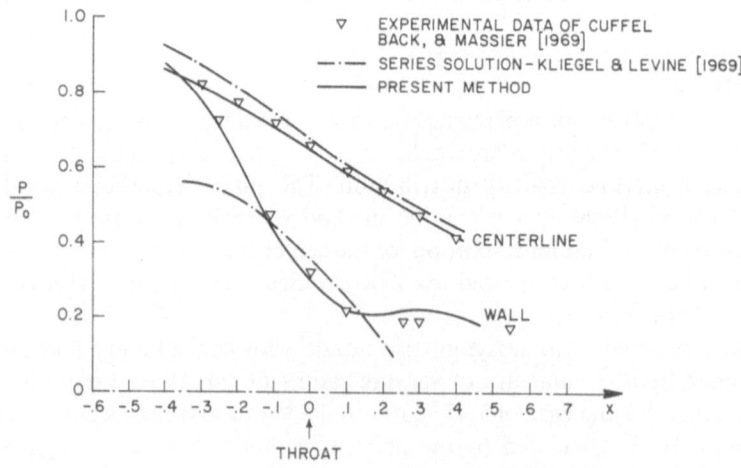

Fig. 6.33   Pressure distribution at wall and centerline for $45° - 15° - 0.625$ axisymmetric nozzle

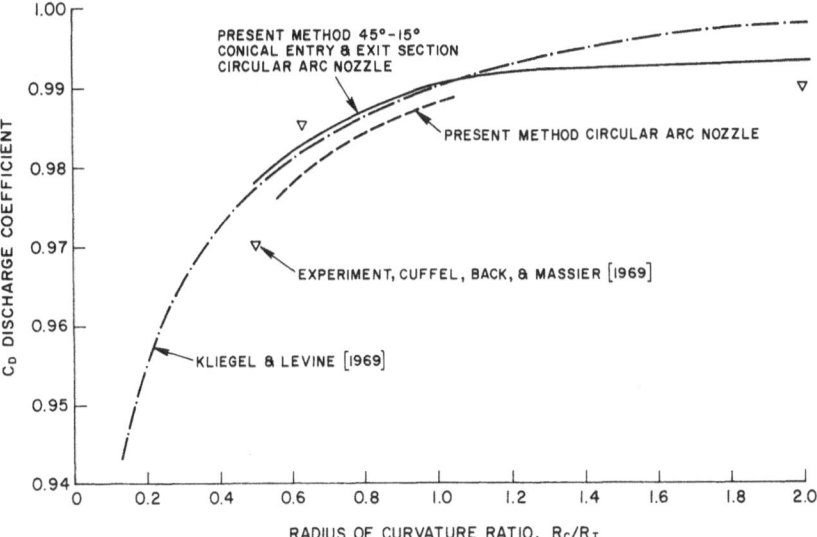

Fig. 6.34   Effect of radius of curvature ratio on discharge coefficient

show that the present method predicts the flow field much better than the method of KLIEGEL and LEVINE, but there are still small disagreements with the experimental data.

The effect of varying the nozzle throat radius of curvature on the discharge coefficient (ratio of the inviscid, adiabatic, two dimensional to the one dimensional mass flow rate) is shown in Fig. 6.34.

The present method has been shown to be an efficient and accurate method for calculating transonic nozzle flows for both the inverse and direct problem.

6.9 Powell's Method Applied to Two Point Boundary Value Problems

In all the applications of Telenin's method and the Method of Lines discussed in Sect. 6.8, problems of Dirichlet or Tricomi type are solved by adjusting values of unknowns on one boundary until prescribed boundary conditions are satisfied on a second boundary. Powell's method is a systematic technique for iterating on two such sets of boundary conditions and is used extensively in the applications to conical and transonic flow problems.

To understand the technique it is sufficient to consider its application (FLETCHER, 1975a) to the yawed cone problem, where parameters defining the conical shock shape must be adjusted to satisfy conditions of zero normal velocity on the cone.

Without loss of generality we suppose that the shock shape can be defined by five angular locations θ_{s_k}, $k=1,...,5$. Then, for given values of these, following the procedure described in Sect. 6.6, we can integrate the conical flow equations from the shock up to the solid cone to determine values of normal velocity components $v_k(\theta_s)$, $k=1,...,5$ at five nodal points on the cone. To determine θ_{s_k} so that $v_k(\theta_s)=0$, $k=1,...,5$ we minimize the sum

$$F(\theta_s)=\sum_{k=1}^{5} v_k(\theta_s)^2 \tag{6.9.1}$$

where

$$\theta_s=(\theta_k), \quad k=1,...,5.$$

If θ_s is an estimate of the shock location vector to minimize F and if $\theta_s+\delta$ is the vector which actually minimizes F, then

$$F(\theta_s+\delta)=0 \tag{6.9.2}$$

and

$$\sum_{k=1}^{5} \frac{\partial v_k(\theta_s+\delta)}{\partial\theta_{s_i}} v_k(\theta_s+\delta)=0, \quad i=1,...,5. \tag{6.9.3}$$

If (6.9.3) is expanded in a Taylor series about θ_s, retaining terms up to first order in δ_i, then

$$\sum_{j=1}^{5}\left\{\sum_{k=1}^{5}\frac{\partial v_k(\theta_s)}{\partial\theta_{s_i}}\frac{\partial v_k(\theta_s)}{\partial\theta_{s_j}}\right\}\delta_j=-\sum_{k=1}^{5}\frac{\partial v_k(\theta_s)}{\partial\theta_{s_i}}v_k(\theta_s), \quad i=1,2,...,5. \tag{6.9.4}$$

This approximation is valid provided that the initial estimates of θ_s are reasonably close to the required values.

To avoid the calculation of the derivatives in (6.9.4) Powell's method replaces the direct calculation of δ_i from (6.9.4) by successive iteration on δ_i, changing only one component at a time.

To initiate the determination of δ five vectors d_i are prescribed which define five independent directions in the θ_{s_i} space. The derivatives of v_k in each of these directions are calculated and their approximate values, denoted by γ_i^k are assumed to satisfy

$$\gamma_i^k=\sum_{j=1}^{5}\frac{\partial v_k}{\partial\theta_{s_j}}(\theta_s)d_i^j, \quad i=1,...,5, \quad k=1,...,5 \tag{6.9.5}$$

where

$$d_i=(d_i^j).$$

We write δ in the form

$$\delta = \sum_{i-1}^{5} q_i d_i \tag{6.9.6}$$

The substitution of (6.9.5) and (6.9.6) into (6.9.4) yields

$$\sum_{j=1}^{5} \left\{ \sum_{k=1}^{5} \gamma_i^k \gamma_j^k \right\} q_j = p_i, \quad i = 1, 2, \dots, 5, \tag{6.9.7}$$

where

$$p_i = -\sum_{k=1}^{5} \gamma_i^k v_k(\boldsymbol{\theta}_s).$$

The vectors d_i are first taken to be in direction of $\boldsymbol{\theta}_s$

$$d_i = (0, \dots, s_i, \dots, 0)$$

where s_i is a scaling factor chosen so that

$$\sum_{k=1}^{5} (\gamma_i^k)^2 = 1$$

in the subsequent calculations. To determine γ_i^k we use the first order difference formula

$$\gamma_i^k = s_i [v_k(\theta_{s_1}, \dots, \theta_{s_i} + \varepsilon_i, \theta_{s_5}) - v_k(\boldsymbol{\theta}_s)]/\varepsilon_i, \quad i = 1, \dots, 5. \tag{6.9.8}$$

We then substitute (6.9.8) in (6.9.7) and solve (6.9.7) for q_i. Eq. (6.9.6) then fixes the first estimate of δ.

With δ known we now evaluate $F(\boldsymbol{\theta}_s + \lambda_n \delta)$ for three values of λ_n (this requires three integrations of the conical flow equations from the shock to the body). A quadratic is fitted through the three F_n and is used to determine the coefficient λ_m minimizing this set of F as described by POWELL (1964). The functional values of $v_k(\boldsymbol{\theta}_s + \lambda_1 \delta)$ and $v_k(\boldsymbol{\theta}_s + \lambda_2 \delta)$, $k = 1, \dots, 5$ which yield the lowest and next lowest values of $F(\boldsymbol{\theta}_s + \lambda \delta)$, during the minimization procedure, are recorded. We can then calculate $\partial v_k / \partial \lambda$ from the approximation

$$\frac{\partial v_k}{\partial \lambda} = \frac{v_k(\boldsymbol{\theta}_s + \lambda_1 \delta) - v_k(\boldsymbol{\theta}_s + \lambda_2 \delta)}{\lambda_1 - \lambda_2}, \quad k = 1, \dots, 5 \tag{6.9.9}$$

and denote this by $u_k(\boldsymbol{\delta})$. Then an improved estimate of $\partial v_k/\partial\lambda$ is given by

$$U_k(\boldsymbol{\delta}) = u_k(\boldsymbol{\delta}) - \mu v_k(\boldsymbol{\theta}_s + \lambda_m\boldsymbol{\delta}) \tag{6.9.10}$$

where $U_m(\delta)$ is the improved $\partial v_k/\partial\delta$ and

$$\mu = \frac{\displaystyle\sum_{k=1}^{5} u_k(\delta) v_k(\boldsymbol{\theta}_s + \lambda_m\boldsymbol{\delta})}{\displaystyle\sum_{k=1}^{5} [v_k(\boldsymbol{\theta}_s + \lambda_m\boldsymbol{\delta})]^2}.$$

The next step is to find the direction, $\boldsymbol{d}_t$, from among $d_1, \ldots, d_5$, such that

$$|p_t \cdot q_t| = \max |p_i \cdot q_i|, \quad i = 1, \ldots, 5.$$

We then replace this particular direction $\boldsymbol{d}$ by $\lambda_m\boldsymbol{\delta}$, γ_i^k by $U_k(\boldsymbol{\delta})$ (leaving the other γ_i^k unchanged) and repeat the cycle starting with (6.9.7).

The procedure converges very rapidly in all applications considered. It is summarized in the attached flow chart.

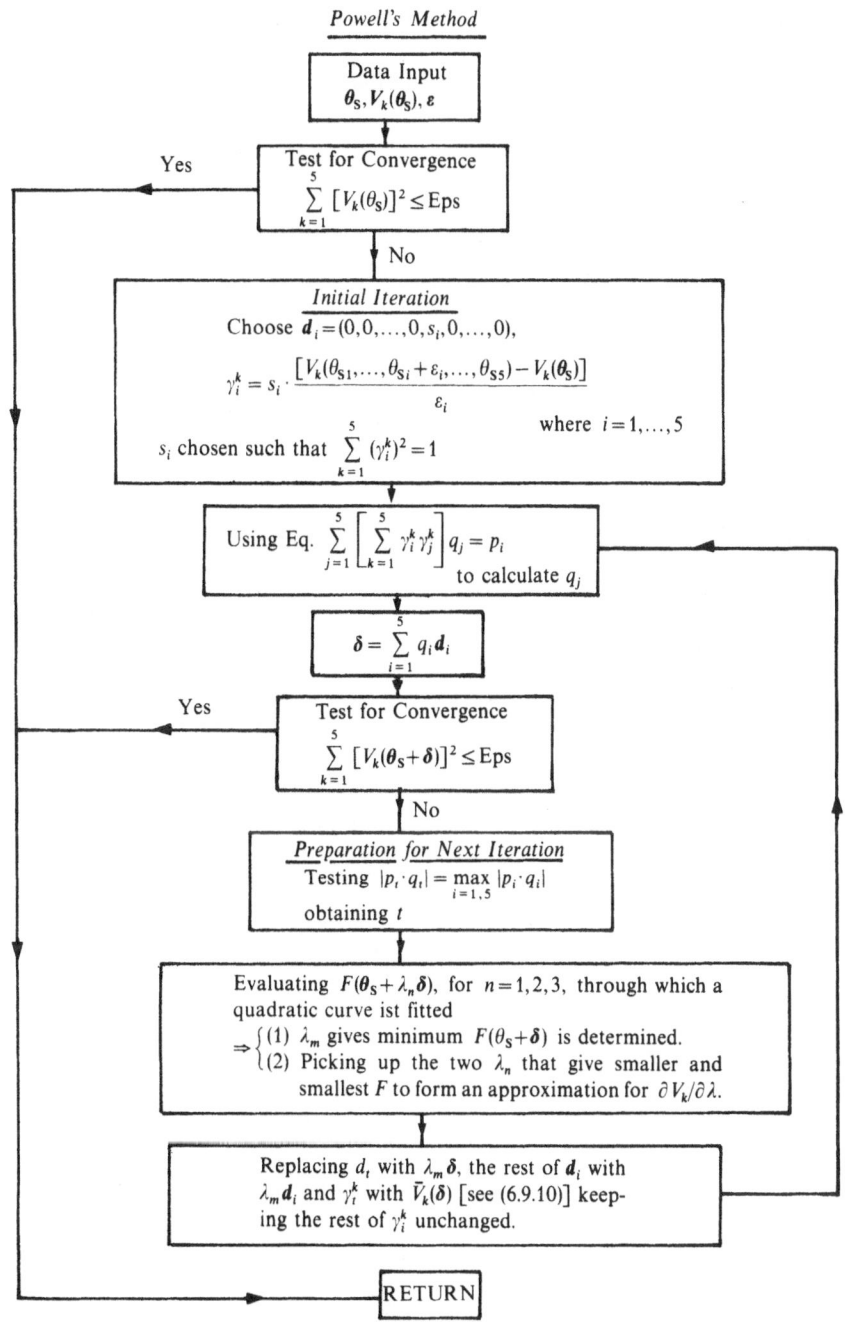

Powell's Method

Data Input
$\theta_S, V_k(\theta_S), \varepsilon$

Test for Convergence
$$\sum_{k=1}^{5} [V_k(\theta_S)]^2 \leq \text{Eps}$$

Yes

No

Initial Iteration
Choose $\boldsymbol{d}_i = (0,0,\ldots,0,s_i,0,\ldots,0)$,

$$\gamma_i^k = s_i \cdot \frac{[V_k(\theta_{S1},\ldots,\theta_{Si}+\varepsilon_i,\ldots,\theta_{S5}) - V_k(\theta_S)]}{\varepsilon_i}$$

s_i chosen such that $\sum_{k=1}^{5} (\gamma_i^k)^2 = 1$ where $i=1,\ldots,5$

Using Eq. $\sum_{j=1}^{5} \left[\sum_{k=1}^{5} \gamma_i^k \gamma_j^k \right] q_j = p_i$
to calculate q_j

$$\delta = \sum_{i=1}^{5} q_i \boldsymbol{d}_i$$

Test for Convergence
$$\sum_{k=1}^{5} [V_k(\theta_S+\boldsymbol{\delta})]^2 \leq \text{Eps}$$

Yes

No

Preparation for Next Iteration
Testing $|p_t \cdot q_t| = \max_{i=1,5} |p_i \cdot q_i|$
obtaining t

Evaluating $F(\theta_S + \lambda_n \boldsymbol{\delta})$, for $n=1,2,3$, through which a quadratic curve ist fitted
$\Rightarrow \begin{cases} (1) \ \lambda_m \text{ gives minimum } F(\theta_S+\boldsymbol{\delta}) \text{ is determined.} \\ (2) \ \text{Picking up the two } \lambda_n \text{ that give smaller and} \\ \quad \text{smallest } F \text{ to form an approximation for } \partial V_k/\partial \lambda. \end{cases}$

Replacing d_t with $\lambda_m \boldsymbol{\delta}$, the rest of $\boldsymbol{d}_i$ with $\lambda_m \boldsymbol{d}_i$ and γ_t^k with $\bar{V}_k(\boldsymbol{\delta})$ [see (6.9.10)] keeping the rest of γ_i^k unchanged.

RETURN

Telenin's Method. Model Problems (G. H. KLOPFER)

(1) Cauchy problem for Laplace's Equation.
Find the solution of Laplace's equation

$$\phi_{yy} + \phi_{xx} = 0 \quad \text{in} \quad 0 \le x \le 1, \quad 0 \le y \le 1$$

with boundary conditions

$$\phi(x, 0) = \sigma(x) = \cos(\pi x)$$

$$\phi_y(x, 0) = \tau(x) = -1$$

$$\phi_x(0, y) = 0$$

The *analytical solution* of this problem, using analytic continuation, is

$$\phi(x, y) = \text{Re} \left\{ \sigma(z) - i \int_0^z \tau(\xi) d\xi \right\}$$

$$= \text{Re} \left\{ \cos(\pi z) + i z \right\}$$

$$\phi(x, y) = \cos(\pi x) \cdot \cosh(\pi y) - y$$

The solution by Telenin's method is given in the listing attached.

73/09/07 UNIVERSITY OF MINNESOTA 7600 FORTRAN COMPILER   BKY 7.0.0 PSR12
        MNF,T.   10 AUG 75 14.14.31

```
      PROGRAM TELENN(INPUT,OUTPUT)
      COMMON/DUMMY     / DUMMY
      DIMENSION XK(20),FY(50),DFY(50),A(20,20),XXI(20,20),XP(20,20)
      DIMENSION B(20,1),SCRTCH(20),ISCRTCH(20,3)
      DOUBLE HPRINT,T(8,20)
      LOGICAL SWPR
      DIMENSION E(20),ER(20)
      COMMON A,XXI,XK,N
      EXTERNAL DERIV
      EXACT(XX,YY) = (.5*COS(PI*XX))*(EXP(PI*YY)+EXP(-PI*YY)) - YY
      SIGMA(XV) = COS(PI*XV)
      TAU(XV) = C
      PI = 4.*ATAN(1.)
    1 CONTINUE
      CALL SECOND(T1)
      READ 2, N,DY,YO,YF,XO,XF,C
      PRINT 4, N, YO,YF,XO,XF,C
    2 FORMAT(I2,/(F10.2))
    4 FORMAT(*1  N = *,I2,*,INITIAL AND FINAL Y = *,2F6.2,*, INITIAL AND
     1 FINAL X = *,2F6.2 ,*,  INITIAL SLOPE = ,*,F6.2//)
      IF (N.LE. 0) GO TO 1
      FN        = N
      DX        = (XF - XO)/FN
      NR        = N + 1
      DO 3  I = 1,NR
      II = I - 1
      FI = II
C     POWER SERIES
      X = XK(I) = FI*DX + XO
      B(I ,1) = 1.
C     INITIAL DATA
      FY(I ) = SIGMA(X)
      IIN     = I + NR
      FY(IIN) = TAU(X)
      XP(I,1) = XXI(I,1) = 1.
      DO 3 J = 2,NR
      JJ = J - 1
      XP(I,J) = X**(2*JJ)
      XXI(I,J) = XP(I,J)
    3 CONTINUE
      DO 5 I = 1,NR
      PRINT 60, (XP(I,L), L = 1,NR)
   60 FORMAT (* MATRIX    *,10G10.4)
    5 CONTINUE
      PRINT 13
C     INVERT MATRIX XP AND CALL IT XXI
      CALL MATINV(XXI,8,20,NR,0,SCRTCH,ISCRTCH,DET)
      DO 6 I = 1,NR
      PRINT 70, (XXI(I,L), L = 1,NR)
   70 FORMAT(  * MATR INV   *,10G10.4)
    6 CONTINUE
      PRINT 13
   13 FORMAT(* *)
COMMENT        COMPUTE COEFFICIENT MATRIX FOR THE 2ND X-DERIVATIVE
```

```
      DO 7 I = 1,NR
      DO 7 K = 1,NR
      A(I,K) = 2.*XXI(2,K)
      DO 7 J = 3,NR
      FJ     = 2*(J-1)*(2*J-3)
      J2     = 2*(J-2)
      IF(XK(I) .GT. 0.) A(I,K) = A(I,K) + FJ*XK(I)**J2*XXI(J,K)
    7 CONTINUE
      DO 8 I = 1,NR
      PRINT 9, (A(I,L),L=1,NR)
    9 FORMAT(* COEF MATR  *,10G10.4)
    8 CONTINUE
C                SET UP INTEGRATION ROUTINES INTO AND INT BY ADAMS-MOULTON,
C                STARTING BY ZONNEVELD
      HPRINT = DBLE(DY)
      YMAX   = YF - DY/2.
      NY = (YF - YO)/DY  + .001
      IF (NY .LE. 0) GO TO 1
      Y = YO
      NEQ    =   2*NR
      PRINT 13
      PRINT 15
   15 FORMAT(10X,*  POWER SERIES APPROXIMATION  *)
      PRINT 13
      PRINT 31,(XK(K),K=1,NR)
   31 FORMAT( 5X,*     X = *,12F10.4)
      PRINT 13
      PRINT 30, N
   30 FORMAT( 5X,*     Y        FUNCTION VALUES PHI(K), K=0,1,...,*I2 ,/
     110X,*            RELATIVE ERRORS  ((APPROX - EXACT)/EXACT)  * ,//)
      SWPR   = .TRUE.
      CALL INTO(NEQ,Y,DERIV,FY,DFY,T,HPRINT)
      DO 20 IJ= 1,5000
      IF(.NOT. SWPR) GO TO 50
      DO 11 I = 1,NR
      E(I) = EXACT(XK(I),Y)
      ER(I) = (FY(I) - E(I))/E(I)
   11 CONTINUE
      PRINT 40,Y,(FY(II), II = 1,NR)
      PRINT 41,(ER(I), I=1,NR)
   40 FORMAT( 5X,12F10.4)
   41 FORMAT(18X,11E10.3)
   50 CONTINUE
      IF(SWPR .AND. (Y .GT. YMAX)) GO TO 100
      CALL INT(Y,DERIV,FY,DFY,T,SWPR)
   20 CONTINUE
  100 CONTINUE
      CALL SECOND(T2)
      TC = T2 - T1
      PRINT 999, TC
  999 FORMAT(//,*  COMPUTATION TIME = *,F6.3,* SECONDS  *,//)
      GO TO 1
      END

      SUBROUTINE DERIV(Y,FY,DFY)
      DIMENSION XK(20),FY(50),DFY(50),A(20,20),XXI(20,20)
      COMMON A,XXI,XK,N
      NR = N + 1
      DO 30 K = 1,NR
      K2 = K + NR
      DFY(K) = FY(K2)
      DFY(K2) = 0.
      DO 30 I = 1,NR
      DFY(K2) = DFY(K2) - A(K,I)*FY(I)
   30 CONTINUE
      RETURN
      END
```

```
      SUBROUTINE INTO(NO,X,DERI ,Y,F,T,HPRO)
      COMMON /INTC/ IPMX,AREF,EMAX,SSSR,HFAC,SWAM,SWEX
      COMMON /INTP/ HPR,XX,N,EUB,ELB,IP,IT,NRKS,SWIN
      DIMENSION Y(50),F(50),T(8,20)
      LOGICAL SWAM,SWEX,SWIN
      INTEGER HFAC
      DOUBLE PRECISION T,HPRO,HPR,XX
      DATA IPMX,AREF,EMAX,SSSR,HFAC,SWAM,SWEX
     $    /1024,1.0,1.0E-6,100.0,2,.TRUE.,.TRUE./
C
      HPR=HPRO
      XX=DBLE(X)
      N=NO
      EUB=EMAX
      ELB=EMAX/SSSR
      IP=1
      IT=0
      NRKS=0
      SWIN=SWEX
      CALL DERI (X,Y,F)
      DO 9 I=1,N
      T(5,I)=DBLE(Y(I))
    9 CONTINUE
      RETURN
      END



      SUBROUTINE INT(X,DERI ,Y,F,T,SWPR
     X )
      COMMON /INTC/ IPMX,AREF,EMAX,SSSR,HFAC,SWAM,SWEX
      COMMON /INTP/ HPR,XX,N,EUB,ELB,IP,IT,NRKS,SWIN
C
      DIMENSION Y(50),F(50),T(8,20)
      LOGICAL SWAM,SWEX,SWIN
      LOGICAL SWPR
      INTEGER HFAC
      DOUBLE PRECISION T,HPR,XX
      DOUBLE PRECISION D,H
 6000 FORMAT (36H0 CANNOT DECREASE H BECAUSE OF HMIN.    ,1PE16.8,I20)
C
  1      CONTINUE
         SWPR=.FALSE.
         TEST=0.0
         H=HPR/DBLE(FLOAT(IP*24))
         IF ((NRKS .LT. 3) .OR. (.NOT. SWAM)) GO TO 200
C
C        ADAMS-MOULTON STEP.
  100    CONTINUE
         DO 109 I=1,N
         D=DBLE(F(I))
         T(4,I)=D
         Y(I)=SNGL(T(5,I)+H*(
     X 55.0D0*D-59.0D0*T(3,I)+37.0D0*T(2,I)- 9.0D0*T(1,I) ))
  109    CONTINUE
         X=SNGL(XX+24.0D0*H)
         CALL DERI (X,Y,F)
         DO 119 I=1,N
         D=DBLE(F(I))
         D=(       T(5,I)+H*(
     X  9.0D0*D+19.0D0)*T(4,I)- 5.0D0*T(3,I)+       T(2,I) ))
         T(6,I)=D
         F=ABS(SNGL(D)-Y(I))/14.0
         TEST=AMAX1(F/AMAX1(AREF,ABS(SNGL(D))),TEST)
  119    CONTINUE
C
```

```
         GO TO 300
C
C        ZCNNEVELD STEP.
  200    CONTINUE
         DO 209 I=1,N
         D=DBLE(F(I))
         T(4,I)=D
C        1
         Y(I)=SNGL(T(5,I)+H*(
     X 12.0D0*D                                                        ))
  209    CONTINUE
         X=SNGL(XX+12.0D0*H)
         CALL DERI (X,Y,F)
         DO 219 I=1,N
         D=DBLE(F(I))
         T(6,I)=D
C        2

         Y(I)=SNGL(T(5,I)+H*(
     X 12.0D0*D                                                        ))
  219    CONTINUE
         CALL DERI (X,Y,F)
         DO 229 I=1,N
         D=DBLE(F(I))
         T(7,I)=D
C        3
         Y(I)=SNGL(T(5,I)+H*(
     X 24.0D0*D                                                        ))
  229    CONTINUE
         X=SNGL(XX+24.0D0*H)
         CALL DERI (X,Y,F)
         DO 239 I=1,N
         D=DBLE(F(I))
         T(8,I)=D
C        4
         Y(I)=SNGL(T(5,I)+H*(
     X 3.75D0*T(4,I)+5.25D0*T(6,I)+9.75D0*T(7,I)-0.75D0*D ))
  239    CONTINUE
         X=SNGL(XX+18.0D0*H)
         CALL DERI (X,Y,F)
         DO 249 I=1,N
         D=DBLE(F(I))
         E=ABS(SNGL(H*(
     X -16.0D0*T(4,I)+48.0D0*T(6,I)+48.0D0*T(7,I)+48.0D0*T(8,I)
     X -128.0D0*D   )))
C        5
         D=(     T(5,I)+H*(
     X 4.0D0*T(4,I)+ 8.0D0*T(6,I)+ 8.0D0*T(7,I)+ 4.0D0*T(8,I)
     X ))
         T(6,I)=D
         TEST=AMAX1(E/AMAX1(AREF,ABS(SNGL(D))),TEST)
  249    CONTINUE
C
C        BOTH ADAMS-MOULTON AND ZONNEVELD METHODS CONTINUE FROM HERE.
  300    CONTINUE
         X=SNGL(XX+24.0D0*H)
         IF (TEST .LE. EUB) GO TO 310
         IF (IP*HFAC .GT. IPMX) GO TO 309
C
C        REPEAT STEP WITH SMALLER H.
         NRKS=0
         IP=IP*HFAC
         IT=IT*HFAC
         DO 305 I=1,N
         Y(I)=SNGL(T(5,I))
         F(I)=SNGL(T(4,I))
```

```
  305    CONTINUE
         GO TO 1
C
C        CANNOT DECREASE H BECAUSE OF HMIN.
  309    CONTINUE
         IF (.NOT. SWIN) GO TO 310
         PRINT 600), X,IPMX
         SWIN=.FALSE.
C
  310    CONTINUE
C
C        ACCEPT CURRENT STEP.
C
C        XX STILL HAS NOT BEEN CHANGED SINCE ENTRY.
C        YY(XX) IS STILL IN T(5, ).
C        F(YY) IS IN T(4, ).
C
         IT=IT+1
         XX=XX+HPR/DBLE(FLOAT(IP))
         NRKS=MIN0(NRKS+1,4)
         DO 319 I=1,N
         D=T(6,I)
         T(5,I)=D
         Y(I)=SNGL(D)
  319    CONTINUE
         X=SNGL(XX)
         CALL DERI (X,Y,F)
         IF (IT .LT. IP) GO TO 320
C
C        X IS A MULTIPLE OF HPRINT.
         SWPR=.TRUE.
         IT=IT-IP
C
  320    CONTINUE
         IF (TEST .GE. ELB) GO TO 330
         IF (MOD(IP,HFAC)+MOD(IT,HFAC) .NE. 0) GO TO 330
C
C        PROCEED TO NEXT STEP WITH LARGER H, USING ZONNEVELD METHOD.
         NRKS=0
         IP=IP/HFAC
         IT=IT/HFAC
         RETURN
C
C
C        PROCEED TO NEXT STEP WITH SAME H.
  330    CONTINUE
         DO 339 I=1,N
         T(1,I)=T(2,I)
         T(2,I)=T(3,I)
         T(3,I)=T(4,I)
  339    CONTINUE
         RETURN
         END
```

```
N =  6,INITIAL AND FINAL Y =      0  1.00,  INITIAL AND FINAL X =
                                              0  1.00,  INITIAL SLOPE =  -1.00

MATRIX      1.000          0          0          0          0          0          0
MATRIX      1.000      .2778E-01  .7716E-03  .2143E-04  .5954E-06  .1654E-07  .4594E-09
MATRIX      1.000      .1111      .1235E-01  .1372E-02  .1524E-03  .1694E-04  .1882E-05
MATRIX      1.000      .2500      .6250E-01  .1563E-01  .3906E-02  .9766E-03  .2441E-03
MATRIX      1.000      .4444      .1975      .8779E-01  .3902E-01  .1734E-01  .7707E-02
MATRIX      1.000      .6944      .4823      .3349      .2326      .1615      .1122
MATRIX      1.000      1.000      1.000      1.000      1.000      1.000      1.000

MATR INV    1.000          0          0          0          0          0          0
MATR INV   -53.69       61.71      -9.643      1.905      -.3214      .3740E-01 -.2165E-02
MATR INV    740.7      -1092.      430.9      -94.65      16.53      -1.954      .1140
MATR INV   -4003.       6412.     -3264.      1032.      -200.9      24.89      -1.489
MATR INV    9730.      -.1618E+05  9217.     -3495.      834.5      -113.9      7.174
MATR INV   -.1061E+05 .1800E+05-.1087E+05 4555.      -1250.      200.0      -13.89
MATR INV    4199.      -7198.      4499.     -2000.      599.9      -109.1      9.089

COEF MATR  -107.4       123.4      -19.29      3.810      -.6429      .7481E-01 -.4329E-02
COEF MATR   58.00      -110.5       59.29     -7.798      1.154      -.1266      .7100E-02
COEF MATR   9.040       30.54      -84.82      49.52     -4.697      .4405      -.2290E-01
COEF MATR  -66.76       112.6      -25.07     -58.80      40.47      -2.509      .1128
COEF MATR   297.1      -527.9      364.4      -147.0      -15.55      29.88      -.9072
COEF MATR  -2018.       3570.     -2466.      1316.      -487.9      68.61      17.07
COEF MATR  .3256E+05-.5723E+05 .3869E+05-.1989E+05 7612.      -2054.      308.5
```

POWER SERIES APPROXIMATION

```
X =          0       .1667      .3333      .5000      .6667      .8333      1.0000
```

Y   FUNCTION VALUES PHI(K), K=0,1,...., 6
          RELATIVE ERRORS ((APPROX - EXACT)/EXACT)

```
   0      1.0000      .8660      .5000     -.0000     -.5000     -.8660     -1.0000
                 0          0          0         -0         -0         -0         -0
 .1000      .9498      .8091      .4249     -.1000     -.6249     -1.0091     -1.1498
           -.145E-08 -.410E-09 -.459E-08 -.537E-07  .429E-07 -.203E-06  .213E-05
 .2000      1.0040      .8427      .4020     -.2000     -.8020     -1.2427     -1.4040
           -.624E-08  .487E-08 -.377E-07 -.220E-06  .202E-06 -.459E-06 -.219E-05
 .3000      1.1780      .9800      .4390     -.3000     -1.0390     -1.5800     -1.7780
           -.297E-07  .390E-07 -.187E-06 -.579E-06  .219E-06  .128E-05 -.155E-04
 .4000      1.4991      1.2447      .5495     -.4000     -1.3495     -2.0447     -2.2991
           -.131E-06  .176E-06 -.574E-06 -.598E-06 -.127E-05  .692E-05 -.102E-04
 .5000      2.0092      1.6730      .7546     -.5000     -1.7546     -2.6730     -3.0093
           -.492E-06  .596E-06 -.102E-05  .301E-05 -.681E-05  .112E-04  .428E-04
 .6000      2.7689      2.3176      1.0845     -.6000     -2.2844     -3.5176     -3.9695
           -.161E-05  .165E-05 -.249E-06  .212E-04 -.173E-04 -.294E-05  .128E-03
 .7000      3.8639      3.2525      1.5820     -.7001     -2.9819     -4.6522     -5.2648
           -.464E-05  .388E-05  .565E-05  .804E-04 -.249E-04 -.577E-04  .165E-03
 .8000      5.4131      4.5808      2.3066     -.8002     -3.9066     -6.1797     -7.0134
           -.121E-04  .791E-05  .255E-04  .227E-03 -.462E-05 -.162E-03  .317E-04
 .9000      7.5804      6.4445      3.3406     -.9005     -5.1408     -8.2421     -9.3771
           -.291E-04  .140E-04  .761E-04  .516E-03  .920E-04 -.284E-03 -.374E-03
1.0000      10.5913      9.0391      4.7969     -1.0009     -6.7982     -11.0352     -12.5786
           -.649E-04  .212E-04  .186E-03  .940E-03  .332E-03 -.337E-03 -.106E-02
```

COMPUTATION TIME =   .157 SECONDS

(2) Cauchy problem for a mixed type equation.
Find the solution of the mixed type equation

$$\phi_{yy} + (1 - x^2)\phi_{xx} = 0 \qquad 0 \le x \le x_c$$
$$0 \le y \le 1/2$$

with boundary conditions

$$\left. \begin{array}{l} \phi_y(x,0) = 0 \\ \phi(x,0) = (1 - x^2) \end{array} \right\} \quad 0 \le x \le x_c$$
$$\phi_x(0,y) = 0 \qquad\qquad 0 \le y \le 1/2$$

where $x_c = \cosh(1/2)$.

The *analytical solution* of this problem is, by separation of variables

$$\phi(x, y) = (1 - x^2)\cosh(\sqrt{2}\, y)$$

The solution by Telenin's method is given in the listing attached.

73/09/07 UNIVERSITY OF MINNESOTA 7600 FORTRAN COMPILER   BKY 7.0.0 PSR12
        MNF,T.   10 AUG 75 14.23.04

```
      PROGRAM TELENN(INPUT,OUTPUT)
      COMMON/DUMMY        / DUMMY
      DIMENSION XK(20),FY(50),DFY(50),A(20,20),XXI(20,20),XP(20,20)
      DIMENSION B(20,1),SCRTCH(20),ISCRTCH(20,3)
      DOUBLE HPRINT,T(8,20)
      LOGICAL SWPR
      DIMENSION E(20),ER(20)
      COMMON A,XXI,XK,N
      EXTERNAL DERIV
      EXACT(XX,YY) = (1.-XX*XX)*(EXP(SQRT(2.)*YY)+EXP(-SQRT(2.)*YY))/2.
      SIGMA(XV) = 1.- XV*XV
      TAU(XV) = C
      PI = 4.*ATAN(1.)
    1 CONTINUE
      CALL SECOND(T1)
      READ 2, N,DY,YO,YF,XO,XF,C
      PRINT 4, N, YO,YF,XO,XF,C
    2 FORMAT(I2,/(F10.2))
    4 FORMAT(*1  N = *,I2,*,INITIAL AND FINAL Y = *,2F6.2,*,  INITIAL AND
     1 FINAL X = *,2F6.2 ,*,   INITIAL SLOPE = *,F6.2//)
      IF (N.LE. 0) GO TO 1
      FN       = N
      DX       = (XF - XO)/FN
      NR       = N + 1
      DO 3  I = 1,NR
      II = I - 1
      FI = II
C     POWER SERIES
      X = XK(I) = FI*DX + XO
      B(I ,1) = 1.
C     INITIAL DATA
      FY(I ) = SIGMA(X)
      IIN     = I + NR
      FY(IIN) = TAU(X)
      XP(I,1) = XXI(I,1) = 1.
      DO 3 J = 2,NR
      JJ = J - 1
      XP(I,J) = X**(2*JJ)
      XXI(I,J) = XP(I,J)
    3 CONTINUE
      DO 5 I = 1,NR
      PRINT 60, (XP(I,L), L = 1,NR)
   60 FORMAT (* MATRIX      *,10G10.4)
    5 CONTINUE
      PRINT 13
C     INVERT MATRIX XP AND CALL IT XXI
      CALL MATINV(XXI,B,20,NR,0,SCRTCH,ISCRTCH,DET)
      DO 6 I = 1,NR
      PRINT 70, (XXI(I,L), L = 1,NR)
   70 FORMAT(  * MATR INV    *,10G10.4)
    6 CONTINUE
      PRINT 13
   13 FORMAT(* *)
COMMENT       COMPUTE COEFFICIENT MATRIX FOR THE 2ND X-DERIVATIVE
```

```
      DO 7 I = 1,NR
      DO 7 K = 1,NR
      A(I,K) = 2.*XXI(2,K)
      DO 7 J = 3,NR
      FJ     = 2*(J-1)*(2*J-3)
      J2     = 2*(J-2)
      IF(XK(I) .GT. 0.) A(I,K) = A(I,K) + FJ*XK(I)**J2*XXI(J,K)
    7 CONTINUE
      DO 8 I = 1,NR
      PRINT 9, (A(I,L),L=1,NR)
    9 FORMAT(* COEF MATR  *,10G10.4)
    8 CONTINUE
C          SET UP INTEGRATION ROUTINES INTO AND INT BY ADAMS-MOULTON,
C          STARTING BY ZONNEVELD
      HPRINT = DBLE(DY)
      YMAX   = YF - DY/2.
      NY = (YF - YO)/DY + .001
      IF (NY .LE. 0) GO TO 1
      Y = YO
      NEQ    = 2*NR
      PRINT 13
      PPINT 15
   15 FORMAT(10X,*  POWER SERIES APPROXIMATION  *)
      PRINT 13
      PRINT 31,(XK(K),K=1,NR)
   31 FORMAT( 5X,*     X =  *,12F10.4)
      PRINT 13
      PRINT 30, N
   30 FORMAT( 5X,*   Y        FUNCTION VALUES PHI(K), K=0,1,...,*I2 ,/
     110X,*          RELATIVE ERRORS  ((APPROX - EXACT)/EXACT)  * ,//)
      SWPR   = .TRUE.
      CALL INTO(NEQ,Y,DERIV,FY,DFY,T,HPRINT)
      DO 20 IJ= 1,5000
      IF(.NOT. SWPR) GO TO 50
      DO 11 I = 1,NR
      E(I) = EXACT(XK(I),Y)
      DENO   = E(I)
      IF(ABS(E(I)) .LE. 1.E-10) DENO = 1.
      ER(I) = (FY(I) - E(I))/DENO
   11 CONTINUE
      PRINT 40,Y,(FY(II), II = 1,NR)
      PRINT 41,(ER(I), I=1,NR)
   40 FORMAT( 5X,12F10.4)
   41 FORMAT(18X,11E10.3)
   50 CONTINUE
      IF(SWPR .AND. (Y .GT. YMAX)) GO TO 100
      CALL INT(Y,DERIV,FY,DFY,T,SWPR)
   20 CONTINUE
  100 CONTINUE
      CALL SECOND(T2)
      TC = T2 - T1
      PRINT 999, TC
  999 FORMAT(//,*  COMPUTATION TIME = *,F6.3,* SECONDS  *,//)
      GO TO 1
      END
```

```
      SUBROUTINE DERIV(Y,FY,DFY)
      DIMENSION XK(20),FY(50),DFY(50),A(20,20),XXI(20,20)
      COMMON A,XXI,XK,N
      NR = N + 1
      DO 30 K = 1,NR
      VAC     = 1. - XK(K)*XK(K)
      K2 = K + NR
      DFY(K) = FY(K2)
      DFY(K2) = 0.
      DO 30 I = 1,NR
      DFY(K2) = DFY(K2) - A(K,I)*FY(I)*VAC
   30 CONTINUE
      RETURN
      END



      SUBROUTINE INTO(NO,X,DERI ,Y,F,T,HPRO)
      COMMON /INTC/ IPMX,AREF,EMAX,SSSR,HFAC,SWAM,SWEX
      COMMON /INTP/ HPR,XX,N,EUB,ELB,IP,IT,NRKS,SWIN
      DIMENSION Y(50),F(50),T(8,20)
      LOGICAL SWAM,SWEX,SWIN
      INTEGER HFAC
      DOUBLE PRECISION T,HPRO,HPR,XX
      DATA IPMX,AREF,EMAX,SSSR,HFAC,SWAM,SWEX
     $    /1024,1.0,1.0E-6,100.0,2,.TRUE.,.TRUE.
C
      HPR=HPRO
      XX=DBLE(X)
      N=NO
      EUB=EMAX
      ELB=EMAX/SSSR
      IP=1
      IT=0
      NRKS=0
      SWIN=SWEX
      CALL DERI (X,Y,F)
      DO 9 I=1,N
      T(5,I)=DBLE(Y(I))
    9 CONTINUE
      RETURN
      END



      SUBROUTINE INT(X,DERI ,Y,F,T,SWPR
     X )
      COMMON /INTC/ IPMX,AREF,EMAX,SSSR,HFAC,SWAM,SWEX
      COMMON /INTP/ HPR,XX,N,EUB,ELB,IP,IT,NRKS,SWIN
C
      DIMENSION Y(50),F(50),T(8,20)
      LOGICAL SWAM,SWEX,SWIN
      LOGICAL SWPR
      INTEGER HFAC
      DOUBLE PRECISION T,HPR,XX
      DOUBLE PRECISION D,H
 6000 FORMAT (36H0 CANNOT DECREASE H BECAUSE OF HMIN.    ,1PE16.8,I20)
C
    1 CONTINUE
      SWPR=.FALSE.
      TEST=0.0
      H=HPR/DBLE(FLOAT(IP*24))
      IF ((NRKS .LT. 3) .OR. (.NOT. SWAM)) GO TO 200
C
C     ADAMS-MOULTON STEP.
  100 CONTINUE
      DO 109 I=1,N
      D=DBLE(F(I))
```

```
        T(4,I)=D
        Y(I)=SNGL(T(5,I)+H*(
     X 55.0D0*D-59.0D0*T(3,I)+37.0D0*T(2,I)- 9.0D0*T(1,I) ))
 109    CONTINUE
        X=SNGL(XX+24.0D0*H)
       CALL DERI (X,Y,F)
        DO 119 I=1,N
        D=DBLE(F(I))
        D=(     T(5,I)+H*(
     X  9.0D0*D+19.0D0*T(4,I)- 5.0D0*T(3,I)+        T(2,I) ))
        T(6,I)=D
        E=ABS(SNGL(D)-Y(I))/14.0
        TEST=AMAX1(E/AMAX1(AREF,ABS(SNGL(D))),TEST)
 119    CONTINUE
C
        GO TO 300
C
C       ZONNEVELD STEP.
 200    CONTINUE
        DO 209 I=1,N
        D=DBLE(F(I))
        T(4,I)=D
C       1
        Y(I)=SNGL(T(5,I)+H*(
     X 12.0D0*D                                              ))
 209    CONTINUE
        X=SNGL(XX+12.0D0*H)
       CALL DERI (X,Y,F)
        DO 219 I=1,N
        D=DBLE(F(I))
        T(6,I)=D
C       2
        Y(I)=SNGL(T(5,I)+H*(
     X 12.0D0*D                                              ))
 219    CONTINUE
       CALL DERI (X,Y,F)
        DO 229 I=1,N
        D=DBLE(F(I))
        T(7,I)=D
C       3
        Y(I)=SNGL(T(5,I)+H*(
     X 24.0D0*D                                              ))
 229    CONTINUE
        X=SNGL(XX+24.0D0*H)
       CALL DERI (X,Y,F)
        DO 239 I=1,N
        D=DBLE(F(I))
        T(8,I)=D
C       4
        Y(I)=SNGL(T(5,I)+H*(
     X 3.75D0*T(4,I)+5.25D0*T(6,I)+9.75D0*T(7,I)-0.75D0*D ))
 239    CONTINUE
        X=SNGL(XX+18.0D0*H)
       CALL DERI (X,Y,F)
        DO 249 I=1,N
        D=DBLE(F(I))
        E=ABS(SNGL(H*(
     X -16.0D0*T(4,I)+48.0D0*T(6,I)+48.0D0*T(7,I)+48.0D0*T(8,I)
     X -128.0D0*D    )))
C       5
        D=(     T(5,I)+H*(
     X 4.0D0*T(4,I)+ 8.0D0*T(6,I)+ 8.0D0*T(7,I)+ 4.0D0*T(8,I)
     X ))
        T(6,I)=D
        TEST=AMAX1(E/AMAX1(AREF,ABS(SNGL(D))),TEST)
```

```
  249    CONTINUE
C
C       BOTH ADAMS-MOULTON AND ZONNEVELD METHODS CONTINUE FROM HERE.
  300    CONTINUE
         X=SNGL(XX+24.0D0*H)
         IF (TEST .LE. EUB) GO TO 310
         IF (IP*HFAC .GT. IPMX) GO TO 309
C
C       REPEAT STEP WITH SMALLER H.
         NRKS=0
         IP=IP*HFAC
         IT=IT*HFAC
         DO 305 I=1,N
         Y(I)=SNGL(T(5,I))
         F(I)=SNGL(T(4,I))
  305    CONTINUE
         GO TO 1
C
C       CANNOT DECREASE H BECAUSE OF HMIN.
  309    CONTINUE
         IF (.NOT. SWIN) GO TO 310
         PRINT 6000, X,IPMX
         SWIN=.FALSE.
C
  310    CONTINUE
C
C
C       ACCEPT CURRENT STEP.
C
C       XX STILL HAS NOT BEEN CHANGED SINCE ENTRY.
C       YY(XX) IS STILL IN T(5, ).
C       F(YY) IS IN T(4, ).
C
         IT=IT+1
         XX=XX+HPR/DBLE(FLOAT(IP))
         NRKS=MIN0(NRKS+1,4)
         DO 319 I=1,N
         D=T(6,I)
         T(5,I)=D
         Y(I)=SNGL(D)
  319    CONTINUE
         X=SNGL(XX)
         CALL DERI (X,Y,F)
         IF (IT .LT. IP) GO TO 320
C
C       X IS A MULTIPLE OF HPRINT.
         SWPR=.TRUE.
         IT=IT-IP
C
  320    CONTINUE
         IF (TEST .GE. ELB) GO TO 330
         IF (MOD(IP,HFAC)+MOD(IT,HFAC) .NE. 0) GO TO 330
C
C       PROCEED TO NEXT STEP WITH LARGER H, USING ZONNEVELD METHOD.
         NRKS=0
         IP=IP/HFAC
         IT=IT/HFAC
         RETURN
C
C
C       PROCEED TO NEXT STEP WITH SAME H.
  330    CONTINUE
         DO 339 I=1,N
         T(1,I)=T(2,I)
         T(2,I)=T(3,I)
         T(3,I)=T(4,I)
  339    CONTINUE
         RETURN
         END
```

```
N =   6; INITIAL AND FINAL Y =        0   .50, INITIAL AND FINAL X =
                                                 0   1.00,  INITIAL SLOPE =          0

MATRIX        1.000             0          0          0          0          0          0
MATRIX        1.000        .2778E-01  .7716E-03  .2143E-04  .5954E-06  .1654E-07  .4594E-09
MATRIX        1.000        .1111      .1235E-01  .1372E-02  .1524E-03  .1694E-04  .1882E-05
MATRIX        1.000        .2500      .6250E-01  .1563E-01  .3906E-02  .9766E-03  .2441E-03
MATRIX        1.000        .4444      .1975      .8779E-01  .3902E-01  .1734E-01  .7707E-02
MATRIX        1.000        .6944      .4823      .3349      .2326      .1615      .1122
MATRIX        1.000        1.000      1.000      1.000      1.000      1.000      1.000

MATR INV      1.000             0          0          0          0          0          0
MATR INV     -53.69        61.71      -9.643      1.905     -.3214      .3740E-01 -.2165E-02
MATR INV      740.7       -1092.       430.9     -94.65      16.53     -1.954      .1140
MATR INV     -4003.        6412.      -3264.      1032.     -200.9      24.89     -1.489
MATR INV      9730.       -.1618E+05  9217.      -3495.      834.5     -113.9      7.174
MATR INV     -.1061E+05   .1800E+05  -.1087E+05  4555.      -1250.      200.0     -13.89
MATR INV      4199.       -7198.      4499.      -2000.      599.9     -109.1      9.089

COEF MATR    -107.4        123.4      -19.29      3.810     -.6429      .7481E-01 -.4329E-02
COEF MATR     58.00       -110.5       59.29     -7.798      1.154     -.1266      .7100E-02
COEF MATR     9.040        30.54      -84.82      49.52     -4.697      .4405     -.2290E-01
COEF MATR    -66.76        112.6      -25.07     -58.80      40.47     -2.509      .1128
COEF MATR     297.1       -527.9       364.4     -147.0     -15.55      29.88     -.9072
COEF MATR    -2019.        3570.      -2466.      1316.     -487.9      68.61      17.07
COEF MATR     .3256E+05  -.5723E+05   .3869E+05 -.1989E+05  7612.      -2054.      308.5

     POWER SERIES APPROXIMATION

X =           0        .1667      .3333      .5000      .6667      .8333     1.0000

Y             FUNCTION VALUES PHI(K), K=0,1,...,6
              RELATIVE ERRORS  ((APPROX - EXACT)/EXACT)


    0         1.0000      .9722      .8889      .7500      .5556      .3056        0
                   0          0          0          0          0          0              0
  .0500       1.0025      .9747      .8911      .7519      .5569      .3063        0
            -.379E-10  -.379E-10  -.379E-10  -.378E-10  -.371E-10  -.341E-10              0
  .1000       1.0100      .9820      .8978      .7575      .5611      .3086        0
            -.270E-09  -.270E-09  -.270E-09  -.270E-09  -.267E-09  -.255E-09              0
  .1500       1.0226      .9942      .9090      .7669      .5681      .3125        0
            -.429E-09  -.429E-09  -.429E-09  -.428E-09  -.422E-09  -.395E-09              0
  .2000       1.0403      1.0114     .9247      .7802      .5779      .3179        0
            -.515E-09  -.515E-09  -.514E-09  -.513E-09  -.503E-09  -.453E-09              0
  .2500       1.0632      1.0336     .9450      .7974      .5906      .3249        0
            -.788E-09  -.788E-09  -.788E-09  -.787E-09  -.772E-09  -.691E-09              0
  .3000       1.0914      1.0610     .9701      .8185      .6063      .3335        0
            -.736E-09  -.736E-09  -.735E-09  -.735E-09  -.715E-09  -.594E-09              0
  .3500       1.1250      1.0938     1.0000     .8438      .6250      .3438        0
            -.111E-08  -.111E-08  -.111E-08  -.111E-08  -.109E-08  -.918E-09              0
  .4000       1.1643      1.1320     1.0349     .8732      .6468      .3558        0
            -.938E-09  -.938E-09  -.937E-09  -.939E-09  -.914E-09  -.675E-09              0
  .4500       1.2094      1.1758     1.0750     .9071      .6719      .3695        0
            -.141E-08  -.141E-08  -.141E-08  -.141E-08  -.139E-08  -.107E-08              0
  .5000       1.2606      1.2256     1.1205     .9454      .7003      .3852        0
            -.113E-08  -.113E-08  -.112E-08  -.113E-08  -.112E-08  -.687E-09              0

COMPUTATION TIME =    .054 SECONDS
```

References

Chattot, J. J.: J. Fluid Mechanics **86**, 161–177 (1978).

Chattot, J. J., Holt, M.: Proc. 3rd. Int. Conf. on Numerical Methods in Fluid Mechanics. *Lecture Notes in Physics No.* **19**, p. 86–91. Berlin-Heidelberg-New York: Springer 1973.

Chushkin, P. I., Shchennikov, V. V.: Zh. Inzh. Fiz. Akad. Nauk SSSR **3**, 88–94 (1962).

Cuffel, R. F., Back, L. H., Massier, P. F.: AIAA J. **7**, 1364–1366 (1969).

Emmons, H. W.: NACA TN 1003, 1946.

Fletcher, C. A. J.: Proc. Fourth Int. Conf. on Numerical Methods in Fluid Dynamics. *Lecture Notes in Physics No.* **35**, p. 161–166. Berlin-Heidelberg-New York: Springer 1975a.

Fletcher, C. A. J.: AIAA J. **13**, No. 8 (1975b).

Friedman, M. P.: J. Aerospace Sci. **29**, 4 (1962).

Gilinskii, S. M., Telenin, G. F., Tinyakov, G. P.: Izv. Akad. Nauk, SSSR Mekh. Mash. **4**, 9–28 (1964) (Translated as NASA TT F297).

Gilinskii, S. M., Telenin, G. F.: Izv. Akad. Nauk SSSR Mekh. Mash. **5**, 148–156 (1964).

Gilinskii, S. M., Lebedev, M. G.: Izv. Akad. Nauk SSSR Mekh. **1** (1965).

Goldstein, S., Ward, G. N.: Aero. Quart. **2**, 39–84 (1950).

Gross, M. B., Holt, M.: *Proc. Symposium Transsonicum II*, Göttingen 1975. pp 369–375. Berlin-Heidelberg-New York: Springer 1976.

Holt, M., Blackie, J.: Aero. Sci. **23**, 931–956 (1956).

Holt, M., Ndefo, D. E.: J. Comp. Phys. **5**, 463–486 (1970).

Hopkins, D. F., Hill, D. E.: AIAA J. **4**, 1337–1343 (1966).

Jones, D. J.: Aero. Rep. LR-507 (NRC No. 10361), Nat. Res. Council, Canada, Ottawa 1968.

Jones, D. J.: Tables of Inviscid Supersonic Flow About Circular Cones at Incidence $\gamma = 1.4$, AGARDograph 137 (1969).

Jones, D. J., South, J. C., Klunker, E. B.: J. Comp. Phys. **9**, 496–527 (1972).

Jones, D. J., South, J. C.: Application of the Method of Lines to the Solution of Elliptic Partial Differential Equations. National Research Council of Canada, National Aeronautical Establishment NRC No. 18021 (1979).

Kliegel, J. R., Levine, J. N.: AIAA J. **7**, 1375–1378 (1969).

Klopfer, G. H., Holt, M.: *Proc. Symposium Transsonicum II*, Göttingen 1975. pp 376–383. Berlin-Heidelberg-New York: Springer 1976.

Klunker, E. B., South, J. C., jr., Davis, R. M.: NASA TR R-374, 1971.

Kopal, Z.: M.I.T. Dept. Elec. Eng. Center of Analysis Tech. Report **4**, 1949.

Kutler, P., Lomax, H.: Proc. 2nd Int. Conf. on Numerical Methods in Fluid Dynamics. *Lecture Notes in Physics No.* **8** (Ed. M. Holt), p. 24–29. Berlin-Heidelberg-New York: Springer 1971.

Li, Kon-ming, Holt, M.: J. Fluid Mech. **114**, 399–418 (1982).

Liskovets, O. A.: The Method of Lines (Review). Differentsial'nye Uravneniya **1**, 1662–1678 (1965).

Meng, J. C. S.: J. Comp. Phys. **15**, 320–344 (1974).

Melnik, R. E.: AIAA J. **5**, 631–637 (1967).

Ndefo, D. E.: Univ. of Calif., Berkeley, Aero. Sci. Rept. AS-68-2, 1968.

Petrovskii, I. G.: *Lectures on Partial Differential Equations*. New York: Interscience 1964.

Pirumov, U. G.: Izv. Akad. Nauk SSSR Mekh. Zh. i G. **2**, 10–22 (1967).

Powell, M. J. D.: Computer J. **7**, 303–307 (1964).

Stone, A. H.: J. Math. Phys. **27**, 67–81 (1948a).

Stone, A. H.: J. Math. Phys. **30**, 200–213 (1948b).

Tinyakov, G. P.: Izv. Akad. Nauk SSSR Mekh. No. **6**, 10–19 (1965).

Subject Index

Springer Series in Computational Physics

Editors: **H. Cabannes, M. Holt, H. B. Keller, J. Killeen, S. A. Orszag**

R. Peyret, T. D. Taylor

Computational Methods for Fluid Flow

1983. 125 figures. X, 358 pages
ISBN 3-540-11147-9

Contents: Numerical Approaches: Introduction and General Equations. Finite-Difference Methods. Integral and Spectral Methods. Relationship Between Numerical Approaches. Specialized Methods. – Incompressible Flows: Finite-Difference Solutions of the Navier-Stokes Equations. Finite-Element Methods Applied to Incompressible Flows. Spectral Method Solutions for Incompressible Flows. Turbulent-Flow Models and Calculations. – Compressible Flows: Inviscid Compressible Flows. Viscous Compressible Flows. – Concluding Remarks. – Appendix A: Stability. – Appendix B: Multiple-Grid Method. – Appendix C: Conjugate-Gradient Method. – Index.

Y. I. Shokin

The Method of Differential Approximation

Translated from the Russian by K. G. Roesner
1983. 75 figures, 12 tables. XIII, 296 pages
ISBN 3-540-12225-7

Contents: Stability Analysis of Difference Schemes by the Method of Differential Approximation. – Investigation of the Artificial Viscosity of Difference Schemes. – Invariant Difference Schemes. – Appendix. – References. – Subject Index.

D. P. Telionis

Unsteady Viscous Flows

1981. 132 figures. XXIII, 408 pages
ISBN 3-540-10481-X

Contents: Introduction. – Basic Concepts. – Numerical Analysis. – Impulsive Motion. – Oscillations with Zero Mean. – Oscillating Flows with Non-Vanishing Mean. – Unsteady Turbulent Flows. – Unsteady Separation. – Index.

F. Thomasset

Implementation of Finite Element Methods for Navier-Stokes Equations

1981. 86 figures. VII, 161 pages
ISBN 3-540-10771-1

Contents: Introduction. – Notations. – Elliptic Equations of Order 2: Some Standard Finite Element Methods. – Upwind Finite Element Schemes. – Numerical Solution of Stokes Equations. – Navier-Stokes Equations: Accuracy Assessments and Numerical Results. – Computational Problems and Bookkeeping. – Appendix 1: The Patch Test of the P1 Nonconforming Triangle: Sketchy Proof of Convergence. – Appendix 2: Numerical Illustration. – Appendix 3: The Zero Divergence Basis for 2-D P1 Nonconforming Elements. – References. – Index.

Finite-Difference Techniques for Vectorized Fluid Dynamics Calculations

Editor: **D. L. Book**
1981. 60 figures. VIII, 226 pages
ISBN 3-540-10482-8

Contents: Introduction. – *D. L. Book, J. P. Boris:* Computational Techniques for Solution of Convective Equations. – *D. L. Book, J. P. Boris, S. T. Zalesak:* Flux-Corrected Transport. – *R. V. Madala:* Efficient Time Integration Schemes for Atmosphere and Ocean Models. – *J. P. Boris:* A One-Dimensional Lagrangian Code for Nearly Incompressible Flow. – *M. J. Fritts:* Two-Dimensional Lagrangian Fluid Dynamics Using Triangular Grids. – *R. V. Madala, B. E. McDonald:* Solution of Elliptic Equations. – *N. K. Winsor:* Vectorization of Fluid Codes. – Appendices A–E. – References. – Index.

F. Bauer, O. Betancourt, P. Garabedian

A Computational Method in Plasma Physics

1978. 22 figures. VIII, 144 pages
ISBN 3-540-08833-4

Contents: Introduction. – The Variational Principle. – The Discrete Equations. – Description of the Computer Code. – Applications. – References. – Listing of the Code with Comment Cards Index.

M. Holt

Numerical Methods in Fluid Dynamics

2nd revised edition. 1984. 114 figures. Approx. 290 pages
ISBN 3-540-12799-2

Contents: General Introduction. – The Godunov Schemes. – The BVLR Method. – The Method of Characteristics for Three-Dimensional Problems in Gas Dynamics. – The Method of Integral Relations. – Telenin's Method and the Method of Lines. – Subject Index.

Springer-Verlag
Berlin
Heidelberg
New York
Tokyo

A new journal

Experiments in Fluids

Experimental Methods and
Their Applications to Fluid Flow

Editors:
W. Merzkirch, Institut für Thermo- und Fluid-
dynamik, Ruhr-Universität Bochum,
Postfach 102148, D-4630 Bochum,
Federal Republic of Germany; **J. H. Whitelaw,**
Mechanical Engineering Department Imperial
College, Exhibition Road, London SW7 2BX,
Great Britain

Experiments in Fluids will publish research papers
and technical notes which describe either:
- The development of new measuring techniques or
  the extension and improvement of existing
  methods for the measurement of flow properties
  necessary for the better understanding of fluid
  flows and their application in science and
  engineering.
- The application of experimental methods to the
  solution of problems of fluid flow.

It is expected that the contributions will encompass a
wide range of applications including aerodynamics,
hydrodynamics, basic fluid dynamics, convective
heat transfer, combustion, chemical, biological, and
geophysical flows, and turbomachinery. Those which
report advances to the analyses of flow problems will
be considered provided they contain a substantial
experimental content.

Subscription information and/or **sample copies** are
available from your bookseller or directly from
Springer-Verlag, Promotion Department,
P.O.Box 105280, D-6900 Heidelberg 1, FRG

Orders from **North America** should be addressed to:
Springer-Verlag New York Inc., Journal Sales Department 175 Fifth Avenue, New York, NY 10010, USA

Springer-Verlag
Berlin
Heidelberg
New York
Tokyo